Electrospun Nanofibres

Electrospinning is a versatile method to synthesize fiber materials. *Electrospun Nanofibres: Materials, Methods, and Applications* explores the technical aspects of electrospinning methods used to derive a wide range of functional fiber materials and their applications in various technical sectors. As electrospinning is a process that can be modified strategically to achieve different fibers of interest, this book covers the wide spectrum of electrospinning methodologies, such as coaxial, triaxial, emulsion, suspension, electrolyte, and gas-assisted spinning processes. It:

- Discusses a broad range of materials, including synthetic polymers, biodegradable polymers, metals and their oxides, hybrid materials, nonpolymers, and more.
- Reviews different electrospinning methods and combined technologies.
- Describes process-related parameters and their influence on material properties and performance.
- Examines modeling of the electrospinning process.
- Highlights applications across different industries.

This book is aimed at researchers, professionals, and advanced students in materials science and engineering.

Electrospun Nanofibres

Materials, Methods, and Applications

Edited by
Chandrasekar Muthukumar,
Senthilkumar Krishnasamy,
Senthil Muthu Kumar Thiagamani,
and Mariyappan Shanmugam

CRC Press is an imprint of the
Taylor & Francis Group, an **informa** business

First edition published 2024
by CRC Press
2385 Executive Center Drive, Suite 320, Boca Raton, FL 33431

and by CRC Press
4 Park Square, Milton Park, Abingdon, Oxon, OX14 4RN

CRC Press is an imprint of Taylor & Francis Group, LLC

ISBN: 978-1-032-35992-2 (hbk)
ISBN: 978-1-032-36798-9 (pbk)
ISBN: 978-1-003-33381-4 (ebk)

DOI: 10.12019781003333814

Typeset in Times New Roman
by Apex CoVantage, LLC

Dedication

The Editor Dr. Senthil Muthu Kumar Thiagamani dedicates this book
to the **Late. "Kalvivallal" Thiru T. Kalasalingam**
Founder Chancellor – Kalasalingam Academy of Research and Education
on his birth centenary

Contents

Figures

Tables

Editor Biographies

Dr. Chandrasekar Muthukumar is presently working as an Associate Professor at the Department of Aeronautical Engineering, Hindustan Institute of Technology & Science, Chennai, India. He graduated with a Bachelor's in Aeronautical Engineering from Kumaraguru College of Technology, Coimbatore, India. He obtained his Master's in Aerospace Engineering from Nanyang Technological University-TUM ASIA, Singapore. He earned his PhD in Aerospace Engineering from Universiti Putra Malaysia (UPM), Malaysia. His PhD was funded through a research grant from the Ministry of Education, Malaysia. During his association with the UPM, he obtained internal research fund of 16,000 and 20,000 MYR from the University . He has five years of teaching and academic research experience. His field of expertise includes fibre metal laminate (FML), natural fibers, biocomposites, aging and their characterization. He has authored and co-authored research articles in reputed SCI Journals, book chapters and in the conference proceedings. He has published edited books with CRC Press, Wiley & Sons, Springer and Elsevier. He is a peer reviewer for Journal of Composite Materials, Polymer Composites, Materials Research Express and Journal of Natural Fibers.

Dr. Senthilkumar Krishnasamy is an Associate Professor in the Department of Mechanical Engineering at PSG Institute of Technology and Applied Research, Coimbatore, Tamil nadu, India. Dr. Krishnasamy graduated with a Bachelor's degree in Mechanical Engineering from Anna University, Chennai, India, in 2005. He then chose to continue his master's studies and graduated with a Master's degree in CAD/CAM from Anna University, Tirunelveli, in 2009. He has obtained his PhD from the Department of Mechanical Engineering- Kalasalingam University (2016). He had been working in the Department of Mechanical Engineering, Kalasalingam Academy of Research and Education (KARE), India, from 2010 (January) to 2018 (October). He has completed his post-doctoral fellowship at Universiti Putra Malaysia, Serdang, Selangor, Malaysia and King Mongkut's University of Technology North Bangkok (KMUTNB) under the research topics of "Experimental investigations on mechanical, morphological, thermal and structural properties of kenaf fibre/mat epoxy composites" and "Sisal composites and Fabrication of Eco-friendly hybrid green composites on tribological properties in a medium-scale application," respectively. His areas of research interest include the modification and treatment of natural fibres, nanocomposites,

3D printing, and hybrid-reinforced polymer composites. He has published research papers in international journals, book chapters, and conferences in the field of natural fibre composites. He also edits books from different publishers.

Dr. Senthil Muthu Kumar Thiagamani is currently working as Associate Professor in the Department of Mechanical Engineering at Kalasalingam Academy of Research and Education (KARE), Tamil Nadu, India. He graduated from Anna University, Chennai with a Bachelor's in Mechanical Engineering and he obtained his Master's in Automotive Engineering from Vellore Institute of Technology, Vellore. He received his Doctor of Philosophy in Mechanical Engineering (specialized in Biocomposites) from KARE. He has also completed his Postdoctoral Research from the Materials and Production Engineering Department at The Sirindhorn International Thai-German Graduate School of Engineering (TGGS), KMUTNB, Thailand. He has over 14 years of teaching and research experience. He is a member of International Association of Advanced Materials. His research interests include biodegradable polymer composites and characterization. He has authored more than 120 research articles published in reputed SCI indexed journals, book chapters, and conference proceedings. He has also edited 10 books in the theme of biocomposites and currently editing 6 books. He is currently guiding 4 PhD students and has completed 1 PhD student. He is also serving as reviewer for several journals published by reputed publishers like Elsevier, Springer, Wiley, Hindawi, Taylor and Francis and De Gruyter etc.

Mariyappan Shanmugam, PhD, is Professor in the Department of Physics at Hindustan Institute of Technology and Science, Chennai, India.

Contributors

Mônica L. Aguiar
Department of Chemical Engineering
UFSCar
São Carlos, Brazil

Annie Aureen Albert
Department of Physics
Hindustan Institute of Technology and Science
Padur, Chennai, India

Abdul Aziz AlGhamdi
Department of Chemical Engineering
Imam Mohammad Ibn Saud Islamic University
Riyadh, Saudi Arabia

Ozan Avinc
Head of Textile Sciences Main Science Branch
Pamukkale University
Engineering Faculty
Textile Engineering Department
Denizli, Turkey

A. Surendra Babu
Department of Food Science and Technology
School of Agricultural Sciences
Malla Reddy University
Hyderabad, Telangana, India

Trishna Bal
Department of Pharmaceutical Sciences & Technology
Birla Institute of Technology
Mesra, Ranchi, India

Paulo A. M. Chagas
Department of Chemical Engineering
UFSCar
São Carlos, Brazil

George G. Chase
Department of Chemical
Biomolecular and Corrosion Engineering
The University of Akron
Akron, Ohio, USA

Jorge Alberto Vieira Costa
Laboratory of Biochemical Engineering
College of Chemistry and Food Engineering
Federal University of Rio Grande (FURG)
Rio Grande, Brazil

Samsur Ali Dafadar
Department of Pharmaceutical Sciences & Technology
Birla Institute of Technology
Mesra, Ranchi, India

Biplob De
Regional Institute of Pharmaceutical Science and Technology
Abhoynagar
Agartala, Government of Tripura
Tripura, India

Hossein Ebrahimnezhad-Khaljiri
Department of Materials Science and Engineering
Faculty of Engineering
University of Zanjan
Zanjan, Iran

Harshal Gade
Logic Technology Department
Process Engineer
Intel Corporation, Hillsboro, Oregon
and
Imam Mohammad Ibn Saud Islamic University
Riyadh, Saudi Arabia

Anand Gobiraman
Department of Mechanical Engineering
Achariya College of Engineering Technology
Puducherry, India

Vádila G. Guerra
Department of Chemical Engineering
UFSCar
São Carlos, Brazil

N. Guruprasad
Department of Food Technology
Hindustan Institute of Technology and Science
Padur, Chennai, India

Syed Mohd Saiful Azwan Syed Hamzah
Faculty of Ocean Engineering Technology and Informatics
Universiti Malaysia Terengganu
Kuala Nerus Terengganu, Malaysia

Pui San Khoo
Centre for Advanced Composite Materials (CACM)
Universiti Teknologi Malaysia
Johor Bahru, Malaysia
and
Faculty of Mechanical Engineering
Universiti Teknologi Malaysia
Johor, Malaysia

Senthilkumar Krishnasamy
Department of Mechanical Engineering
PSG Institute of Technology and Applied Research
Coimbatore, Tamilnadu, India

Sedat Kumartasli
Research and Development Centre
Polyteks Textile Company
Bursa, Turkey

Suelen Goettems Kuntzler
Laboratory of Microbiology and Biochemistry
College of Chemistry and Food Engineering
Federal University of Rio Grande (FURG)
Rio Grande, Brazil

S. Mahalakshmi
Department of Physics, Anand Institute of Higher Technology
Kazhipattur, Chennai, India

Adrika Maji
Department of Pharmaceutical Sciences & Technology
Birla Institute of Technology
Mesra, Ranchi, India

Gustavo C. Mata
Department of Chemical Engineering
UFSCar
São Carlos, Brazil

Sauvik Mazumdar
Department of Pharmaceutical Sciences & Technology
Birla Institute of Technology
Mesra, Ranchi, India

Gabriela B. Medeiros
Department of Chemical Engineering
UFSCar
São Carlos, Brazil

Michele Greque de Morais
Laboratory of Microbiology and Biochemistry
College of Chemistry and Food Engineering
Federal University of Rio Grande (FURG)
Rio Grande, Brazil

Sirlene Morais
Faculty of Pharmaceutical Sciences of Ribeirão Preto
USP
São Paulo, Brazil

Juliana Botelho Moreira
Laboratory of Microbiology and Biochemistry
Federal University of Rio Grande (FURG)
Rio Grande, Brazil

Zaleha Mustafa
Fakulti Kejuruteraan Pembuatan
Universiti Teknikal Malaysia Melaka
Durian Tunggal, Melaka, Malaysia

M. Muthukrishnan
Department of Mechanical Engineering
KIT-Kalaignarkarunanidhi Institute of Technology
Coimbatore, Tamil Nadu, India

Anant Nag
Department of Pharmaceutical Sciences & Technology
Birla Institute of Technology
Mesra, Ranchi, India

Lin Feng Ng
Centre for Advanced Composite Materials (CACM)
Universiti Teknologi Malaysia
Johor Bahru, Johor, Malaysia.
and
Faculty of Mechanical Engineering
Universiti Teknologi Malaysia
Johor Bahru, Johor, Malaysia

Wanderley P. Oliveira
Faculty of Pharmaceutical Sciences of Ribeirão Preto
University of São Paulo
Ribeirão Preto, Brazil

Anima Pandey
Department of Pharmaceutical Sciences & Technology
Birla Institute of Technology
Mesra, Ranchi, India

V. Parthasarathy
Department of Physics
Rajalakshmi Institute of Technology
Chennai, Tamil Nadu, India

Mrinal Kanti Pradhan
Department of Pharmaceutical Sciences & Technology
Birla Institute of Technology
Mesra, Ranchi, India

Aditya Dev Rajora
Department of Pharmaceutical Sciences & Technology
Birla Institute of Technology
Mesra, Ranchi, India

M. Ramesh
Department of Mechanical Engineering
KIT-Kalaignar Karunanidhi Institute of Technology
Coimbatore, Tamil Nadu, India

A. O. Adeyeye Samuel
Department of Food Technology, Hindustan Institute of Technology and Science
Padur, Chennai, India

P. Sankarganesh
Department of Food Technology
Hindustan Institute of Technology and Science
Padur, Chennai, India

N. Santhosh
Department of Mechanical Engineering
MVJ College of Engineering
Whitefield, Bangalore, India

A. Saravanan
Department of Chemical Engineering
Hindustan Institute of Technology and Science
Padur, Tamil Nadu, India

Daiane Angelica Schmatz
Laboratory of Microbiology and Biochemistry
College of Chemistry and Food Engineering
Federal University of Rio Grande (FURG)
Rio Grande, Brazil

Edilton N. Silva
Department of Chemical Engineering
UFSCar
São Carlos, Brazil

Ana Luiza Machado Terra
Laboratory of Microbiology and Biochemistry
College of Chemistry and Food Engineering
Federal University of Rio Grande (FURG)
Rio Grande, Brazil

Lívia da Silva Uebel
Laboratory of Microbiology and Biochemistry
College of Chemistry and Food Engineering
Federal University of Rio Grande (FURG)
Rio Grande, Brazil

Bruna da Silva Vaz
Laboratory of Microbiology and Biochemistry
College of Chemistry and Food Engineering
Federal University of Rio Grande (FURG)
Rio Grande, Brazil

N. Vigneshwari
Department of Biomedical Engineering
KIT-Kalaignarkarunanidhi Institute of Technology
Coimbatore, Tamil Nadu, India

S. Vishvanathperumal
Department of Mechanical Engineering
S.A. Engineering College
Thiruverkadu, Tamilnadu, India

Mohd Yazid Yahya
Centre for Advanced Composite Materials (CACM)
Universiti Teknologi Malaysia, Johor Bahru, Malaysia
and
Faculty of Mechanical Engineering
Universiti Teknologi Malaysia
Johor, Malaysia

1 Characterization of the Metal-Organic Framework Nanofibers Prepared via Electrospinning

M. Ramesh and M. Muthukrishnan

1.1 INTRODUCTION

Reticular chemistry relates to the linking of organic and inorganic molecules by strong bonds creating crystalline frameworks called MOFs. Omar M. Yaghi pioneered reticular chemistry and popularized the word MOF in 1995 [1]. MOFs are considered a new genre of porous hybrid crystallized organic-inorganic constituents which are held together in an organized manner by metal ions and clusters. MOFs are developed through mesh synthesis i.e. template guided assembly which offers flexibility in developing tailor-made geometry and performance for specific applications [2]. MOFs when compared to conventional porous materials have a specific characteristic of tunable porosity at the molecular level by modifying MOF geometry and organic ligands and metal combination structures with ultrahigh surface areas that generally range from 1,000 to 10,000 m^2g^{-1}. MOF as a concept arises from coordination chemistry which involves coordination polymers [3]. These coordination polymers are created by joining metal ions, which act as connectors and linkers, respectively.

Many researchers [4–8] done significant improvements in developing various ranges of crystalline MOFs with better porous capacity, gas sorption, and catalytic properties, and because of these significant characteristics, they are widely used as thin films, and membranes in various fields like gas purification, carbon dioxide capture, rechargeable batteries, nano-reactors, biomedical imaging, chemical catalysis, bio-luminescence, drug delivery, etc. As of today, based on different functionalities, geometry, and constituents there are more than 20,000 MOFs currently available. MOFs synthesis involves explicated building blocks with defined knowledge of various possible MOF geometries, and performances of various organic linkers with appropriate metal coordinations. Some of the other aspects that are to be considered for synthesizing the MOFs are i) source material availability and cost; ii) reaction temperature (room temperature, solvothermal, and non-solvothermal); iii) fail safe procedure for synthesis; iv) molecular arrangement of polymers; and v) lattice arrangement of MOFs. Some of the prominent MOFs commercially available are HKUST-1 commercially available as Basolite C 300, AlMIL-53 is

DOI: 10.1201/9781003333814-1

offered in commercial form as Basolite A100, FeBTC is commercially available as Basolite F300, $Mg(O_2CH)_2$ is commercially available as Basosiv M050, IRMOF-8 is an organic solvent with 15% Zn yield [9]. Many recent studies have attempted to improvise the physical and chemical properties of these MOFs for tailor-made applications at the mesoscopic/macroscopic scale by altering the pathways of MOF structures. MOFs are mostly fragile and can be disintegrated into fine powder due to their intrinsic physical structure and thus exhibit low MOF efficiency due to poor mechanical and thermal stability. Processing of MOFs into various shapes is improvised as pellets by pressing MOF crystals, in the form of granules, substrates, and electrospun nanofibers [10–12].

MOF combined with polymer forms an effective structure where polymers give superior flexibility mechanically and ensures chemical stability. The polymer acts as a binder and effective methods like spray drying, dip coating, hard or soft templates, lithography, and printing are used for structuring polymer-MOFs. Due to the varying physical and chemical characteristics of these materials, polymer and MOFs have poor interfacial compatibility, which induces non-selective voids to form between the particles and polymer. The open and interconnected web structure of MOF-polymer nanofibers is excellent in allowing the free flow of liquid and gases to overcome transport limitations [13, 14].

1.2 ELECTROSPINNING PROCESS

The electrospinning process is an electro-hydro spinning process used for producing micro to macro nanofibers that are in random or aligned orientations. It is one of the fabrication methods that use polymers to produce extended ultrafine fibers. John Cooley and William Morton each received an electrospinning patent in 1902 that detailed a working prototype of the technology [15, 16]. Soviets in 1938 were the first to bring the practical application of electrospun nanofibers with the commercial name "Petryanov filters" for air filters to capture aerosol particles. As a result, more gas masks with electrospun nanofiber mats are commercially produced. The fibers can be produced as nonwoven mats, yarns, etc. that are in the diameter range of 2 nanometers to micrometers. MOF polymer nanofibers are distinguished for their high surface area to volume, light weight, high aspect ratio, high flexibility, and multi-scale porosity. Currently, electrospun nanofibers are widely employed in both industrial and consumer devices (such as automobile filters) for water and air filtration.

The electrospinning setup (Figure 1.1) comprises of high voltage power supply, spinneret, and collector plates. A high voltage source is used to introduce a specific polarity of charge into a polymer solution. The electrospinning accelerates the charge toward a collector with the opposite polarity. When the more repulsive charge has accumulated and the repulsive force is proportional to the surface tension, the drop surface on the conducting tube begins to build a cone known as a Taylor cone. Thus, the charged polymer fluid droplets under the influence of electrostatic forces overcome the surface tension, and the electrostatic repulsion deformed from a rounded meniscus to a conical shape called the Taylor cone [17].

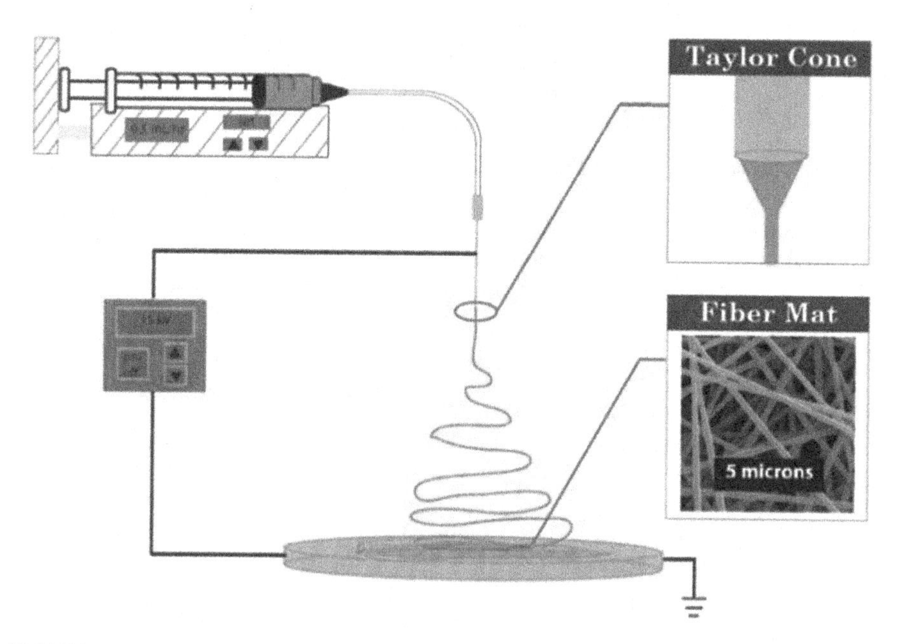

FIGURE 1.1 Electrospinning process setup [18].

The discharged polymer increases in size due to the turbulence in jet flow, which minimizes the diameter of the extruded fiber. The electrified fluid jet passing through the atmosphere gets solidified and collected as nanofibers in the collector. The temperature inside the chamber, the solution vapor pressure, and the distance between tip and collector are some of the parameters affecting the solvent evaporation. Since both electrospinning and electrospraying rely on high voltage to expel continuous jet streams to get the desired result, electrospinning can be considered to be a form of electrospraying technology. Currently, horizontal and vertical are two types of electrospinning processes used for the extraction of MOF polymer nanofibers.

1.2.1 Process Parameters Affecting the Electrospinning Process

Though the setup for the electrospinning process looks simple, the process is quite complex where it involves three main parameters that can be manipulated to get a stable nanofiber. They are solution parameters, process parameters, and ambient parameters [19] as shown in Figure 1.2.

1.2.2 Solution Parameters

The morphology and the diameter of the nanofiber can be modified by the solution parameters like molecular weight, polymer conductivity, surface tension, viscosity, and solution concentration.

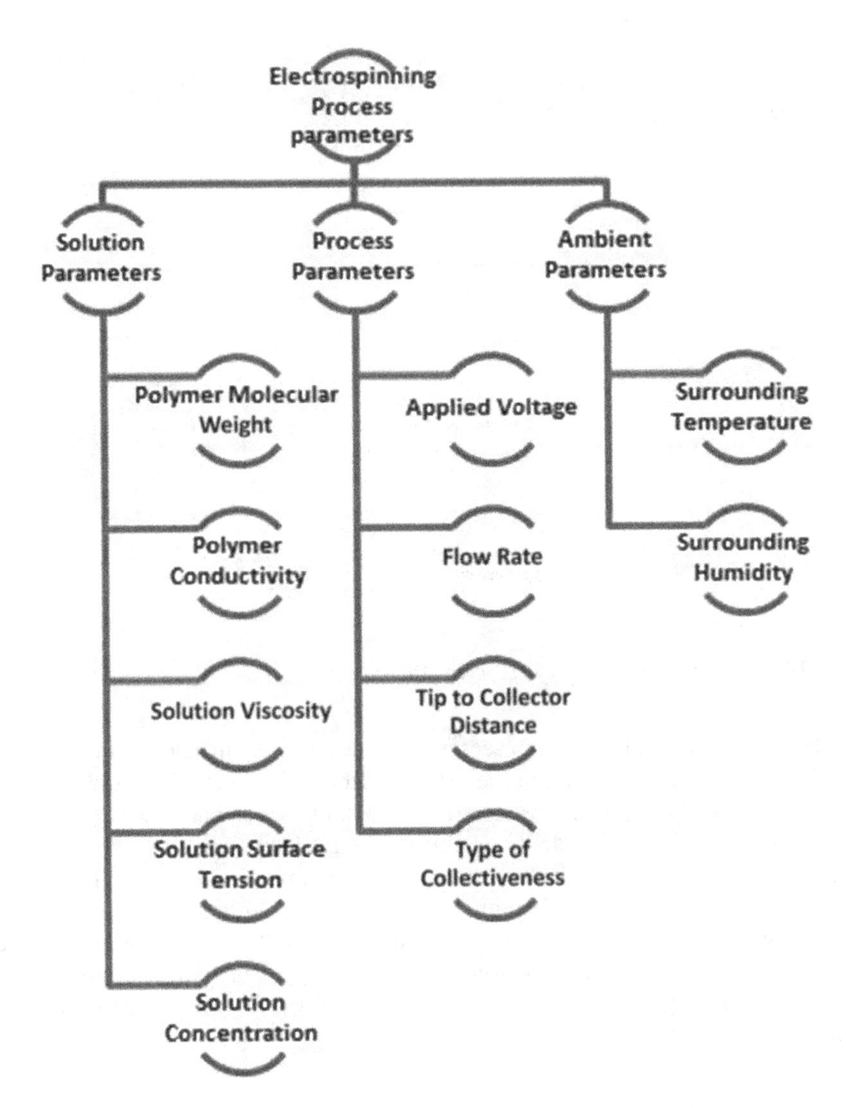

FIGURE 1.2 Process parameters affecting the electrospinning process.

1.2.2.1 Molecular Weight

Polymer molecular weight is one of the important parameters in determining the morphology of nanofibers. Decreasing molecular weight induces bead-like structural formation in the case of polyvinyl alcohol. On the other hand, smooth and long fibers followed by ribbon-like fibers are formed with the increase in molecular weight [20]. Polymers like polyacrylamide are categorized as ultrahigh molecular weight (9×10^6 g/mol) and exhibit wider variations in the morphologies of fiber even with small changes in the concentration (0.3 to 0.7%) [21]. Similarly, studies on polypropylene under variable molecular weights show an incremental trend in the fiber diameter with an increase in molecular weight. In the case of poly(vinyl alcohol), the formation of bead-like structures results in a decrease in molecular weight. Viscosity is another variable that has a significant role in the pore size and fiber diameter and the viscosity of the polymer is greatly influenced by the molecular weight. On the other hand higher molecular weight results in the formation of smooth elongated fibers with reduced fiber diameters [22]. In the case of polyacrylamide, smooth fibers formed when the viscosity of the solution is 0.3 to 0.7 wt. %. Whereas, smooth and ribbon-like structures were formed for the solutions with viscosities around 0.7 to 2 wt. %. At higher viscosities above 2 wt. %, helical ribbons with triangular beads are prominently found [23]. Entanglement is another factor that is influenced by molecular weight. With the higher molecular weight of the solution, the fibers are prone to higher entanglement which in turn affects the efficiency of the electric field in forming a thin fiber [24].

1.2.2.2 Polymer Conductivity

The amount of electric charges on the polymer solution's surface is indicated by its conductivity, which affects whether the solution tends to form nanofibers or nano drops. Polymer solutions having poor conductivity are not suitable for electrospinning. The electrical charge transfer used in the electrospinning process occurs from the electrode to the polymer droplet located at the injection needle's tip. The conductivity may decrease with the increasing concentration of the solution. The conductivity of the solution will be substantially greater if the polymer itself exhibits polyelectrolyte-type ionic properties [25]. Inorganic ions such as NaCl or ionic organic chemicals such as pyridiniumformate palladium diacetate trialcoholbenzyl ammonium chloride can be used to increase the conductivity of the polymer solution.

1.2.2.3 Solution Viscosity

An important variable that can be used to change the dimensional properties of manufactured nanofibers is the viscosity of the solution. Each polymer has a significant band of viscosity that should be regulated for proper solution preparation. In the influence of surface tension, low viscosity results in the formation of fiber beads due to coagulation of the solution. As a result, the mechanical strength and overall surface area of the nanofibers are reduced because of this bead formation. An increase in viscosity results in an increase in the concentration of fibers over the smaller area resulting in long and smooth nanofibers. Ideally, the viscosity of above 6 wt. % is suitable for producing good quality fibers [26, 27]. On the other hand, the viscosity of more than 20 poises prevents the electrospinning process resulting in larger solution beads, an increase in distance between the beads, and larger diameter fibers formed. Therefore, the 0.5 ml/hr of fluid flow is adjusted to be proportional to the feed rate in the electrospinning process.

1.2.2.4 Surface Tension

Surface tension is attributed to the cohesive property of water molecules that resists the effect of external forces over the surface of the liquid. In contrast to surface tension, electrospinning jets stretch to increase surface area; meanwhile surface tension lessens the amount of surface area per mass. More surface tension on the solution liquid results in the congregation of the polymers, and the electrospinning process setup has to exert more power to overcome this surface resistance. Also, by adding surfactants to the solution, the surface tension in the solution is reduced in general; electrospinning solutions with viscosities of 1 to 20 poise and surface tensions of 35 to 55 days/cm are considered to be suitable for electrospinning [28]. Polymer jets are created when an electric field is applied. Polymer solutions with lower surface tension will become unstable when the charges exceed the solution's surface tension, resulting in drops as the extruded filament tends to break due to the reduced surface tension [29].

1.2.2.5 Effect of Solution Concentration

The homogeneity and evenness of the diameter of the nanowebs and nanofibers are strongly influenced by the solution concentration, which is a crucial parameter in the formation of bead-free and non-branched fibers [30]. The critical concentration range depends on the polymer and solvent properties. Electrospinnability apart from the concentration of the solution liquid also depends on the molecular weight of the polymers. Electrospinning of PA-6/formic acid solution under variable concentrations (5, 15, 25 wt. %) keeping other variables constant like the applied voltage, distance, needle gauge, needle length, and electrospinning time reveal that i) bead formations are impossible to spin at low concentration of 5 wt. %, ii) at 15 wt. %, fiber formation is affected by branching, and iii) the branching and uneven web fiber formations disappear at the concentration of 25 wt. %.

It can be seen from Figure 1.3 that higher concentrations of polymer solutions are necessary for the formation of uniform, homogenous, bead, and branch-free

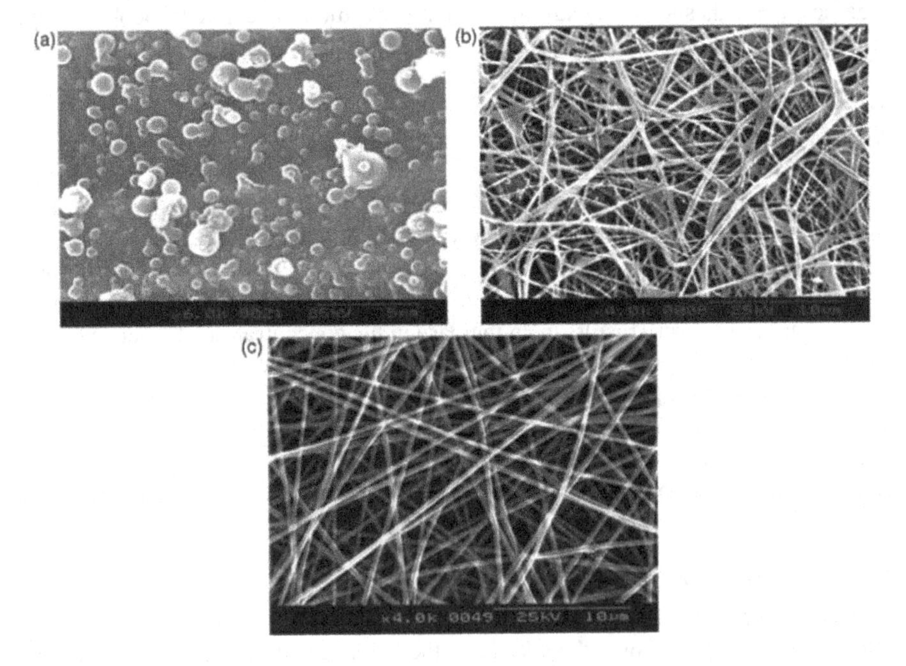

FIGURE 1.3 Electrospinning process of PA-6/formic acid solution under variable concentrations a) 5 wt. %, b) 15 wt. %, and c) 25 wt. % [31].

TABLE 1.1
Polymers and Their Suitable Solvents with Concentration [19]

S.No	Polymer	Solvent	Concentration (wt. %)
1	Polystyrene PS	Dimethyl formamide	18–35
2	Polylactic acid		15
3	Polyethylene-co-vinyl acetate		14
4	Polyacrylonitrile		15
5	Polyvinylchloride	Tetrahydrofuran/dimethylformamide	10–15
6	Polyamide	Dimethylacetamide	4
7	Poly(vinylidene fluoride)	Dimethylformamide: dimethylacetamide	20
8	Polyvinylcarbazole	Polyvinylcarbazole	7.5
9	Polyether imide	Hexafluoro-2-propanol	10
10	Polycarboate	Dimethyl formamide: tetrahydrofuran (1:1)	10

nanofibers. An increase in concentrations increases the entanglement of the fibers. Cohesion between polymer chains and solvents reduces the bead formation and breaking. Thus, stabilization of the Taylor cone during the electrospinning process takes place. Table 1.1 lists the concentrations in wt. % of some of the polymer–solvent combinations.

1.2.3 Effect of Process Parameters

The process parameters involve applied voltage, flow rate, tip to collector distance, and type of collectiveness.

1.2.3.1 Applied Voltage

DC voltage is applied during electrospinning, which has a greater influence on fiber formation. Increased applied voltage will allow the needle tip to draw more solution. In the case of electrospinning of polyacrylonitrile (PAN) fibers, a significant rise in voltage results in stretching of the solution and thinning of the fiber diameter initially but a further increase of voltage results in increasing of the fiber diameter [32, 33].

Under constant flow rate, increasing the voltage results in an unstable Taylor cone which recedes into the needle (Figure 1.4). Similarly, secondary jets are prone to occur for the combination of higher voltage and lower viscosity [35]. Critical voltage (V_c) applied to the solution results in a convex shape on the drop surface that begins the initiation of the electrospinning. Also, several studies have reported that at lower voltage i.e. below the V_c, the solution has more flight duration which enables the drop to be stretched before deposition on the collector [36–38].

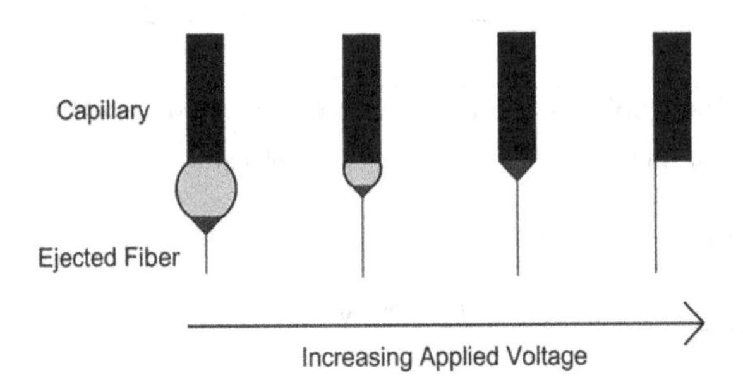

FIGURE 1.4 Taylor cone formation under varying voltages [34].

1.2.3.2 Effect of Flow Rate

At a higher flow rate, the lesser time is available for the solvent solution for stretching and drying and therefore, as more fibers aggregate, greater fiber diameter and web-like structures are formed [39]. Additionally, the electric field strength required to stretch the ejected solvent solution at a greater flow rate at constant voltage is insufficient. As the solvent solution requires more time to evaporate, a low flow rate from the needle tip is desirable [40].

A unique strategy to reduce the effects of the solution's viscosity during the process is to use limited vibration of the polymer solution [41]. The threshold value for the flow rate via the needle must therefore be determined. Variations in the flow rate affect the formation of the Taylor cone and breakage of the fiber strands occurs. It results in increasing the shear stress between the polymer liquid and the wall of the needle tip. The breaking of the flow involves more drawing forces to overcome the shear force and thus new Taylor cone is formed every time. Shamim et al. has identified the ideal flow rate in the case of Nylon-6 nanofiber. Among the tested flow rates of 0.1 ml/hr, 0.5 ml/hr, 1 ml/hr, and 1.5 ml/hr, a uniform Taylor cone was formed at 0.5 ml/hr which results in uniform fibers with narrow distribution being produced. Figure 1.5 shows the SEM images of the electrospun fiber flow at various flow rates. It can be seen that improper flow rates result in defects in fibers like a blob, splitting, or branched fibers due to less flight time and insufficient time for evaporation. Figure 1.5(b) at 0.5 ml/hr shows the uniform distribution of fibers devoid of any defects.

1.2.3.3 The Impact of the Needle Tip on the Collector Distance

In the nanofiber formation process, the distance between the needle tip and collector plays a significant role in solvent evaporation. According to several studies, the more bending and whipping that occurs during Taylor cone creation, the smaller the fiber diameter becomes as the needle tip to collector distance increases. On the other hand, evaporation of the solvent is independent of varying the distance [42,43]. The effective spinning distance between the capillary–collector is identified to be around 15–20 cm. Increasing the capillary collector distance results in a decrease in the fiber diameter [44].

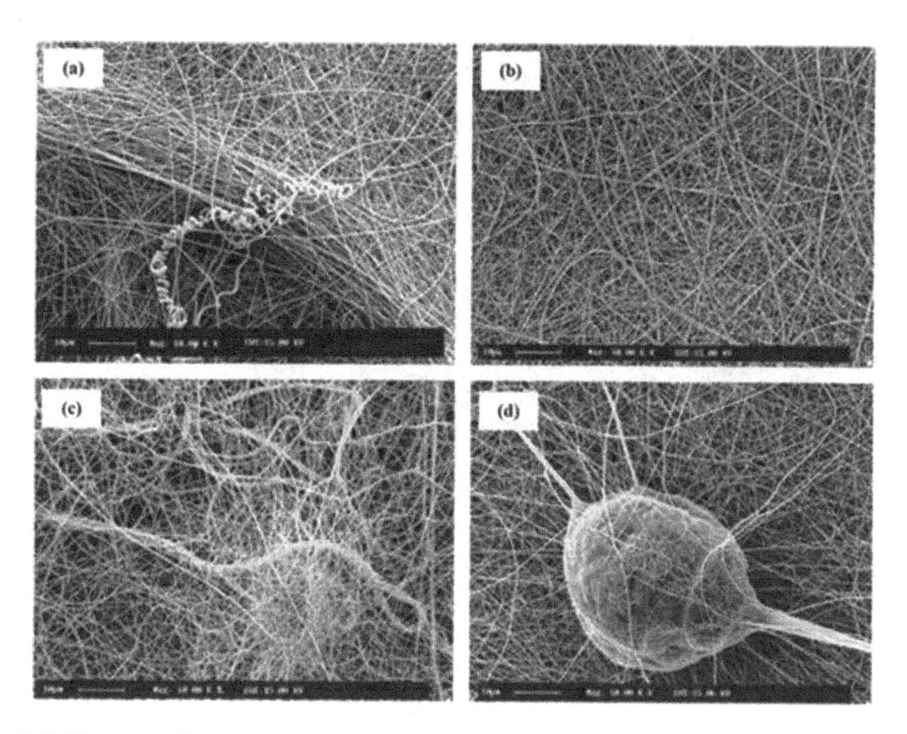

FIGURE 1.5 Effect of variable flow rates on electrospun nylon-6 nanofibers: a) 0.1 ml/hr, b) 0.5 ml/hr, c) 1 ml/hr, and d) 1.5 ml/hr [40].

1.2.4 Effect of Ambient Conditions

The morphology of the fiber is influenced by ambient conditions such as temperature, humidity, and gas composition during electrospinning. At higher temperatures, the viscosity of the fluid is decreased and as a result, fibers with thin diameters are produced in the case of polyacrylonitrile/dimethylformamide produced. A temperature-dependent effect is also observed on the properties of viscosity, surface tension, and solution conductivity. Fibers of 65–85 nm were formed under the higher temperature of 88.7 °C. On the other hand, at room temperature, the formation of larger fiber diameter of 190–240 nm is developed. At lower temperatures, the evaporation rate is reduced exponentially, thus taking more time for the solvent to evaporate. At higher temperatures, the rigidity of the polymer chains is affected which results in free movements. This had a direct effect on viscosity and greater stretching of fibers occurs. It is the reason for the thinner fiber diameters during electrospinning at elevated temperatures. Generally, the appropriate process temperature for MOFs depends on the melting point of the polymer which dictates the viscosity and the subsequent stretching of the fiber during the electrospinning process.

Another important factor in the electrospinning process is humidity where the increase in humidity results in bead formation during electrospinning. Poly(vinylpyrrolidone) ethanol 10 wt. % solutions under varying relative humidity

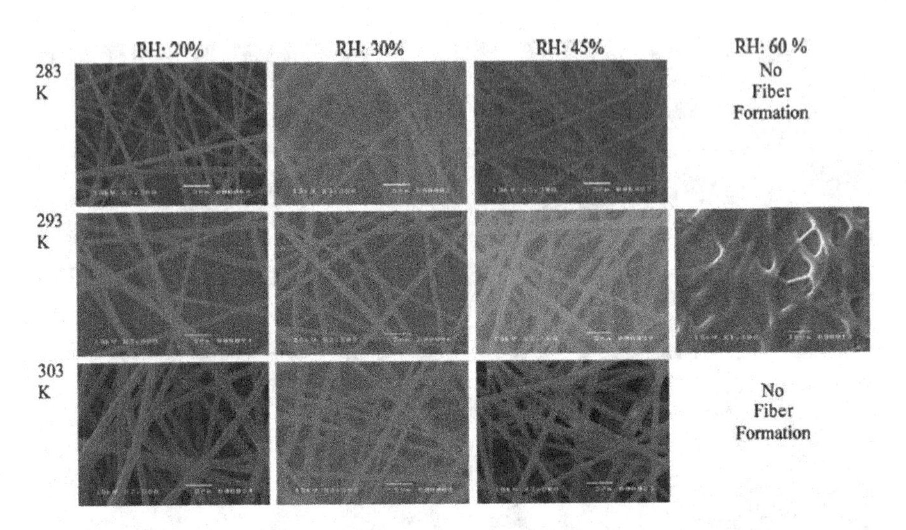

FIGURE 1.6 Effect of temperature and relative humidity in the fiber formation during electrospinning of poly(vinylpyrrolidone) ethanol 10 wt. % solutions [45].

conditions of 20%, 30%, 45%, and 60% were undergone electrospinning. It was found that thinning of fibers occurs at increasing humidity and no well-defined fiber formation takes place at around 60% humidity due to higher absorption of water at high humidity. In addition, higher humidity prevents the solvent solution from being evaporated properly, resulting in homogenous fibers (Figure 1.6).

There are diverse processes, including wet, dry, melt, and gel spinning, for developing fibers from synthetic polymers [46]. In wet spinning, when a polymer solution is extruded from a spinneret submerged in the polymer solution of a chemical bath, the polymer is precipitated out due to dilution or chemical reaction, resulting in the formation of fibers. On the other hand, in dry spinning, fibers are produced as a result of solvent absorption aided by hot air streams, and the polymer solution is extruded into the air. A polymer melt is extruded from a spinneret and then cooled to produce fibers by melt spinning. In gel spinning, polymers are spun in a gel state, dried outside, and cooled in a liquid bath to generate fibers with high mechanical strength or other characteristics. When electrospun fibers have diameters less than 500 nm, they are frequently referred to as nanofibers [47].

1.3 SYNTHESIS OF MOF NANOFIBERS

MOFs/electrospun nanofibers are the 3D porous framework that facilitates various exclusive applications in wider fields like: i) energy conservation and storage, ii) gas storage and separation, iii) air pollutant filtration, iv) water treatment, and v) heterogeneous catalysis. Electrospinning can thus create new materials with multifunctional properties with tailor-made architectures, and nanoscale structures with variable chemical functionality [48, 49]. Some of the prominent MOFs that have been synthesized in large amounts are MOF-5 [52], MOF-74, HKUST-1 (Basolite C 300) [51], Basolite F300 [49], and ZIF-8 (Basolite Z1200) [50–52].

The manufacture of MOFs as well as their intrinsic features provide difficulties for their use in industrial applications. Polymer-MOFs suffer from dispersion and compatibility issues due to their different physical and chemical properties, which is a hurdle in some applications. High porosity, thermal and chemical stability, as well sustainability, are demands that must be met by MOFs. Because of the low interfacial compatibility, there would be spontaneous voids between MOF particles and polymers as a result of molecular particle aggregation [53]. To find novel MOFs, laboratory syntheses that were already established must also be modified. As a result, the following factors must be considered: synthesis circumstances (low temperature, ambient pressure), availability and cost of the raw components, workup technique, activation process, etc. In recent studies, electrospinning is an efficient method for developing hybrid materials with multiscale porosity. Ostermann et al. [54] synthesized MOF-NFs the polyvinylpyrrolidone (PVP) and zeolitic imidazolate framework-8 (ZIF-8) in 2011.

1.3.1 Direct Spinning Process

During direct electrospinning, MOF particles are electrospun together with polymer solutions into composite nanofibers. The generated MOFs are then encapsulated in a matrix of polymeric nanofibers. In most cases, the polymer matrix covers the pores of the MOF particles which restricts the MOF's accessibility and activity in most applications. The synthesis of MOF-nanofibers using a direct spinning process involves three stages: i) synthesis of MOF powder particles, ii) preparation of MOF-polymer slurry, and iii) directing the slurry into the syringe for electrospinning. Polymer MOFs-nanofibers of HKUST-1and MIL-100 (Fe) is a typical example of the direct spinning process [55]. Figure 1.7 represents the variation of nanofibers like PAN(polyacrylonitrile) and PS (polystyrene) with MOFs (ZIF-8, UiO-66-NH_2, MOF-199, Mg-MOF-74) and their associated SEM images.

López-Maya et al. developed a novel method for self-cleaning air filtration materials using an electrospinning process by doping MOF nanoparticles into silk fibroin fabric fibers. The MOFs used in this process are UiO-66 (UiO for the University of Oslo) with the composition of (Zr_6O_4 $(OH)_4$ $(bdc)_6$] (bdc = benzene-1,4-dicarboxylate) system [57]. Dai et al. [58] attempted the electrospinning of the PLA/ZIF-8 (Polylactic acid/Zeoliticimidazolate frameworks) nanofibers to prepare the nanocomposite electrospun membrane. However, the probability of the polymer PLA covering the ZIF-8 surface is high and has a negative effect on the properties of the micropores. The ZIF-8 was dispersed in the PLA solution using highly volatile CH_2Cl_2 as the solvent. In PLA/ZIF-8 nanofibers created using CH_2Cl_2, rapid evaporation of ZIF-8 at a substantial number of microparticles with high pore area ratios are produced. Modifying the loading fraction can also be used to regulate the nanofibers' diameter of MOF-ZIF-8. The resultant composite material has improved mechanical property and increased wettability [59–61]. Thus nanoparticles like ZIF-8 can be applied to minimize surface energy, which also reduces nanofiber

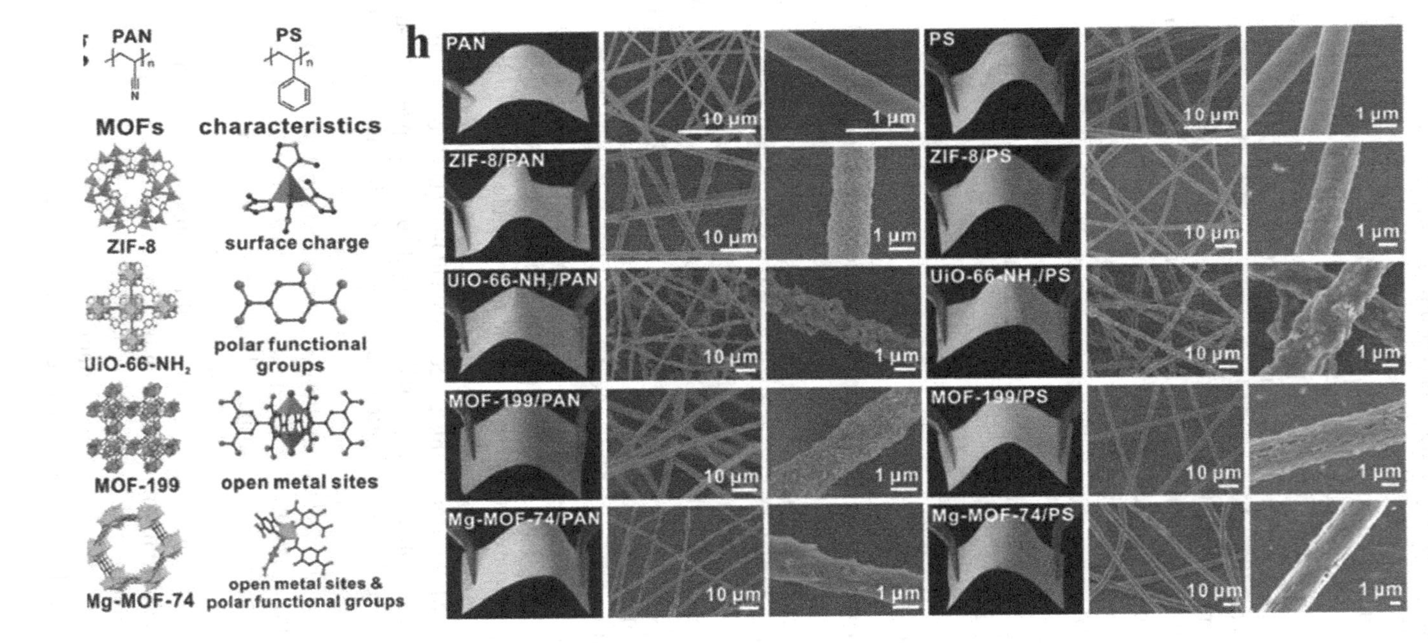

FIGURE 1.7 SEM images of MOFs and nanofibers combinations in the electrospinning process [56].

FIGURE 1.8 Crystal structure of ZIF-8 MOF and the composite of PLA/ZIF-8 [58].

diameter. As a result, the average pore size, pore area fraction, and pore perimeter all increase dramatically as shown in Figure 1.8 [58].

Polyethylene oxide and polyamide are used to form porous MOF nanofibers that are independent of the distribution of nanoparticles. The mixture is then washed with methanol to remove PEO. Thus, 50% of PEO is washed which enables an increase in N_2 adsorption and increase in pore channels in the polymer [62]. A drawback of this method is that MOF particles do not disperse well in polymers, so MOF pores do not open, or mechanical properties are reduced due to the MOF loading, such as high brittleness.

1.3.2 Surface Decoration of Nanofibers with MOF Particles

The fundamental benefit of this approach is that the pristine MOF crystals' surfaces and interior pores are completely accessible because they are not shielded by the polymer. It involves a two-step process where in the first step, polymer nanofibers are developed using electrospinning and in the following step, MOF particles are grown at the open porosity of the nanofibers layer [63]. Surface decoration is different from direct spinning where there is a controlled thickness of the deposition of MOF layers on the nanofiber surface. Several methods are used to deposit the material including layer-by-layer assembly, sol-gel processes, physical deposition, microwave heating, etc. The general methods for fabricating MOF fibers for surface decoration are given as follows: i) in situ growth and ii) phase transformation [64].

1.3.2.1 In Situ Growth MOFs on Polymer Nanofibers

The MOF-polymer nanofiber composite is formed by dissolved MOF precursors crystallizing on the surface of polymer nanofibers. On the nanofibers, MOF crystals often form, grow, and inter-grow all at the same time. MOF nanofiber structure can consist of either supported layers or independent layers that are grown in situ on the nanofiber. Compared with supported MOFs, independent nanofiber layers are more flexible and have a larger surface area. MOF growth on nanofibers involves the following: i) a seed-assisted growth technique involves embedding MOF seeds inside the polymer during electrospinning, initiating MOF growth by dipping the

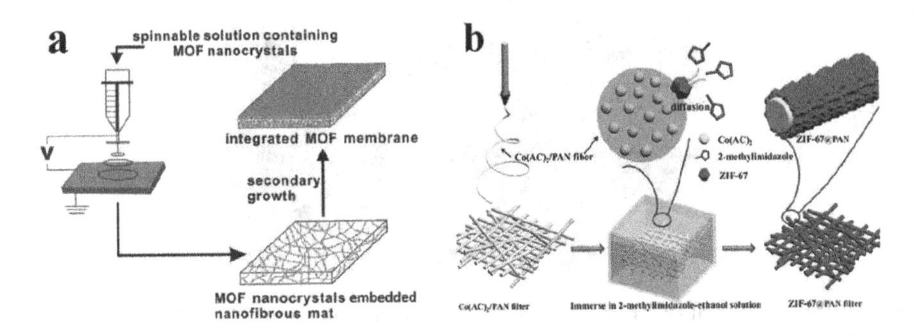

FIGURE 1.9 In situ growth of MOF crystals on polymer: a) growth of MOF seeds on polymer nanofibers [65] and b) incorporation of metals ions into polymer nanofibers [66].

polymer into a precursor solution for MOFs. ii) An alternative approach is to add MOF precursors like metals, organic ligands, etc. to polymers during electrospinning. MOF growth is initiated by immersion of MOF precursor-polymer nanofiber into another precursor solution. iii) The process of hydrothermal growth involves immersing polymer into a MOF precursor solution and allowing MOF crystals to grow directly on the surface.

Wu et al. attempted seeding of MOFs like HKUST-1, ZIF-8, MIL-101(Fe), and Zn_2(bpdc) 2(bpee) in nanofibers (Figure 1.9a) and demonstrated the MOF growth in the surface and inside of the nanofibers. In the second method, electrospinning involves a mixture of polymers and MOFs (metal salts) in a suitable solvent to produce a nanofiber. To initiate the local formation of a MOF, a nanofiber is dipped into a solution of an organic linker compatible with the fiber. An electrospun nanofiber of MOF($Co(AC)_2$) and polymer (PAN) is submerged in a mixture of ethanol and the organic linker methylimidazole (HmIM). It results in the formation of ZIF-67 nanocrystals with homogenous coating on the surfaces of the PAN nanofibers (Figure 1.9b). Liu et al. [67] reported that hybrid nanofibers like PVA/PAA/SiO_2 hybrid nanofiber surfaces exhibit excellent affinity for the in situ growth of various types of MOFs (HKUST-1, MIL-53(Al), ZIF-8, and MIL-88B(Fe)). The affinity is attributed to the availability of active surface hydroxyl and carboxyl groups for anchoring the MOF precursors.

To facilitate the formation of MOF structures, Zhao et al. [68] used the atomic layer deposition (ALD) technique to cover polymer nanofibers with metal oxides. A very thin layer of TiO_2 was applied to polyamide-6 nanofibers to improve the next heterogeneous nucleation of Zr-MOFs on the surface of TiO_2 (Figure 1.10).

1.3.2.2 Layer-by-layer

Another feasible technique for forming nanolayers on nanofiber surfaces is a layer-by-layer (LBL) assembly. LBL technique is a multilayer technique and is a time-consuming process. For example, crosslinked poly(acrylic acid)/PVA (PAA/PVA) nanofibers were employed with MOFs HKUST-1 using an LBL assembly technique [69].

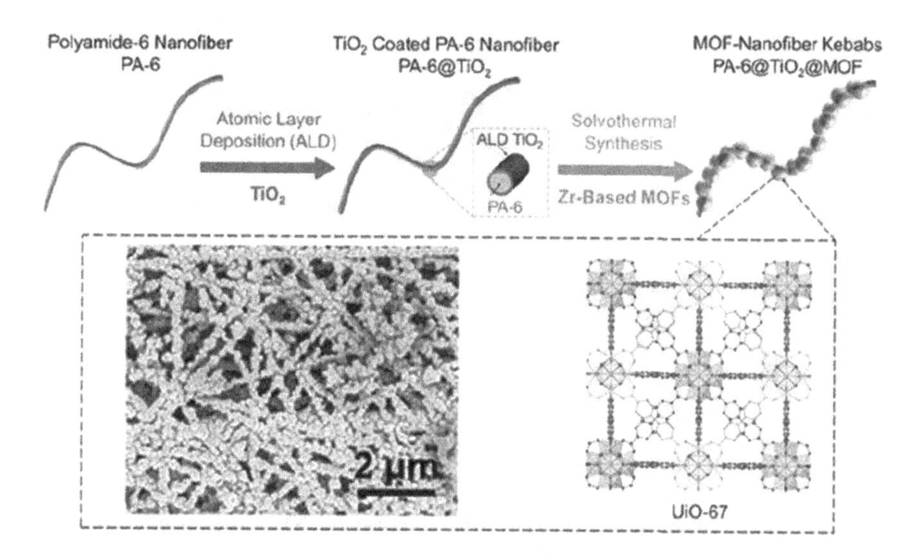

FIGURE 1.10 SEM photos of the resulting PA-6@TiO_2@UiO-66-NH_2 nanofibers show the core-shell structure with an ultrathin ALD coating of TiO_2 on PA-6 nanofibers to permit the formation of Zr-MOF [68].

1.3.2.3 Solvothermal Method

It is another method of creating MOFs. In this method, MOFs crystallization is developed by dissolving polymer nanofibers like PVP, PAN, or PVA in a solvent of dimethylformamide (DMF) solution. Sometimes the crystallization growth of MOF demands higher processing temperatures. Solvothermal growth of MOFs (UiO-66-NH_2) took place on PAN nanofibers in the acetone solvent. Acetone is preferred over DMF owing to its stability [70].

1.3.2.4 Phase Transformation

This is another method for surface decoration of polymer with MOFs. It begins with a precursor (often a metal oxide or hydroxide) that is integrated into the surface of the polymer nanofiber. In a subsequent phase, the precursor is transformed into the MOF structure during in situ growth. The phase transformation is made possible by a kinetic match between the precursor and the targeted MOF structures.

In Liang et al. [71] through phase conversion of metal oxide nanofibers, an improved hydro/solvothermal synthesis method was described for creating self-sustaining flexible MOF nanofiber mats (Figure 1.11). Electrospinning is used to create nanofibers of the metal oxide, which are then subjected to calcination. In hydrothermal conditions, conversion of the metal oxides on the nanofibers into their corresponding MOF nanofiber structures occurs through phase transformation, producing uniform growth of MOF crystallites on the nanofiber surface. The surface morphology of MOF on the polymer nanofibers relies on the reaction type and MOF

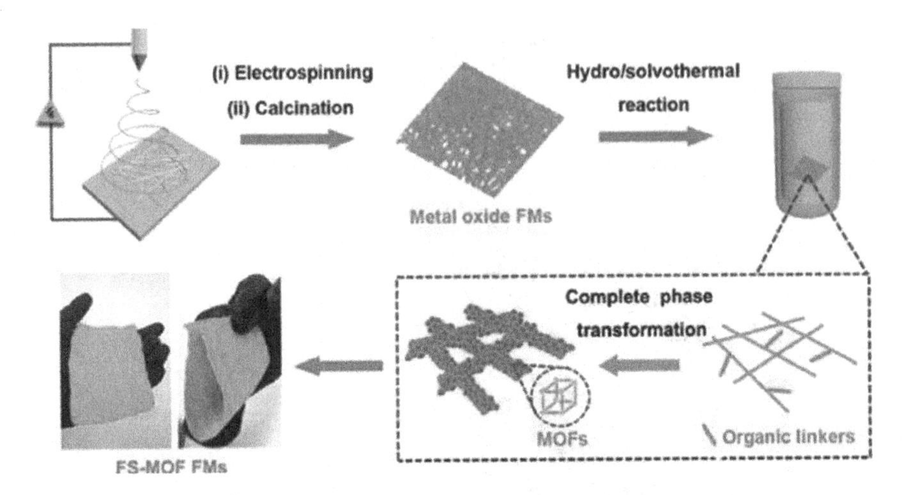

FIGURE 1.11 Solvothermal conversion of metal oxide nanofibers into self-supporting nanofiber mats of MOF in stages [71].

properties. Figure 1.12 displays the different types of phase-converted MOF nanofibers made from a variety of metal oxides Co_3O_4, Fe_2O_3 ZnO, ZrO_2, and CuO, to ZIF-8, ZIF-67, UiO-66, and MIL-88B(Fe), HKUST-1 respectively and the matching SEM photos of the microstructures.

1.4 APPLICATIONS FOR MOF NANOFIBERS

Numerous uses for nanofibers exist, including those in purification, biotechnology, drug delivery, wound healing, micro-electronics, ecological safety, and energy harvesting owing to their substantial surface-to-volume ratio, surface functionalities in addition to superior mechanical performance [72]. Some of the areas considered for the discussion are filters, energy, sensors, biotech, and smart materials.

1.4.1 Air Filtration

The effects of air pollution on the ecosystem and public health are long-lasting. The main elements of air pollutants include ozone, COx, NOx, SO_2, and PMs in liquid or solid form. An airborne combination of microscopic particles and liquid droplets known as particulate matter (PM) is made up of both inorganic substances in the form of dust, smoke, soot, pollen, and other hazardous liquid droplets. Particularly, PM 2.5 and PM 10 (that is, particles with aerodynamic equivalent size less than 2.5 and 10 m, respectively) can easily infiltrate the human body, leading to lung cancer and pneumonia, congestive heart failure, and coronary artery blockage, among other health problems [91].

There are three major problems with conventional air filters: i) poor filtration performance, particularly for PM 1.0; (ii) reduced air pressure over the filter by a

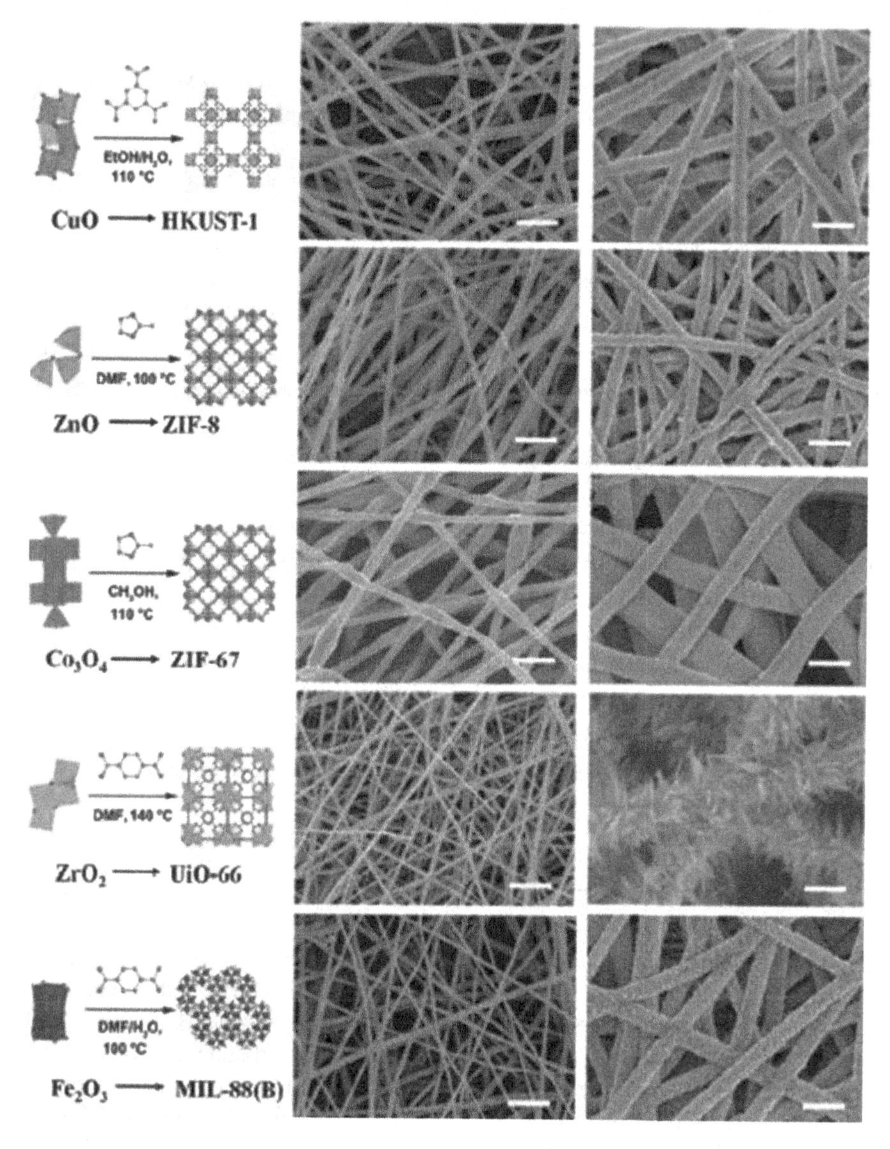

FIGURE 1.12 Phase conversion of different types of MOF nanofibers and their respective SEM images [71].

significant amount; and (iii) unsafe conditions like fire and dust explosion. An ideal filter requires high filtration efficiency and little pressure loss, which are considered typical requirements for high-performance air filters. The electrospun fiber efficiency also depends on the flexibility and pore diameter size of the fiber. An effective filtration system displays high filtration performance with the least air resistance [92]. Also, electrospun fibers are effective in filtering up to 0.3 μm of

TABLE 1.2
Applications of Electrospun Polymer Derivatives and Their Solvent

S. No.	Electrospun polymer derivatives/ solvent	Applications	Ref.
1.	Polyurethane/dimethylformamide	Filter and protective cloths	[73, 74]
2.	Polycarbonate/dimethylformamide and tetrahydrofuran	Filter, sensor, protective cloths	[75]
3.	Polylactic acid/dimethylformamide	Drug delivery system, filter, sensor	[76]
4.	Polyethylene oxide/distilled water/ethanol	Microelectronics, filter	[77]
5.	Polyaniline/chloroform	Conductive fibers	[78]
6.	Polystyrene/tetrahydrofuran, toluene	Enzymatic biotransformation, catalyst, filter	[79, 80]
7.	Polyamide/dimethylacetamide	Glass fiber filter	[81]
8.	Polyvinyl phenol/tetrahydrofuran	Antimicrobial agent	[82]
9.	Cellulose acetate/acetone	Membrane, fabrics	[83]
10.	Polyvinylidene fluoride/ dimethylformamide	Fiber ribbon	[84]
11.	Chitin/chitosan/dimethylacetamide	Wound dressing	[85]
12.	Polyacrylic acid/water/ethanol	Gas sensor	[86, 87]
13.	Polyacrylonitrile/dimethylformamide	Hydrogen storage	[88, 89]
14.	Polyacrylonitrile/nanofillers/ dimethylformamide	Hydrogen storage, photovoltaic applications	[90]
15.	Polycaprolactone/chloroform	Scaffold and tissue engineering, packaging	[91]

adverse tiny contaminants. Despite their lack of durability and mechanical strength, electrospun membranes have been used as a layer inside filters that enhances filtration in general without a substantial pressure change drop [93]. Another study showed that while layers of thicker fibers are stronger and support the filter, layers of thinner fibers promote high flux and efficient filtration [94]. The nanofibers membranes are used for reverse osmosis of 7 MPa along with high pressure.

Su et al. worked (Figure 1.13) by coating cellulose fibrous mats with a ZIF nanofiller (CFs@ZIF-8). Nanocrystals ZIF-8 enhance the filter's interactions with PMs by increasing its specific surface area. ZIF-8 nanocrystals offer plenty of cavities for gas adsorption. The effectiveness of CFs@ZIF-8's filtration has achieved an extremely high degree of 99.99% and nitrogen gas absorption of CFs@ZIF-8 is 200 times more than that of the conventional cellulose-based filter (Figure 1.14). There are many open metal and organic adsorption sites on ZIF-8 nanocrystal, which makes it extremely capable of gas adsorption. Also, a ZIF-8 nanocrystal facilitates improved porous structure and thus porous volume and surface area to the fiber network.

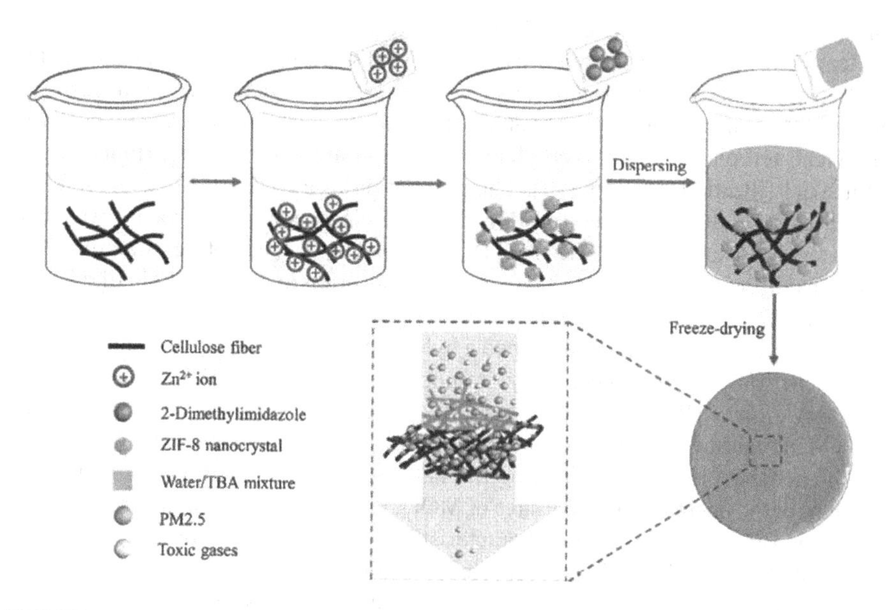

FIGURE 1.13 Schematic representation of absorption of PMs and toxic gases by CFs@ZIF-8's [95].

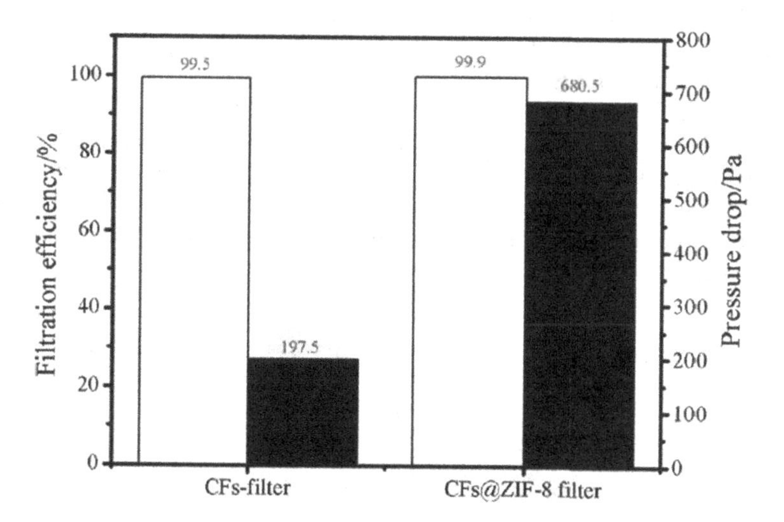

FIGURE 1.14 Filtration efficiency of absorption of PMs and toxic gases by CFs@ZIF-8's compared with CFs-filter [96].

1.4.2 Water Treatment

MOF nanofilters are effectively used in the removal of contaminants like heavy metals by adsorption. Before the standard reverse osmosis process, wastewater is first treated with nanofiltration. Adsorption is thought to be one of the most promising methods of waste removal among the other techniques due to its high efficacy, low cost, and adaptability. Since MOFs are naturally nano-distributed and enhance active functional sites, they have been widely used as adsorbents to remove contaminants from water. The fibrous systems polypropylene, poly (ethylene terephthalate), polyethylene, and nylon are of great importance to the industrial filtration industry [97]. The production of PVA/La-TBC, PVA/Sr-TBC, and PVA/La-TBC nanofibers produced by the electrospinning process was also reported by Shooto et al., where TBC is benzene 1,2,4,5-tetracarboxylic. The modified hybrid nanofibers are effective at adsorption of heavy metals like Pb(II) from tainted water [98].

Guo et al., in their study, developed a hybrid membrane comprising stable MOFs by electrospinning to successfully immobilize a Zr-based MOF (UiO-66) into polyacrylonitrile (PAN) substrate that is water-stable. The hybrid membranes with 10% MOF loading are highly effective in the adsorption of arsenic, and MOF loading higher than 15% results in a significant decrease in the adsorption efficiency because of the decrease of active sites of MOFs/PAN. The fibers can be used without sacrificing the MOFs' inherent characteristics. After use, the hybrid membrane is simply collected and sustainably recycled (Figure 1.15).

Another important factor of adsorption is pH which may harm the structures and the charges of the target. The UiO-66/PAN exhibited excellent arsenate adsorption efficiency across a broad pH range of 2–6 but decreases with high pH values above 7. It is found that the maximum adsorption efficiency is around the neutral pH conditions. Similarly, Fe-MOF/PAN membranes are prepared by Efome et al. for the removal of lead and mercury ions from an aqueous solution. The MOF-polymer nanofiber membrane comprises iron-based Materials of Institut Lavoisier (MIL100-Fe) and zirconium-based MOF (MOF-80) structures embedded in polyacrylonitrile (PAN) and polyvinylidene fluoride polymer is effective in removing heavy metals like Pb(II) by competitive ion exchange. The ions of heavy metals are removed by electrostatic interactions between polymer or MOF and binding it with the open metal sites [100].

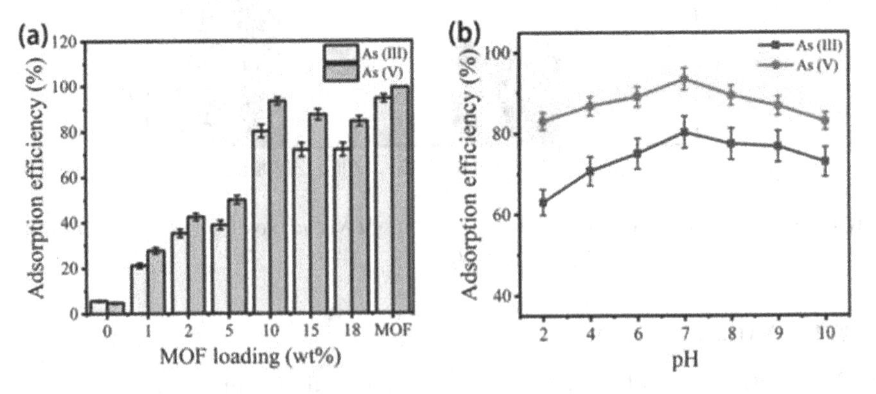

FIGURE 1.15 Effect of a) MOF loading in adsorption efficiency over arsenic removal; b) pH in adsorption efficiency over As(III) and As samples [99].

1.4.3 Energy Applications

The use of electrospinning in the creation of porous fiber mats for electronic components has received more attention. The fiber mats are used extensively as g separators and electrodes. It has also found its place in tools like supercapacitors and batteries. Electrospun fiber mats have the potential to enhance battery performance, capacitor energy density, and fuel cell and solar cell efficiency.

1.4.3.1 Battery

The performance of the battery is directly impacted by the physical differences between electrode materials. Two factors are mostly responsible for poor battery performance: i) its very poor intrinsic electrical conductivity; and ii) the material's volume expansion during charging and discharging, which results in spalling and comminution of the particles [101]. Rechargeable lithium-ion batteries are viewed as the most practical response to the rising need for high-energy-density electro-chemical power sources among the many energy storage options because of their high energy density, extended cycle lifetimes, and flexible design. This demand has encouraged research towards the creation of high-capacity alternative electrode materials with enhanced safety, lower carbon footprints, longer life cycles, and lower costs.

Ji et al. [102] created a range of composite materials like copper/carbon composite and polyacrylonitrile/polypyrrole composite nanofibers using electrospinning for improving the performance of the lithium-ion batteries. High reversible capacity, increased cycle performance, and the ability to keep the fibrous shape after 50 cycles were all obtained by this specific anode material. The unusual surface characteristics and distinctive structural features of these composite nanofibers, which magnify surface area and extensive intermingling between electrode and electrolyte phases over small length scales, are responsible for their improved electrochemical performance. This results in fast kinetics and short pathways for both Li-ions and electrons.

Recently, ultrafine fibers for polymer electrolyte membranes have been developed from polyvinylidene fluoride (PVDF), polyvinylidene fluoride-co-hexa fluoropropylene, and polyacrylonitrile. However, due to polymer degradation and leakage, electrospun fibers made from pure polymers have a limited ability to stabilize the battery at high discharge rates. Similarly, silica (SiO_2)/PAN composite electrolyte membranes where the processing parameters were optimized for smaller pore diameter, high porosity, and large surface area. Numerous studies have concentrated on the production of metal oxide/carbon nanofiber composites as electrode materials for batteries using the electrospinning of polymer and MOF nanoparticles together.

1.4.3.2 Fuel Cells

Two devices, namely fuel cells and batteries, produce electricity using similar principles, but one is a free standing device, while the other requires a continuous supply of chemicals. Typically, electrolyte type or operating temperature is used to categorize fuel cells. Fuel cells can be divided into three categories based on their operating temperatures: solid oxide fuel cells (700–1,000 °C), polymer exchange membrane fuel cells (60–80 °C), and molten carbonate fuel cells (600 °C). The fuel cell employs membranes for effective proton exchange that improves the efficiency. Thus, proton exchange membranes (PEMs) for fuel cells are in considerable research interest due to their high efficiency in energy conversion and low operating temperature. Efforts are being focused on creating membranes with improved proton conductivity, low gas penetration, and high thermal stability. The PEMs fuel cells are made up of seal rings, collector plates, and membrane electrode assemblies (MEAs). Fuel cell performance is directly influenced by MEAs, which convert chemical energy into electrical energy. An MEA usually contains a PEM, gas diffusion layer, and catalytic layer. Thus, composite ion-exchange membranes for fuel cells like Nafion/PVA [103], and SiO_2/PEI-coated PI are nonwoven [104]. Commercially available PEMs are Nafion membrane from DuPont and other alternative sources of PEMs like sulfonated poly (aryl ether) have also proven to be effective in the conductivity [105].

Quaniyi et al. analyzed PEM by doping with various MOFs like UiO, ZIF, and MIL series for improved proton conductivity. UiO-series MOFs especially the UiO-66 exhibit excellent thermal stability under high temperatures, and chemical stability when working with various solvents like water and chemicals. They also exhibit high working capacity and high regenerability [106]. The large surface area of the MOFs facilitates more proton migration for conductivity, and it is found that the addition of UIO-66 to PEMs has improved the performance of conduction by 10–15% [107]. According to Donnadio et al., the Nafion matrix series has been improved by 30% by adding UiO-66 as filler material [108]. Some researchers [109–112] have increased the proton conductivity to the order of four by associating UiO-66 MOF with sulphanic acid groups (SO_xH). A sulphanic acid group provides more proton transfer pathways and improves the conduction sites by increasing proton transfer pathways.

MIL-series have distinctive benefits of porosity, and thermal and water stability and can add different functional sites within the channels or frameworks to tightly control how subjects and objects interact. Incorporating MIL-series MOFs will undoubtedly increase the composite membranes' proton conductivity. By hydrolyzing one of the several coordinatively unsaturated metal sites in MIL-101, an abundant -OH group can be produced. In the proton exchange membrane, these -OH groups can create hydrogen-bond networks that facilitate proton conduction. Li et al. [113] have increased the proton transition channels by adding the MOF with the sulphanic acid group increasing the proton conductivity by 2 to 3 times.

As one of the subclasses of MOFs, zeolite imidazolate frameworks (ZIF) consist of four imidazolic acid rings coordinated to tetrahedral metal cations (CO^{2+} and Zn^{2+}). The chemical formula for ZIF is $Zn_2(C_2O_4)(C_2N_4H_3)_2(H_2O)_{0.5}$ (ZCCH). ZIF exhibits excellent chemical stability in aqueous solutions containing water, organic solvents, and alkali solutions. Microporosity and cavities with many proton carriers enhance the conductivity of PEM. Similarly, PBI@ZIF-8, PBI@ZIF-67, and PBI@ZIF-mix have successfully been created as hybrid membranes using ZIF MOFs [114]. At high temperatures, ZIF-8 added to polyvinylphosphonic acid (PVPA) increased proton conductivity 3 times that of pure PVPA membranes [115].

1.4.4 Biomedical Applications

Medical and drug-related research can benefit from electrospinning since it is simple, inexpensive, and provides a wide range of material options, as well as high loading capacities and encapsulation efficiencies. The general requirements for materials used in drug delivery systems must, among other things, have mass transfer qualities, be biodegradable, allow for drug loading, and be biocompatible. Antibiotics, cancer treatments, and biomolecules including DNA, RNA, and proteins are just a few examples of the drugs that can be incorporated into polymeric nanofibers [115]. In the polymeric solution for electrospinning, there are several ways to load the medication. They are blending, surface modification, emulsion, multidrug delivery, and multilayer coating. Blending involves the dispersion of drugs directly into the polymer and the resulting solution is subsequently electrospinning. Some of the factors to be considered in this method for efficient drug delivery are compatibility of the physico-chemical properties of the drug and the solution, distribution of the drug in the fiber solution, and kinetic of the drug during release and delivery (Figure 1.16a).

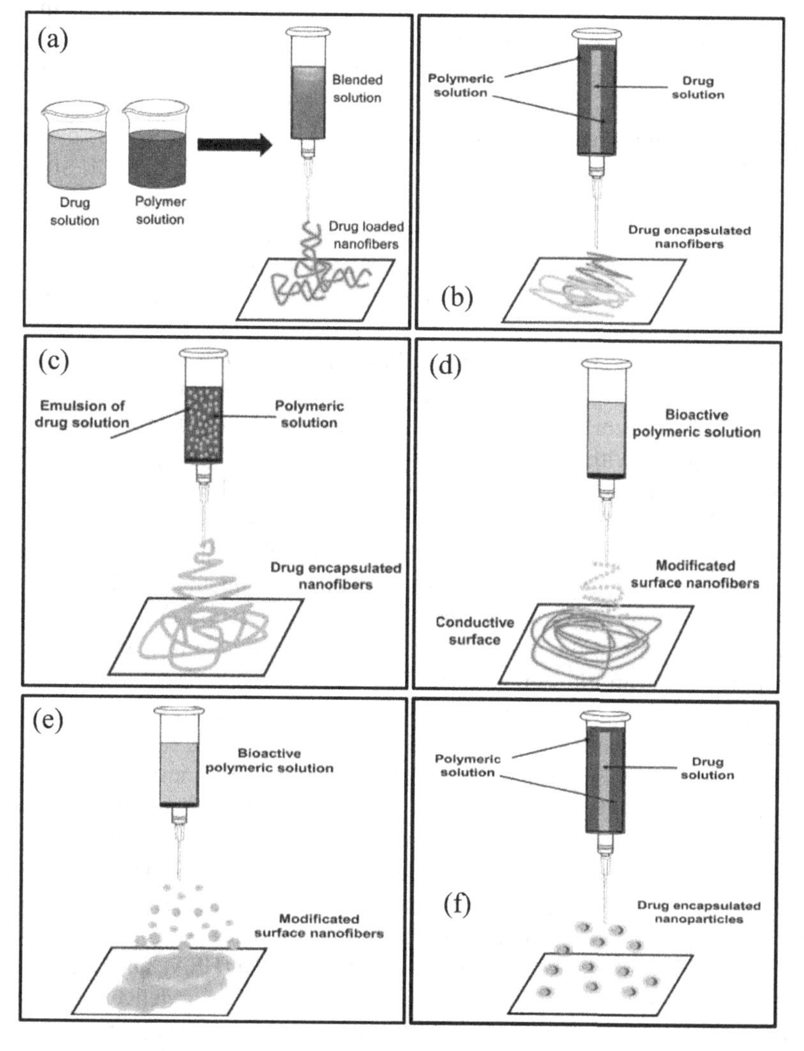

FIGURE 1.16 Drug incorporation techniques using electrospinning [116].

The blending process is more suitable for hydrophilic polymers like PEG and PVA, which are used for dissolving hydrophilic drugs such as doxorubicin. On the other hand, hydrophobic drugs like polyester have better compatibility with hydrophilic drugs like gelatin, paclitaxet, and rifampicin [117]. The drawback of this method is that they work well for only a single phase and it is not possible to produce fibers with a core-shell structure. Coaxial electrospinning overcomes the drawbacks of blended electrospinning, and this method facilitates high surface area and 3D networks with the sustained and regulated release of the drugs that are embedded in the protective fiber core-shell structures as shown in Figure 1.16(b). The core-shell structure protects the drugs from the external environment, and they are perfectly suitable for loading proteins, antibodies, and other biological agents.

Emulsion electrospinning enables direct encapsulation of hydrophilic or hydrophobic chemicals into core-shell fibers possible using the simultaneous spinning of two immiscible that creates core-shell nanofibers (Figure 1.16c). In this method, either hydrophilic compounds or hydrophobic compounds are used for water-in-oil or oil-in-water emulsions. Sanchez et al. [118] fabricated core-shell nanofibers by loading Lidocaine hydrochloride (LidHCl) into a polymer matrix (PVA) using the emulsion electrospinning technique. PVA in the aqueous phase enables controlled phase delivery of the drug into the system by encapsulating the LidHCl. Similarly, ultrafine nanofibers produced using poly(ethylene glycol)-poly(l-lactic acid) matrix in the oil phase is used to dissolve anticancer medications (doxorubicin hydrochloride), in aqueous solution (Figure 1.16d). The tests on mice glioma cells (C6 line cells) prove that the medication doxorubicin could be released from the nanofiber without losing its toxicity [119].

Surface modification electrospinning aims at modifying the surface of the nanofibers to imbibe desirable functions like hydrophilic or hydrophobic to the original polymer matrix (Figure 1.16e). Ladan et al. investigated the effect of coating polyethersulfone (PES) with polyvinylidene fluoride (PVDF) hydrophobic nanofibers for water filtration [120]. Similarly, Park et al. developed a nanofibrous scaffold by chemically modifying the surfaces of electrospun PGA, PLLA, and PLGA with oxygen plasma treatment and in situ grafting of hydrophilic acrylic acid. The resulting scaffold shows significant improvement in cell attachment [121]. Drug delivery using electrospray involves the utilization of a strong electric field between a coaxial capillary needle and the ground. Compared to electrospinning, electrospraying uses a lower concentration of polymer. The electrodynamic forces increase with the increase in the applied electric field, and it results in the stretching of the liquid to form the Taylor cone at the threshold limit, it destabilizes the jet, and the liquid breaks into droplets and initiates the electrospraying phenomenon which favors the formation of nanoparticles and microparticles. These electrosprayed nanoparticles are suitable for biomedical applications owing to dimensionless nature, ease of reproduction, higher encapsulation efficiency, and scalability. It has been reported that the therapy of retinoblastoma and other tumors can be achieved by administering cyclophosphamide-loaded natural gliadin nanoparticles [122]. One of the most prevalent plasma proteins is called human serum albumin (HSA), and due to its low immunogenicity, it has received substantial research as a drug delivery vehicle. At physiological pH (7.4), albumin has a high solubility in water, making it a desirable candidate to transport a wide range of medicines [123].

Coaxial electrospray (Figure 1.16f) works on a similar principle to conventional electrospray, but it enables the deposition of multi-layer particles with diameters ranging from 10 to 100 m. Here the jet on the threshold limit breaks into multi-layer droplets and ensures nano-encapsulation of the drugs to be carried for vital medical applications. Another water-soluble protein called sericin is non-immunogenic and biocompatible and is widely used as a coating material in drug delivery. Electrosprayed nanoparticles of sericin proteins are used for the delivery of drugs for antioxidants and the treatment of cancerous tumors. Polymer solutions of PCL, PLA, PLGA, PMMA, and polyethylene are found to be useful in the fabrication of hardshell multilayer nanoparticles [124]. The important process parameters that are to be considered for effective electrospraying are voltage, viscosity, flow rate, and tip to collector distance [125].

1.4.5 Application in Sensors

Sensors are known for their shorter time responses, small hysteresis, higher sensing properties, higher stability, and reproducibility. The performance of the sensors can be increased by effective control over the composition morphology and surface area of the sensing materials. Electrospinning of composite nanofibers provides wider opportunities in sensor applications owing to its variable morphologies, inherent large surface area, and high porosity. When a fiber's diameter decreases from a micrometer to a nanometer, the electrospun nanofibrous membranes' specific surface area rises in tandem. This creates a huge number of regions and channels that strengthen the interaction and boost sensitivity. Many types of research have been carried out on the effective use of conducting polymers like polyaniline, polypyrrole, and polythiophene in the fabrication of sensor materials.

By suspension polymerization aniline with electrospun Mn_3O_4/TiO_2 nanofibers, Li et al. have synthesized PANI/TiO_2 composite nanofibers. In this case, the sensitiveness of PANI increases with the addition of TiO_2. PANI is highly insoluble in most solvents but possesses excellent electrical conductivity and thermal stability [126]. Thus, in order to fabricate PANI as solid-state material for sensor applications, PANI and polyamide-66 (PA-66) were combined by Wen et al. and then dissolved in formic acid to create a mixed solution for electrospinning. As a result, nanofiber membranes possess an even distribution of PANI. Similarly, the electrical conductivity of the PANI is enhanced by doping PANI with polyacrylonitrile (PAN) and silver nanowires [127].

PANI is also used in surface acoustic wave gas sensors. Polyaniline/tri-tungsten oxide complex nanofibers with an integrated layer of ZnO are used for sensing hydrogen leakage. In the presence of H_2, the conductivity of the PANI/tri-tungsten oxide nanofibers increases, resulting in the decrease of surface acoustic wave velocity, and thus the frequency is decreased. It can be functionalized to improve the target sites for detection, and it depends on electrostatic forces, hydrogen bonding, and van der Waals forces for effective interaction between probe molecules. Wang et al. improved the sensitivity and response of NH_3 detection by using TiO_2/ZnO coated with polypyrrole (PPy) over the surface [129]. Similarly, in the chitosan platform, polyethyleneimine is coated to detect formaldehyde [130]. To directly adorn Au nanoclusters on a polysulfone surface, Senthamizhan et al. utilized dip-coating. They

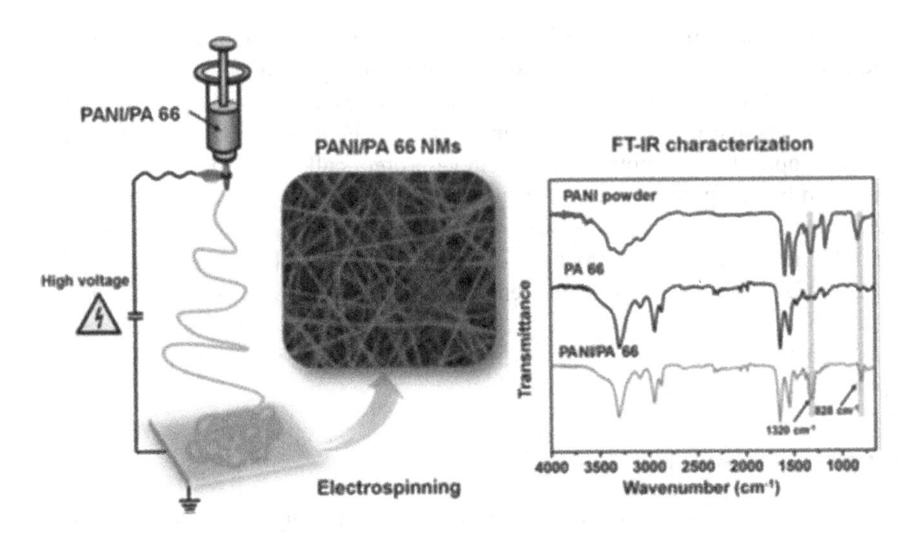

FIGURE 1.17 Fabrication of polyaniline/polyamide-66 (PANI/PA 66) composites via electrospinning [128].

subsequently used the decorated nanoclusters to detect gases like H_2O_2. Tin-doped indium oxide electrospun nanowires with increased conductivity by 10^7 times the conventional ones are beneficial for designing compact, extremely sensitive chemical and biological sensors [131].

1.5 CONCLUSION

In this chapter MOF nanofibers-based composite materials have been discussed. From the literature the following conclusions have been arrived.

(i) In comparison to its MOF nanoparticle constituents, the material's activity and surface area decrease with an increase in MOF nanofibers.
(ii) When MOF particles are loaded through electrospinning, they are less controllable, and in situ growth of MOF particles on fibers is difficult.
(iii) Heat treatment drastically decreases the mechanical properties of electrospun MOF-derived nanomaterials, preventing continued application.
(iv) A closer look at the relationship between performance and structure is warranted. It is suggested that electrospinning of MOF nanofibers and their derivatives could be used for a wide range of novel applications.

REFERENCES

[1] Yaghi, O., Li, G., Li, H. Selective binding and removal of guests in a microporous metal-organic framework. Nature 1995, 378, 703–706.
[2] Ramesh, M., Deepa, C. Metal-organic frameworks and their composites. Met. Org. Fram. Chem. React. 2021, 1–18. https://doi.org/10.1016/B978-0-12-822099-3.00001-0.

[3] Ramesh, M., Kuppusamy, N., Praveen, S. Metal-organic frameworks for batteries and supercapacitors. Met. Org. Fram. Chem. React. 2021, 19–35. https://doi.org/10.1016/B978-0-12-822099-3.00002-2.
[4] Herm, Z. R., Bloch, E. D., Long, J. R. Hydrocarbon separations in metal-organic frameworks. Chem. Mater. 2014, 26, 323.
[5] Dou, Y., Zhang, W., Kaiser, A. Electrospinning of metal-organic frameworks for energy and environmental applications. Adv. Sci. 2020, 7(3), 1902590.
[6] Lee, K., Krishnaraj, C., Verhoeven, L., et al. Catalytic carpets: Pt@MIL-101@electrospun PCL, a surprisingly active and robust hydrogenation catalyst. J. Catal. 2018, 360, 81.
[7] Liu, Y., Lin, S., Liu, Y., et al. Super-stable, highly efficient, and recyclable fibrous metal-organic framework membranes for precious metal recovery from strong acidic solutions. Small 2019, 15, 1805242.
[8] Gu, Y. F., Wu, Y. N., Li, L. C., et al. Controllable modular growth of hierarchical MOF-on-MOF architectures. Angew. Chem. Int. Ed. 2017, 56, 15658–15662.
[9] Stock, N., Biswas, S. Synthesis of metal-organic frameworks (MOFs): Routes to various MOF topologies, morphologies, and composites. Chem. Rev. 2012, 112, 933–969.
[10] Akhtar, F., Andersson, L, Ogunwumi, S, et al. Structuring adsorbents and catalysts by the processing of porous powders. J. Eur. Ceram. Soc. 2014, 34, 1643–1666.
[11] Wang, H., Wang, Q., Teat, S. J., et al. Synthesis, structure, and selective gas adsorption of a single-crystalline zirconium based microporous metal-organic framework. Cryst. Grow. Des. 2017, 17(4), 2034–2040.
[12] Jeong, S., Kim, D., Park, J., et al. Topology conversions of non-interpenetrated metal-organic frameworks to doubly interpenetrated metal-organic frameworks. Chem. Mater. 2017, 29, 3899–3907.
[13] Ramesh, M., Muthukrishnan, M. Bio-Based magnetic metal-organic framework nanocomposites. Met. Org. Fram. Nanocomp. 2020, 167–190.
[14] Ramesh, M., Muthukrishnan, M., Khan, A. Metal-organic frameworks and permeable natural polymers for reasonable carbon dioxide fixation. Met. Org. Fram. Chem. React. 2021, 417–440. https://doi.org/10.1016/B978-0-12-822099-3.00017-4.
[15] Cooley, J. F. Apparatus for Electrically Dispersing Fluids U.S. Pat. 692, 631, 1902.
[16] Morton, W. J. Method of Dispersing Fluid U.S. Pat. 705,691, 1902.
[17] Sill, T. J., von Rectum, H. A. Electrospinning: Applications in drug delivery and tissue engineering. Biomaterials 2008, 29, 1989–2006.
[18] Zagho, M. M., Elzatahry, A. Recent trends in electrospinning of polymer nanofibers and their applications as templates for metal oxide nanofibers preparation. Intec Open. DOI: 10.5772/65900, 2016.
[19] Ramesh Kumar, P., Khan, N. S., Vivekanandhan, N., et al. Nanofibers: Effective generation by electrospinning and their applications. J. Nanosci. Nanotechnol. 2012, 12, 1–25.
[20] Koski, A., Yim, K., Shivkumar, S. Effect of molecular weight on fibrous PVA produced by electrospinning. Mater. Lett. 2004, 58, 493–497.
[21] Zhao, Y., Yang, Q., Lu, X.-F., et al. Study on the correlation of morphology of electrospun products of polyacrylamide with ultrahigh molecular weight. J. Polym. Sci. Pt B Polym. Phys. 2005, 43, 2190–2195.
[22] Lee, J. S., Choi, K. H., Do Ghim, H., et al. Role of molecular weight of atactic poly (vinyl alcohol)(PVA) in the structure and properties of PVA nano fabric prepared by electrospinning. J. Appl. Polym. Sci. 2004, 93, 1638–1646.
[23] Kim, J. R., Choi, S. W., Jo, S. M., et al. Characterization and properties of P(VdF-HFP)-based fibrous polymer electrolyte membrane prepared by electrospinning. J. Electrochem. Soc. 2005, 152, A295.
[24] Gupta, P., Elkins, C., Long, T. E., et al. Electrospinning of linear homopolymers of poly(methyl methacrylate): Exploring relationships between fiber formation, viscosity, molecular weight and concentration in a good solvent. Polymer 2005, 46, 4799.

[25] Hristian, L., Ostafe, M. M., Manea, L. R., et al. The study about the use of the natural fibers in composite materials. IOP Conf. Series: Mater. Sci. Eng. 2016, 145, 032004.

[26] Schreuder-Gibson, H., Gibson, P., Senecal, K., Sennett, M., Walker, J., Yeomans, W., Ziegler, D., Tsai, P.P.J. Protective textile materials based on electrospun nanofibers. J. Adv. Mater. 2002, 34, 44–55.

[27] Huang, Z. M., Zhang, Y. Z., Kotaki, M., et al. A review on polymer nanofibers by electrospinning and their applications in nanocomposites. Compos. Sci. Technol. 2003, 63(15), 2223–2253.

[28] Amariei, N., Manea, L. R., Bertea, A. P., et al. The influence of polymer solution on the properties of electrospun 3D nanostructures. IOP Conf. Ser. Mater. Sci. Eng. 2017, 209, 012092.

[29] Ramesh, M., Muthukrishnan, M. Polymer electrocatalysis. In *Methods of Electrocatalysis: Advanced Materials and Allied Applications*, Springer-Nature, 2020, 125–147.

[30] Deitzel, J. M., Kleinmeyer, J, Harris, D, et al. The effect of processing variables on the morphology of electrospun nanofibers and textiles. Polymer 2001, 42, 261–272.

[31] Hekmati, A. H., Rashidi, A., Ghazisaeidi, R., et al. Effect of needle length, electrospinning distance, and solution concentration on morphological properties of polyamide-6 electrospun nanowebs. Textile Research Journal. DOI: 10.1177/0040517512471746.

[32] Heikkilä, P., Harlin, A. Electrospinning of polyacrylonitrile (PAN) solution: Effect of conductive additive and filler on the process. Polym. Lett. 2009, 3, 437–445.

[33] Bhattarai, R. S., Bachu, R. D., Boddu, S. H. S., et al. Biomedical applications of electrospun nanofibers: Drug and nanoparticle delivery. Pharmaceutics 2019, 11(1), 5. DOI: 10.3390/pharmaceutics11010005.

[34] Habeeb, S. Fabrication and characterization of tubular polymeric nanofibrous composite, 2013. DOI: 10.13140/RG.2.2.15658.93126.

[35] Bakar, S. S. S., Fong, K. C., Eleyas, A., et al. Effect of voltage and flow rate electrospinning parameters on polyacrylonitrile electrospun fibers. IOP Conf. Ser. Mater. Sci. Eng. 2018, 318, 012076.

[36] Lee, J. S., Choi, K. H., Ghim, H. D., et al. Role of molecular weight of atactic poly(vinyl alcohol) (PVA) in the structure and properties of PVA nano fabric prepared by electrospinning. J. Appl. Polym. Sci. 2004, 93, 1638.

[37] Hao, S. L., Wu, X. H., Wang, L. G., et al. Electrospinning of ethyl-cyanoethyl cellulose/tetrahydrofuran solutions. J. Appl. Polym. Sci. 2004, 91, 242.

[38] Zhong, X. H., Kim, K. S., Fang, D. F., et al. Structure and process relationship of electrospun bioabsorbable nanofiber membranes. Polymer 2002, 43, 4403

[39] Rutledge, G. C., Li, Y., Fridrikh, S., et al. Electrostatic spinning and properties of ultrafine fibers. Nat. Text. Center 2000, 2001.

[40] Zargham, S., Bazgir, S., Tavakoli, A., et al. The effect of flow rate on morphology and deposition area of electrospun nylon 6 nanofiber. J. Eng. Fib. Fabr. 2012, 7, 42–49.

[41] Wan, Y.-Q., He, J.-H., Wu, Y., Yu, J.-Y. Vibrorheological effect on electrospun polyacrylonitrile (PAN) nanofibers. Mater. Lett. 2006, 60(27), 3296–3300.

[42] Buchko, C.J., Chen, L. C., Shen, Y, et al. Processing and microstructural characterization of porous biocompatible protein polymer thin films. Polymer 1999, 40, 7397–7407.

[43] Li, D., Xia, Y. Electrospinning of nanofibers: Reinventing the wheel. Adv. Mater. 2004, 16, 1151–1170.

[44] Lyons, J., Li, C., Ko, F. Melt-electrospinning part I: Processing parameters and geometric properties. Polymer 2004, 45, 7597–7603.

[45] De Vrieze, S., Van Camp, T., Nelvig, A., et al. The effect of temperature and humidity on electrospinning. Mater. Sci. 2009, 44, 1357–1362.

[46] Xue, J., Wu, T., Dai, Y., et al. Electrospinning and electrospun nanofibers: methods, materials, and applications. Chem Rev. 2019, 119(8), 5298–5415.

[47] Jian, S., Zhu, J., Jiang, S., et al. Nanofibers with diameter below one nanometer from electrospinning. RSC Adv. 2018, 8, 4794–4802.
[48] Li, W. J., Laurencin, C. T., Caterson, E. J., et al. Electrospun nanofibrous structure: A novel scaffold for tissue engineering. J. Biomed. Mater. Res. 2002, 60, 613–621.
[49] Bhardwaj, N., Kundu, S. C. Electrospinning: A fascinating fiber fabrication technique. Biotechnol. Adv. 2010, 28, 325–347.
[50] Czaja, A. U., Trukhan, N., Muller, U. Industrial applications of metal–organic frameworks. Chem. Soc. Rev. 2009, 38, 1284.
[51] Wang, Q. M., Shen, D., B€ulow, M., et al. Semanscin. Metallo-organic molecular sieve for gas separation and purification. J. Micropor. Mesopor. Mater. 2002, 55, 217.
[52] Muller, U., Schubert, M., Teich, F., et al. Metal-organic frameworks—prospective industrial applications. Mater. Chem. 2006, 16, 626.
[53] Erucar, I., Yilmaz, G., Keskin, S. Recent advances in metal-organic framework-based mixed matrix membranes. Chem. Asian J. 2013, 8, 1692.
[54] Ostermann, R., Cravillon, J., Weidmann, C., et al. Smarsly. Metal–organic framework nanofibers via electrospinning. Chem. Commun., 2011, 47, 442.
[55] Rose, M., Böhringer, B., Jolly, M., et al. MOF processing by electrospinning for functional textiles. Adv. Eng. Mater. 2011, 13, 356.
[56] Zhang, Y., Yuan, S., Feng, X., et al. Preparation of nanofibrous metal-organic framework filters for efficient air pollution control. J. Am. Chem. Soc. 2016, 138, 5785.
[57] López-Maya, C., Montoro, L. M., Rodríguez-Albelo, S. D., et al. Textile/metal-organic-framework composites as self-detoxifying filters for chemical-warfare agents Angew. Chem. Int. Ed. 2015, 54, 6790–6794.
[58] Dai, X., Cao, Y., Shi, X. W., et al. The PLA/ZIF-8 Nanocomposite Membranes: The diameter and surface roughness adjustment by ZIF-8 nanoparticles, high wettability, improved mechanical property, and efficient oil/water separation. Adv. Mater. Interf. 2016, 3, 1600725.
[59] Haase, M. F. Stebe, K. J., Lee, D. Continuous fabrication of hierarchical and asymmetric Bijelmicroparticles, fibers, and membranes by solvent transfer-induced phase separation (STRIPS). Adv. Mater. 2015, 27, 7013–7013.
[60] Thompson, C. J., Chase, G. G., Yarin, A. L., Reneker, D. H. Effects of parameters on nanofiber diameter determined from electrospinning model. Polymer 2007, 48, 6913–6922.
[61] Kim, J. Y., Kim, S. H., Kang, S. W., et al. Crystallization and melting behavior of silica nanoparticles and poly(ethylene 2,6-naphthalate) hybrid nanocomposites. Macromol. Res. 2006, 14, 146–154.
[62] Armstrong, M. R., Shan, B., Maringanti, S. V., et al. Hierarchical pore structures and high ZIF-8 loading on matrimid electrospun fibers by additive removal from a blended polymer precursor. Ind. Eng. Chem. Res. 2016, 55(37), 9944–9951.
[63] Liang, H., Yao, A., Jiao, X., et al. Fast and sustained degradation of chemical warfare agent simulants using flexible self-supported metal-organic framework filters. ACS Appl. Mater. Interf. 2018, 10, 20396.
[64] Zhao, J., Gong, B., Nunn, W. T., et al. Conformal and highly adsorptive metal-organic framework thin films via layer-by-layer growth on ALD-coated fiber mats. J. Mater. Chem. 2015, 3, 1458.
[65] Wu, Y., Li, F., Liu, H., et al. Electrospun fibrous mats as skeletons to produce free-standing MOF membranes. J. Mater. Chem. 2012, 22, 16971.
[66] Bian, Y., Wang, R., Wang, S., et al. Metal-organic framework-based nanofiber filters for effective indoor air quality control. Mater. Chem. A 2018, 6, 15807.
[67] Liu, C., Wu, Y., Morlay, C., et al. General deposition of metal-organic frameworks on highly adaptive organic-inorganic hybrid electrospun fibrous substrates. ACS Appl. Mater. Interf. 2016, 8, 2552.

[68] Zhao, J., Lee, D. T., Yaga, R. W., et al. Ultra-fast degradation of chemical warfare agents using MOF–nanofiber kebabs. Angew. Chem. Int. Ed. 2016, 55, 13224.
[69] Shangguan, J., Bai, L., Li, Y., et al. Layer-by-layer decoration of MOFs on electrospun nanofibers. RSC Adv. 2018, 8, 10509.
[70] Lu, A. X., Ploskonka, A. M., Tovar, T. M., et al. Direct surface growth of UIO-66-NH_2 on polyacrylonitrile nanofibers for efficient toxic chemical removal. Ind. Eng. Chem. Res. 2017, 56, 14502.
[71] Liang, H., Jiao, X., Li, C., et al. Flexible self-supported metal-organic framework mats with exceptionally high porosity for enhanced separation and catalysis. J. Mater. Chem. A 2018, 6, 334.
[72] Hosseini Ravandi, S. A., Gandhimathi, C., Valizadeh, M., et al. Application of electrospun natural biopolymer nanofibers. Curr. Nanosci. 2013, 9, 423–433.
[73] Sathish Kumar, Y, Unnithan, A. R, Sen, D., et al. Microgravity biosynthesized penicillin loaded electrospun polyurethane-dextran nanofibrous mats for biomedical applications. Coll. Surf. A Physicochem. Eng. 2015, 477, 77–83.
[74] Sheng, J., Li, Y., Wang, X., et al. Thermal inter-fiber adhesion of the polyacrylonitrile/fluorinated polyurethane nanofibrous membranes with enhanced waterproof-breathable performance. Sep. Purif. Technol. 2016, 158, 53–61.
[75] Kim, S. J., Nam, Y. S., Rhee, D. M, et al. Preparation and characterization of antimicrobial polycarbonate nanofibrous membrane. Eur. Polym. J. 2007, 43, 3146–3152.
[76] Nobeshima, T, Ishii, Y, Sakai, H., et al. Actuation behavior of polylactic acid fiber films prepared by electrospinning. J. Nanosci. Nanotechnol. 2016, 16, 3343–3348.
[77] Kazemi Pilehrood, M., Dilamian, M, Mirian, M., et al. Nanofibrous chitosan-polyethylene oxide engineered scaffolds: A comparative study between simulated structural characteristics and cell viability. Biomed. Res. Int. 2014, 99, 99.
[78] Qavamnia, S. S., Nasouri, K. Conductive polyacrylonitrile/polyaniline nanofibers prepared by electrospinning process. Polym. Sci. Ser. A 2015, 57, 343–349.
[79] Alayande, O., Olatubosun, S, Adedoyin, O, et al. Porous and non-porous electrospun fibers from discarded expanded polystyrene. Int. J. Phys. Sci. 2012, 7, 1832–1836.
[80] Wu, J., Lu, X., Shan, F., et al. Polydiacetylene-embedded supramolecular electrospun fibers for a colorimetric sensor of organic amine vapor. RSC Adv. 2013, 3, 22841–22844.
[81] Cai, Y., Zong, X., Ban, H., et al. Fabrication, structural morphology and thermal energy storage/retrieval of ultrafine phase change fibers consisting of polyethylene glycol and polyamide 6 by electrospinning. Polym. Polym. Compos. 2013, 21, 525–532.
[82] Kenawy, E.-R., Abdel-Hay, F., El-Newehy, M. H, et al. Processing of polymer nanofibers through electrospinning as drug delivery systems. Mater. Chem. Phys. 2009, 113:296–302.
[83] Anitha, S., Brabu, B., John Thiruvadigal, D., et al. Optical, bactericidal and water repellent properties of electrospun nano-composite membranes of cellulose acetate and ZnO. Carbohydr. Polym. 2013, 97, 856–863.
[84] Liao, Y., Wang, R., Tian, M., et al. Fabrication of polyvinylidene fluoride (PVDF) nanofiber membranes by electro-spinning for direct contact membrane distillation. J. Membr. Sci. 2013, 425–426, 30–39.
[85] Dobrovolskaya, I. P., Lebedeva, I. O., Yudin, V. E., et al. Electrospinning of composite nanofibers based on chitosan, poly(ethylene oxide), and chitin nanofibrils. Polym. Sci. Ser. A 2016, 58, 246–254.
[86] Lin, Q., Li, Y, Yang, M. Polyaniline nanofiber humidity sensor prepared by electrospinning. Sens. Actuators B Chem. 2012, 161, 967–972.
[87] Oktay, B., Kayaman-Apohan, N., Erdem-Kuruca, S., et al. Fabrication of collagen immobilized electrospun poly (vinyl alcohol) scaffolds. Polym. Adv Technol. 2015, 26, 978–987.

[88] Shen, L., Li, X., Yang, Y., et al. High flux PEO doped chitosan ultrafiltration composite membrane based on PAN nanofibrous substrate. Mater. Res. Innov. 2014, 184, 808–811.
[89] Wang, P., Wang, H., Liu, J., et al. Montmorillonite@chitosan-poly (ethylene oxide) nanofibrous membrane enhancing poly (vinyl alcohol-co-ethylene) composite film. Carbohydr. Polym. 2018, 181, 885–892.
[90] Hu, Y., Yu, H., Yan, Z., et al. The surface chemical composition effect of a polyacrylic acid/polyvinyl alcohol nanofiber/quartz crystal microbalance sensor on ammonia sensing behavior. RSC Adv. 2018, 8, 8747–8754.
[91] Zhang, L., Aboagye, A., Kelkar, A., et al. A review: Carbon nanofibers from electrospun polyacrylonitrile and their applications. J. Mater. Sci. 2014, 49, 463–480.
[92] Geise, G. M., Lee, H. S., Miller, D. J., et al. Water purification by membranes: the role of polymer science. J. Polym. Sci. Pt. B Polym. Phys. 2010, 48, 1685–1718.
[93] Shutov, A. A. Composite fluoroplastic fibrous filtration membranes. Tech. Phys. Lett. 2005, 31, 1026–1028.
[94] Ramesh, M., Ramkumar, C., Ravanan, A. Fibre selections in polymer matrices for protective structures. In *Fibre-reinforced Polymers: Processes and Applications*, Nova Science Publishers, 2021, 211–236.
[95] Su, Z., Zhang, M., Lu, Z., et al. Functionalization of cellulose fiber by in situ growth of zeolitic imidazolate framework-8 (ZIF-8) nanocrystals for preparing a cellulose-based air filter with gas adsorption ability. Cellulose 2018, 25, 1997.
[96] Furukawa, H., Cordova, K. E., O'Keeffe, M., et al. The chemistry and applications of metal-organic frameworks. Science 2013, 341, 1230444.
[97] Shooto, N. D., Dikio, C. W., Wankasi, D., et al. Novel PVA/MOF nanofibres: Fabrication, evaluation, and adsorption of lead ions from aqueous solution. Nanoscale Res. Lett. 2016, 11, 414.
[98] Guo, Q., Li, Y., Wei, X.-Y., et al. Electrospun metal-organic frameworks hybrid nanofiber membrane for efficient removal of As(III) and As(V) from water. Ecotoxicol. Environ. Saf. 2021, 228, 112990.
[99] Efome, J. E., Rana, D., Matsuura, T., et al. Insight studies on metal-organic framework nanofibrous membrane adsorption and activation for heavy metal ions removal from aqueous solution. ACS Appl. Mater. Interf. 2018, 10, 22, 18619–18629.
[100] Lavoie, N., Malenfant, P. R., Courtel, F. M., et al. High gravimetric capacity and long cycle life in Mn_3O_4/graphene platelet/LiCMC composite lithium-ion battery anodes. J. Power Sour. 2012, 213, 249–254.
[101] Ji, L. W., Yao, Y. F., Toprakci, O., et al. Fabrication of carbon nanofiber-driven electrodes from electrospun polyacrylonitrile/polypyrrole bicomponents for high-performance rechargeable lithium-ion batteries. J. Power Sour. 2010, 195, 2050–2056.
[102] Bose, A. B., Gopu, S., Li, W. Enhancement of proton exchange membrane fuel cells performance at elevated temperatures and lower humidities by incorporating immobilized phosphotungstic acid in electrodes. J. Power Sour. 2014, 263, 217–222.
[103] Molla, S., Compan, V., E. Gimenez, V., et al. Novel ultrathin composite membranes of Nafion/PVA for PEMFCs. Int. J. Hydr. Ener. 2011, 36, 9886–9895.
[104] Lee, N., Kim, Y., Lee, M. S., et al. SiO_2-coated polyimide nonwoven/Nafion composite membranes for proton exchange membrane fuel cells. J. Member. Sci. 2011, 367, 265–272.
[105] Hooshyari, K., Moradi, M., Salarizadeh, P. Novel nanocomposite membranes based on PBI and doped-perovskite nanoparticles as a strategy for improving PEMFC performance at high temperatures. Intern. J. Energy Res. 2020, 44, 2617–2633.
[106] Devautour-Vinot, S., Maurin, G., Serre, C., et al. Structure and dynamics of the functionalized MOF type UiO-66(Zr): NMR and dielectric relaxation spectroscopies coupled with DFT calculations. Chem. Mater. 2012, 24, 2168–2177.

[107] He, T., Xu, X., Ni, B., et al. Fast and scalable synthesis of uniform zirconium, hafnium-based metal-organic framework nanocrystals. Nanoscale 2017, 9, 19209–19215.
[108] Yang, D., Bernales, V., Islamoglu, T., et al. Tuning the surface chemistry of metal organic framework nodes: protontopology of the metal-oxide-like Zr6 nodes of UiO-66 and NU-1000. J. Am. Chem. Soc. 2016, 138, 15189–15196.
[109] Yang, F., Huang, H., Wang, X., et al. Proton conductivities in functionalized UiO-66: tuned properties, thermogravimetry mass, and molecular simulation analyses. Cryst. Grow. Des. 2015, 15, 5827–5833.
[110] Yang, L., Tang, B., Wu, P. Metal–organic framework–grapheneoxide composites: a facile method to highly improve the proton conductivity of PEMs operated under low humidity. J. Mater. Chem. A 2015, 3, 15838–15842.
[111] Yang, Q., Wiersum, A. D., Llewellyn, P. L., et al. Functionalizing porous zirconium terephthalate UiO-66(Zr) for natural gas upgrading: A computational exploration. Chem. Commun. 2011, 47, 9603–9605.
[112] Li, Z., He, G., Zhao, Y., et al. Enhanced proton conductivity of proton exchange membranes by incorporating sulfonated metal-organic frameworks. J. Power Sour. 2014, 262, 372–379.
[113] Shi, G. M., Yang, T., Chung, T. S. Polybenzimidazole (PBI)/zeolitic imidazolate frameworks (ZIF-8) mixed matrix membranes for pervaporation dehydration of alcohols. J. Memb. Sci. 2012, 415–416, 577–586.
[114] Sen, U., Erkartal, M., Kung, C. W., et al. Proton conducting self-assembled metal-organic framework/polyelectrolyte, hollow hybrid nanostructures. ACS Appl. Mater. Interf. 2016, 8, 23015–23021.
[115] Sill, T. J, von Recum, H. A. Electrospinning: Applications in drug delivery and tissue engineering. Biomater. 2008, 29(13), 1989–2006.
[116] Manuel, C. B. J., Jesús, V. G. L., Aracely, S. Electrospinning for drug delivery systems: Drug incorporation techniques. In S. Haider, A. Haider (eds.), *Electrospinning—Material, Techniques, and Biomedical Applications*, IntechOpen, 2016. DOI: 10.5772/65939.
[117] Ravi Kumar, R. M. V. *Hand Journal of Polyester Drug Delivery Systems*. 1st ed. CRC Press, 2016, 1–738.
[118] Sanchez, M. A., Rodriguez, A. P., Monsalve, L. N., et al. Emulsion electrospinning for drug delivery: Two encapsulation methods. Austin J. Pharmacol. Ther. 2020, 8(1), 1117.
[119] Xu, X., Yang, L., Xu, X., et al. Ultrafine medicated fibers electrospun from W/O emulsions. J Control Release. 2005, 108(1), 33–42.
[120] Zoka, L., Narbaitz, R., Matsuura, T. Effect of surface modification with electrospun nanofibers on the performance of an ultrafiltration membrane. J. Membr. Sci. Res. 2020, 6(4), 351–358.
[121] Park, K., Ju, Y. M., Son, J. S., et al. Surface modification of biodegradable electrospun nanofiber scaffolds and their interaction with fibroblasts. J. Biomater. Sci. Polym Ed. 2007, 18(4), 369–382.
[122] Sridhar, R., Ramakrishna, S. Electrosprayed nanoparticles for drug delivery and pharmaceutical applications. Biomater. 2013, 3(3), 242811–242813.
[123] Kratz, F. Albumin as a drug carrier: Design of prodrugs, drug conjugates, and nanoparticles. J. Control. Rel. 2008, 132, 171–183.
[124] Zhang, L., Huang, J., Si, T., et al. Coaxial electrospray of microparticles and nanoparticles for biomedical applications. Exp. Rev. Med. Dev. 2012, 9(6), 595–612.
[125] Hazari, N., Tavana, H., Moradi, A. R. Production and properties of electrosprayed sericin nanopowder. Sci. Technol. Adv. Mat. 2012, 13, 035010.
[126] Sen, T., Mishra, S., Shimpi, N. G., 2016. Synthesis and sensing applications of polyaniline nanocomposites: A review. RSC Adv. 6, 42196–42222.

[127] Rezaei, B., Ghani, M., Shoushtari, A. M., et al. Electrochemical biosensors based on nanofibres for cardiac biomarker detection: A comprehensive review. Biosen. Bioelect. 2016, 78, 513–523.
[128] Wen, J., Tan, X., Hu, Y., et al. Filtration and electrochemical disinfection performance of PAN/PANI/AgNWs-CC composite nanofiber membrane. Environ. Sci. Technol. 2017, 51, 6395–6403.
[129] Wang, Y., Jia, W., Strout, T., et al. Ammonia gas sensor using polypyrrole-coated TiO_2/ZnO nanofibers. Electroanalysis 2009, 21, 1432–1438.
[130] Wang, X., Yang, Y., Gao, H. A novel solid-state electrochemiluminescence quenching sensor for detection of aniline based on luminescent composite nanofibers. J. Luminesc. 2014, 156, 229–234.
[131] Lin, D., Wu, H., Zhang, R., et al. Preparation and electrical properties of electrospun tin-doped indium oxide nanowires. Nanotechnol. 2007, 18, 465301.

2 Impact of Storage Methods on Charge Decay of Electrospun PVDF Fibers

Harshal Gade, George G. Chase, and Abdul Aziz AlGhamdi

2.1 WHAT ARE ELECTRETS?

Electrets are known to carry quasi-permanent electrical charge on them. Electrets are mostly dielectric materials which can include various forms of nanostructured materials, ceramics, polymer composites and natural materials such as bees wax and gum [1]. Electrospun polymeric nanofibers are said to carry electrical charge. Electrospinning can induce dipoles in nanofiber structure giving them additional charge in addition to inherent piezoelectricity the polymer structure exhibits. PVDF, a well-known polymer, is said to have very high dipole moment due to its inherent spontaneous polarization. Beta-phase, known to be the most polar phase of PVDF polymer, is said to be inducing electrical properties such as piezo, pyro and ferroelectricity making it a polymer finding usage in myriad of applications such as semi-conductors, electronic materials, drug delivery, filtrations and separations such as aerosol filtrations and water purification. Electrospun nanofibers can be polarized to enhance their electric charge using different functionalization techniques [2–8].

There are lot of applications where life span of charge matters. Charge on nanofibers can be preserved by storing in proper storage methods so that their shelf life is enhanced. Lolla et al showed the difference in the performance of as-spun and polarized PVDF fibers in aerosol filtration. It was observed that over a period of one year, efficiency of polarized fibers was better than as compared to as-spun fibers, and those could be used over a long period of time. In this study, electrostatic voltameter was used to measure surface potential of fiber mats as no other precise instrument was available. A systematic study of storage methods was not done [9].

According to principle of electrostatics, the charged ions in air can interact with charged ions on the surface of nanofiber mat sample and that is how charge can decay or neutralize over the period of time and the charged nanofiber sample can no longer be of use. The literature does not have an instrument accurate and sensitive enough accurate and sensitive instrument to measure charge of electrospun

DOI: 10.1201978100333814-2

nanofibers. In this chapter, efforts have been made to store as-spun and polarized PVDF nanofibers in various storage methods as described in abstract and maintain a systematic study of charge decay with different storage methods.

2.2 MATERIALS USED AND CHARACTERIZATION TECHNIQUES FOR NANOFIBERS

2.2.1 Materials Used

Electrospinning of PVDF nanofibers was carried out by making a homogenous solution of Kynar® 761 grade PVDF powder (Arkema Inc, King of Prussia, PA, USA) (MW of about 250,000, melt viscosity of 35 cP, melting point of 165–172°C, as per material safety data sheet) and solvents. PVDF powder concentration used was 18% by weight. Solution making involved mixing of the co-solvents N-N-Dimethylformamide (DMF) and Acetone (Sigma Aldrich, St. Louis, MO, USA) in a ratio of 50/50 wt%, simultaneous heating and stirring at 70°C for 30mins.

2.2.2 Electrospinning as Technique for Nanofiber Production

Electrospinning of fiber mats was carried out using an electrospinning setup as shown in Figure 2.1. The setup makes use of high voltage power supply to create an electric field between the droplet at the needle tip and collector. The electrospinning conditions used are listed in Table 2.1.

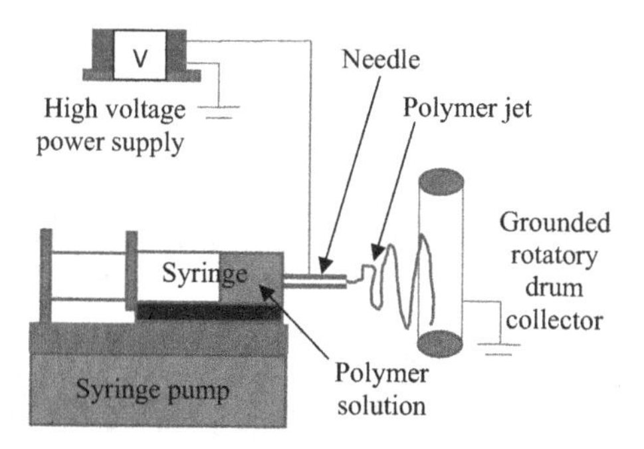

FIGURE 2.1 Schematic of electrospinning setup [2].

TABLE 2.1
Electrospinning Conditions and Fiber Diameter Data

Concentration of PVDF (wt%)	DMF:Acetone mass ratio	Tip to collector distance (cm)	Voltage (kV)	Flow rate (mL/hr)	Ave fiber diameter (nm)	Standard deviation (nm)	Rotations per min (rpm)
18	1:1	20	27	5	1239	512	30

2.2.3 Polarization Procedure

Figure 2.2 a) shows the end view of the polarizing setup used to polarize as-spun PVDF fibers. The setup was fabricated out from Teflon. Figure 2.2 b) shows the sample holder inside of a Fischer Scientific iso-temp oven. The sample holder was used to perform all polarization treatments including simultaneous heating, stretching and poling on electrospun PVDF fiber mats and yarns [10–11].

2.2.4 Characterization Methods

FTIR analysis was performed on Thermo Scientific FTIR (Thermo Nicolet 380, Waltham, MA, USA) to study the IR absorption spectra of different crystalline phases exhibited by PVDF molecules. IR spectra in the 600–1400 cm^{-1} wave number range with a resolution of 0.5cm^{-1} were obtained. All samples were run for 32 scans [10].

Electric charge of fiber samples was measured using a custom-made Faraday bucket which works on the principle of electrostatics. Details regarding working mechanism, measurement techniques, procedures and applications have been explained well in paper [12].

2.2.5 β-Phase Calculation and Analysis

FTIR spectra for the α and β-phases have been studied extensively in literature and equations below have been used to calculate fraction of β-phase led by Gregario and Cestari in 1994. IR absorption spectra was assumed to be following Beer-Lambert's law.

$$A_\alpha = \log \frac{I_\alpha^0}{I_\alpha} = K_\alpha C X_\alpha L \qquad (1.1)\ [13]$$

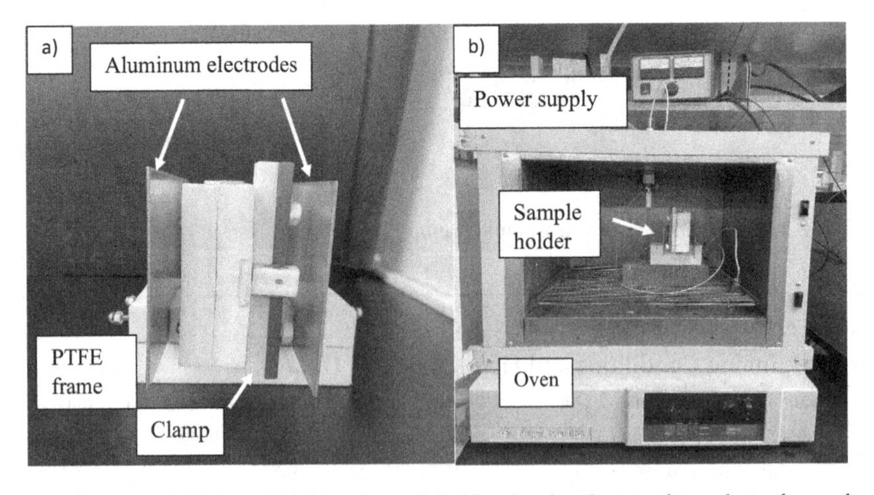

FIGURE 2.2 a) End view of fabricated sample holder showing the two planar electrodes used to apply the electric field for poling the sample. b) Photo of sample holder inside of oven and high voltage power supply for charging the electrode above the oven [7].

$$A_\beta = \log \frac{I_\beta^0}{I_\beta} = K_\beta C X_\beta L \qquad (1.2)\ [13]$$

where I and I^0 are the incident and transmitted radiation intensities respectively, L is the sample thickness and C average total monomer concentration (0.0305 mole/cm^3) and X is the degree of crystallinity of each phase and absorption coefficient at corresponding wave number for each phase ($K\alpha$=6.1 × 10^4 and K_β=7.7 × 10^{4}cm^2/mole). Thus, for a system containing β-phase and α-phase, the relative fraction of β-phase, F(β), can be calculated by (1.3) [11]:

$$F(\beta) = \frac{x_\beta}{X_{\alpha + x_\beta}} = \frac{A_\beta}{\left(\frac{K_2}{K_\alpha}\right) A_\alpha + A_\beta} = \frac{A_\beta}{1.26 A_\alpha + A_\beta} \qquad (1.3)\ [13]$$

2.3 RESULTS AND DISCUSSIONS

2.3.1 Exposed to Open Air, Light and Darkness

Figure 2.3 b) shows pictures of samples stored in open light and air, and Figure 2.3 c) shows mats when stored inside the drawer. The drawer was shut once samples were stored and only opened when the samples were to be tested. There was a small amount of air inside the drawer which also contributed to the charge decay. Figure 2.3 a) shows how the surface charge decays with time for polar and non-polar PVDF fiber mats when stored in open air and light as well as in darkness. They were stored in lab conditions which had a temperature of around 25°C and relative humidity of around 50%. Charge decay was fastest for non-polar samples which were open to light and air. These can be considered to be the harshest conditions amongst which were tested as the mats got exposed to all sorts of external factors such as moisture, electrons and variable lights of various wavelengths. For these samples, charge was lost within a week. Polar samples always showed more initial charge as well as charge retention for the same set of conditions. The polar samples, when stored in darkness and not exposed to light, retained charge for over four weeks. Samples stored measure charge as well as beta-phase were taken from same electrospun sheet in order to maintain consistency of properties. Figure 2.3 d) shows beta-phase decay in stored samples with time. It was found out that the polar samples had slightly more beta-phase as compared to non-polar samples. But with time, beta-phase decay was not very significant for all the samples. It did not change much when exposed to external atmospheric elements with time. Polar samples did show slightly higher beta-phase but were not significantly higher than non-polar samples. This helped in concluding that storage methods didn't have much effect on molecular structure but only paved the path for dissipation of free ions which were induced into fiber due to electrospinning and external polarization treatments which included uniaxial stretching, heating and electrical poling. The charge values of any of the mats did not go down to complete zero as there was some residual charge always remaining at the end.

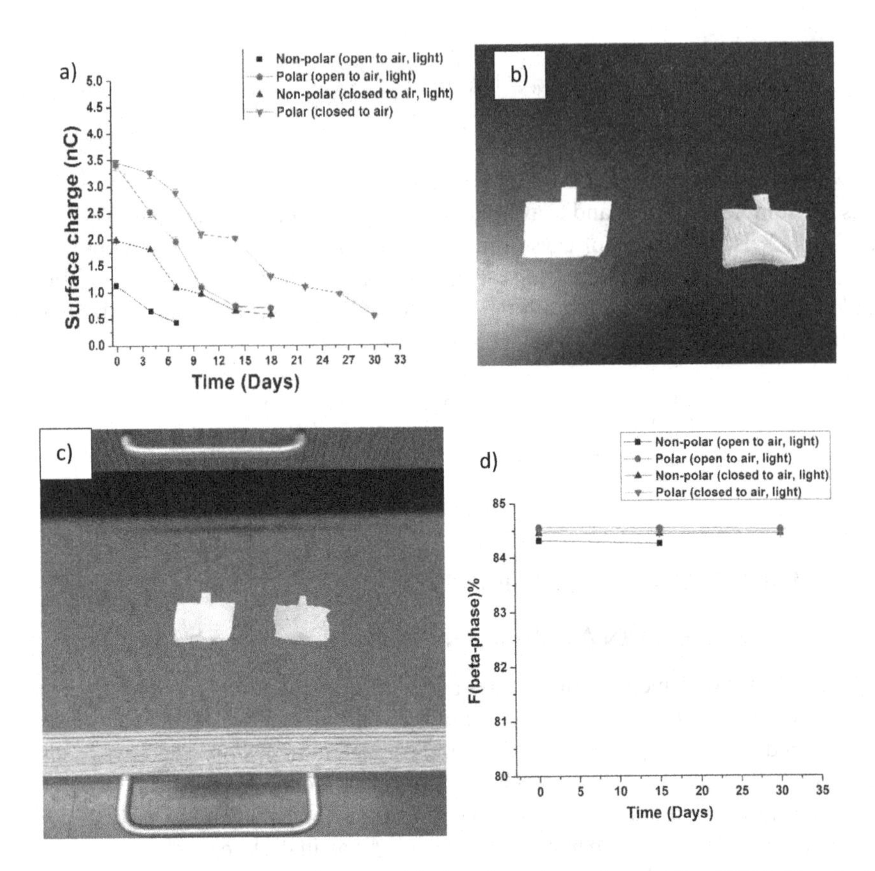

FIGURE 2.3 a) Surface charge decay with time for polar and non-polar PVDF fiber mats when stored in open air, light and darkness. b) Pics of mats stored in open light and air. c) Mats stored in darkness inside a drawer. d) Beta-phase decay for all samples plotted in a).

2.3.2 Samples Covered with Aluminum Foil

Figure 2.4 b) shows pictures of samples stored in open light and air, and Figure 2.4 c) shows mats when stored inside the drawer. All samples were stored inside an aluminum foil. The drawer was shut once samples were stored and was only opened when the samples were to be tested. There was a small amount of air inside the drawer which also contributed to the charge decay. Figure 2.4 a) shows how the surface charge decays with time for polar and non-polar PVDF fiber mats when stored in open air and light as well as in darkness when stored inside the aluminum foil. They were stored at lab conditions which had a temperature of around 25°C and relative humidity of around 50%. It was observed that use of aluminum foil helped prolong the shelf life of all mats irrespective of conditions. As a result, the polar and non-polar mats which were exposed to air and light but inside an aluminum foil could increase their shelf life by around three weeks and one week respectively as compared to those without foil. The polar mats stored in an aluminum foil and stored inside the drawer in darkness retained the charge for the longest times for around 105 days which was around 75 days more than those polar mats stored

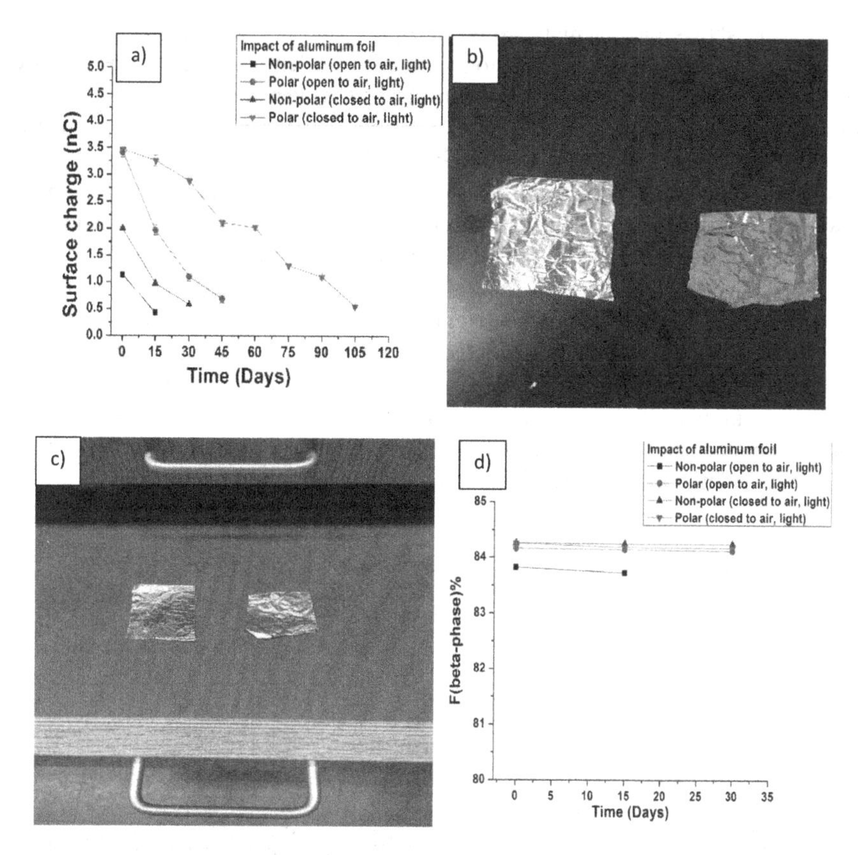

FIGURE 2.4 a) Surface charge decay with time for polar and non-polar PVDF fiber mats when stored in open air, light and darkness when stored inside an aluminum foil. b) Pics of mats stored in open light and air inside an aluminum foil. c) Mats stored in darkness inside a drawer inside an aluminum foil. d) Beta-phase decay for all samples plotted in a).

in darkness without aluminum foil. The primary reason for the increase in shelf life was that use of aluminum foil helped reduce the exposure of mats to outside elements such as moisture, dust and other electrons in air and lights of variable wavelengths. Figure 2.4 d) shows beta-phase decay in stored samples with time. A similar trend was found, that the polar samples had slightly more beta-phase as compared to non-polar ones, but with time, beta-phase decay was not very significant for all the samples. Beta-phase content was not recorded for further storage methods as these storage methods which were considered to be most harsh did not seem to have any effect on beta-phase decay with time.

2.3.3 Sample Stored in Zipped and Non-Zipped Storage Bags without Foil

Figure 2.5 a) shows surface charge decay with time for polar and non-polar PVDF fiber mats when stored in zipped and non-zipped plastic bags and static shielding bags. Figure 2.5 b) shows pics of mats stored in zipped and non-zipped plastic bags.

Figure 2.5 c) shows pics of mats stored in zipped and non-zipped static shielding bags. They were stored at lab conditions which had a temperature of around 25°C and relative humidity of around 50%. Similar trends were observed such as polar samples showed higher charge as well as retention in all types of combinations of storages. Plastic bags are commonly used in labs, but static shielding bags are unique and silverish in color. They prevent the build-up of static electricity but also do not allow electrostatic discharge. It was observed that static shielding bags helped to preserve the charge for a longer time as compared to plastic bags for both types of samples. The non-polar samples stored in non-zipped plastic bags and static shielding bags showed the lowest charge retention which was around 2 and 3.5 weeks respectively. The non-polar samples stored in zipped plastic and static shielding bags showed higher charge retention which was around 63 days and 125 days respectively. Moisture accumulation can be said to be the biggest factor for faster charge decay in both the non-zipped bags. The highest charge retention was found to be in polar samples stored in zipped static shielding bags which was close to around 212 days.

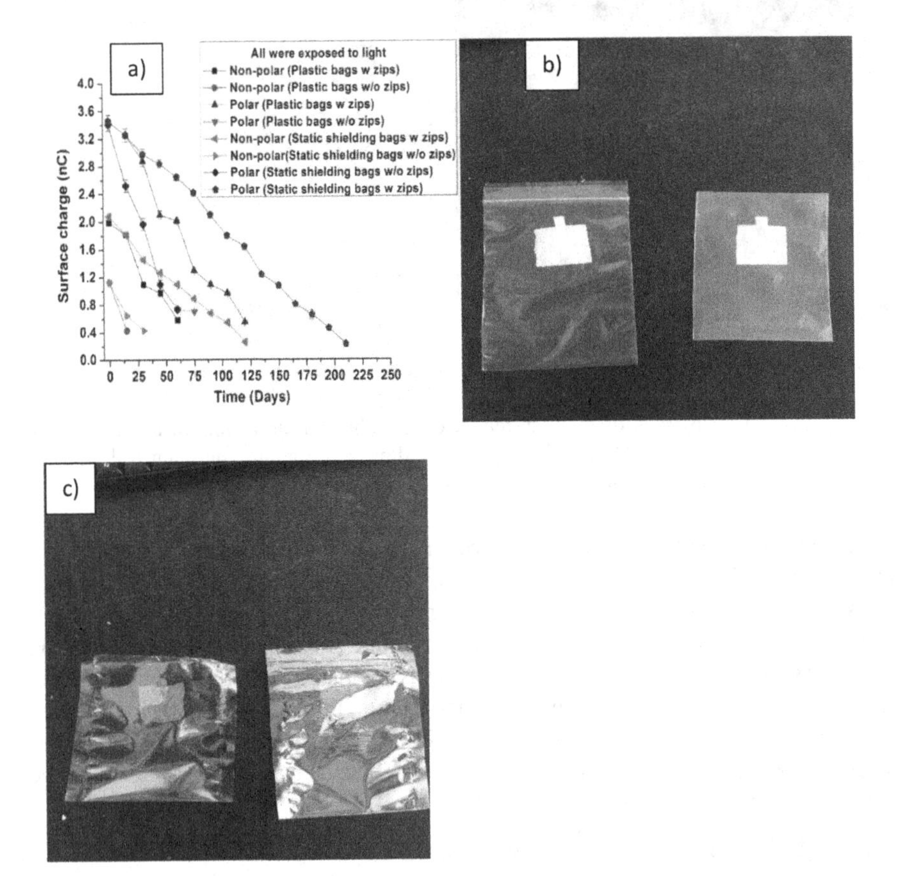

FIGURE 2.5 a) Surface charge decay with time for polar and non-polar PVDF fiber mats when stored in zipped and non-zipped plastic bags and static shielding bags. b) Pics of mats stored in zipped and non-zipped plastic. c) Pics of mats stored in zipped and non-zipped static shielding bags.

2.3.4 Sample Stored in Zipped and Non-Zipped Storage Bags with Foil

Figure 2.6 a) shows surface charge decay with time for polar and non-polar PVDF fiber mats when stored in zipped and non-zipped plastic bags and static shielding bags covered in aluminum foil. Figure 2.6 b) shows pics of mats stored in zipped and non-zipped static shielding bags covered in foil. Figure 2.6 c) shows pics of mats

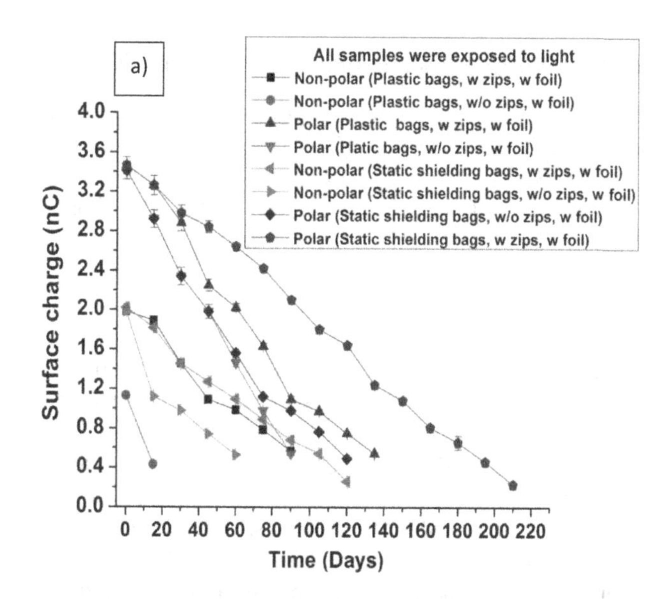

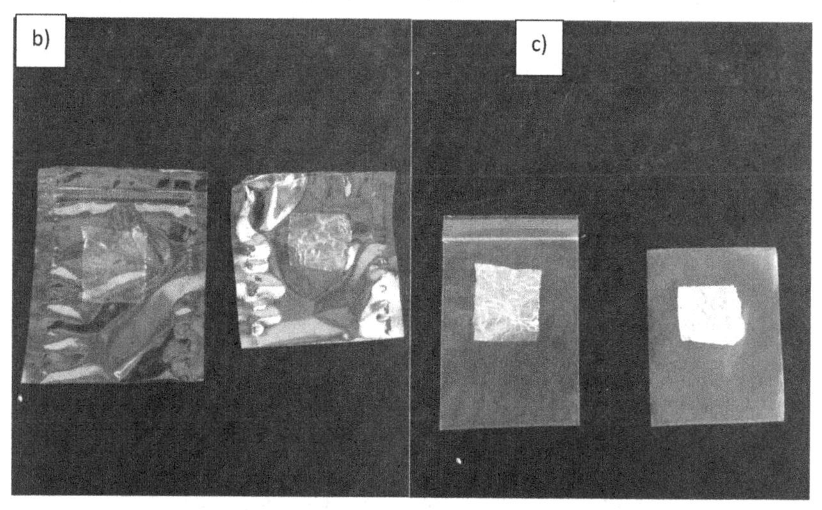

FIGURE 2.6 a) Surface charge decay with time for polar and non-polar PVDF fiber mats when stored in zipped and non-zipped plastic bags and static shielding and covered with aluminum foil bags. b) Pics of mats stored in zipped and non-zipped static shielding bags covered in foil. c) Pics of mats stored in zipped and non-zipped plastic bags covered in foil.

stored in zipped and non-zipped plastic bags covered in foil. They were stored at lab conditions which had temperature of around 25°C and relative humidity of around 50%. Similar trends were observed regarding the use of aluminum foil which helped to prolong charge life. The differences in surface charge decay after storing in aluminum foil were not more significant than without when stored in zipped and non-zipped plastic and static shielding bags. In some cases, the charge decay was marginally delayed by two to three weeks.

2.3.5 Samples Stored in Various Temperature and Humidity Values

Figure 2.7 shows surface charge decay with time for polar and non-polar PVDF fiber mats when stored without any covering in various temperatures and humidities. As the temperature and humidity decreased, the charge decay accelerated and was fastest at lowest conditions. One of the primary reasons that can be thought of is deposition of more moisture on sample leading to rapid charge deposition.

2.4 SUMMARY

Charge decay was studied for polar and non-polar samples after storing them in various storage methods such as open to air, light and darkness and storing them in various zipped and non-zipped bags such as plastic bags and static shielding bags. Polar samples retained charge for a longer time than non-polar samples for almost all types of storages. When stored in open exposed to air and light, non-polar samples lost charge in a week whereas polar samples retained for three weeks. When stored in aluminum foil, they retained charge for a much longer time. It was concluded that samples could retain charge for longer time by use of aluminum foil. Amongst bags, zipped bags

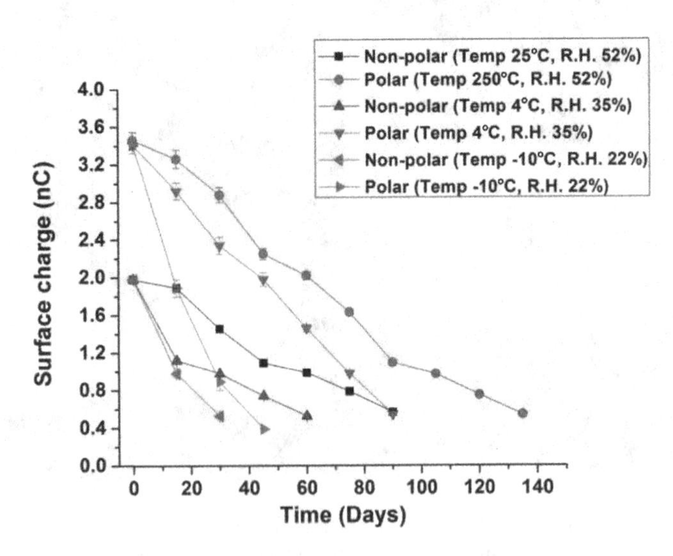

FIGURE 2.7 Surface charge decay with time for polar and non-polar PVDF fiber mats when stored without any covering in various temperature and humidity values.

always helped samples to retain charge for a longer time. Non-zipped bags tended to collect moisture within which allowed charge to dissipate at a faster rate. Static shielding bags with zips and foil on samples were found to be best performers as polar samples could withhold charge for around 230 days. Beta-phase content was also measured alongside for the storage methods which were the harshest as in when samples were exposed to light and all atmospheric elements. It was found that beta-phase content did not show any significant reduction with time under any storage method. The charge dissipation was inverse in proportion to temperature and humidity levels. The charge dissipation was fastest for lowest temperature and humidity levels.

ACKNOWLEDGMENTS

This work was funded by Coalescence Filtration Fibers Consortium (CFNC) comprising of major filtration companies including Parker Hannifin, Cummins Filtration, Hollingsworth & Vose, Donaldson, Alhstrom. We appreciate all the technical assistance provided by technicians Steve Roberts and William Imes for the building up of a Faraday bucket.

REFERENCES

[1] https://en.wikipedia.org/wiki/Electret Last Accessed Sept 11th, 2022.

[2] Lolla D.; Pan L.; Gade H.; Chase G.G.; *Functionalized Polyvinylidene Fluoride Electrospun Fibers and Applications, in Electrospinning Method Used to Create Functional Nanocomposite Films* (Ed. T Tański), Volume 8. London: IntechOpen Limited, 2018. DOI: 10.5772/intechopen.76261

[3] Kumar B.; ed. *Nanofibers—Synthesis, Properties and Applications*. London: IntechOpen, 2021. DOI: 10.5772/intechopen.92484

[4] Gade H.; Bokka S.; Chase G.G.; Polarization of Electrospun PVDF Fiber Mats and Fiber Yarns. In *Nanofibers—Synthesis, Properties and Applications*. London: IntechOpen, 2021. DOI: 10.5772/intechopen.96305

[5] Kasbe P.; Gade H.; Liu S.; Chase G.G.; Xu W.; Ultrathin Polydopamine-Graphene Oxide Hybrid Coatings on Polymer Filters with Improved Filtration Performance and Functionalities. *ACS Applied BioMaterials*, 2021. https://doi.org/10.1021/acsabm.1c00367

[6] Kantenwein R.; Investigation of Surface Charge using Faraday Bucket Measurements: Analysis on Electrospun Polyvinylidene Fluoride Fiber Mats. *Williams Honors College, Honors Research Projects*, 2021, 1280. https://ideaexchange.uakron.edu/honors_research_projects/1280/

[7] Chase G.G.; Reneker D.H.; Lolla D.; Gade H.; Atomic Scale Engineering of Polymer Fibers for Enhanced Filtration Performance. In *8th World Congress on Particle Technology*. AIChE, 2018. https://aiche.confex.com/aiche/wcpt18/meetingapp.cgi/Paper/506343

[8] Tomasz T.; Jarka P.; and Matysiak W.; eds. *Electrospinning Method Used to Create Functional Nanocomposites Films*. BoD–Books on Demand, 2018. DOI: 10.5772/intechopen.70984

[9] Lolla D.; Lolla M.; Abutaleb A.; Shin H.U.; Reneker D.H.; Chase G.G.; Fabrication, Polarization of Electro-spun Polyvinylidene Fluoride Electret Fibers and Effect on Capturing Nanoscale Solid Aerosols. *Materials*, 2016, 9, 671. https://doi.org/10.3390/ma9080671

[10] Gade H.; Bokka S.; Chase G.G.; Polarization Treatments of Electrospun PVDF Nanofibers. *Polymer*, 2020, 212,123152. https://doi.org/10.1016/j.polymer.2020.123152

[11] Gade H.; Nikam S.; Chase G.G.; Reneker D.H.; Effect of Electrospinning Conditions on β-phase and Surface Charge Potential of PVDF Fibers. *Polymer*, 2021, 228, 123902. https://doi.org/10.1016/j.polymer.2021.123902
[12] Gade H.; Parsa N.; Chase G.G.; Renekar D.H.; Roberts O.S.; Charge Measurement of Electrospun Polyvinylidene Fluoride Fibers using a Custom-made Faraday Bucket. *Review of Scientific Instruments*, 2020, 91, 075107. https://doi.org/10.1063/1.5142386
[13] Gregorio Jr G.; Uneo E.M.; Effect of Crystalline Phase, Orientation and Temperature on the Dielectric Properties of Poly (vinylidene fluoride). *Journal of Material Sciences*, 1999, 34, 4489.

3 Effects of Polarization on Filter Properties and Performance of PVDF Fibers in Aerosol Filtration

Harshal Gade, George G. Chase, and Abdul Aziz AlGhamdi

3.1 USE OF NANOFIBERS IN AEROSOL FILTRATION

Nano-aerosols are defined to as airborne solid particles around 100nm or less. Nano-aerosols can be produced because of combustion of fossil fuel from various sources such as vehicles, industries and power plants. Health risks associated with these are huge as they have the ability to penetrate deep inside lungs leading to cardio and pulmonary repairs [1].

Textiles, yarns and fibers and various nanostructured materials have been widely used in a wide range of applications such as water desalination, semi-conductors and catalysis [2,3]. Electrospun polymeric nanofibers are widely used in various filtrations and separations. One of the applications where they find use is aerosol filtration [4,5]. Several materials have been used in literature in aerosol filtration. HEPA and ULPA filters conventionally can get rid of 99.97% and 99.99% of particles in an air stream containing particles of size 100nm and beyond [6]. Polymeric electrospun nanofibers are also being used in this particular application due to their properties of high surface areas and high porosities. There are several methods to manufacture nanofibers, but electrospinning is one such well-documented and established method. Ahn et al synthesized electrospun Nylon 6 fibers lower than 200nm and used it to conduct aerosol filtration experiments and found that Nylon 6 fibrous media performed better than common HEPA filters [7]. Yun et al used PAN electrospun fibers as filter media to study capture efficiency of NaCl nanoparticles from air and inferred that the penetration was strongly dependent on filter thickness while quality factor and fiber efficiency were dependent on fiber size [8]. Gopal et al used electrospun PVDF nanofibrous membranes to segregate contaminants from water to achieve 90% efficiency [9]. There are a variety of polymeric nanofiber materials especially PVDF which is well known for its electrical properties that can work as electrets in such applications [10–12].

DOI: 10.1201978100333814-3

Aerosol capture on fibers predominantly takes places via three different capture mechanisms which are interception, inertial impaction and Brownian diffusion. Electrostatic attraction is another mechanism which comes into picture when electrets or electrically charged filters are used [13]. Several researchers have used electrets in aerosol filtration and carried out different types of studies including theoretical and experimental validation, material development and so on. A few used various functionalization techniques to increase charge on polymeric fibers and then used them to improve filtration efficiencies. Dinesh et al showed that polarized PVDF nanofibers captured particles more efficiently and had lower pressure drops as compared to as-spun media. A customized lab-based setup used simultaneous heating, stretching and poling method to charge as-spun fibers [14].

Certain filter performance properties such as thickness and charge values can affect the theoretical and experimental validation experiments and their interpretations. Thickness of filter has been predominantly measured by thickness gauges and micrometers. Surface charge has been measured predominantly by electrostatic voltmeters. Literature does not report highly accurate instruments. Efforts have been made to use custom-made lab-based instruments for measuring charge in this manuscript. The studies show effects of polarization on basis weights and in turn their performance in aerosol capture have been covered in this chapter.

3.2 MATERIALS USED AND METHODS NEEDED FOR NANOFIBERS

3.2.1 Materials

PVDF nanofibers were synthesized using an electrospinning technique. In the first step, polymer solution was created using Kynar® 761 grade PVDF powder (Arkema Inc, King of Prussia, PA, USA). PVDF powder concentration used was 18% by weight. Solvents used were N-N- Dimethylformamide (DMF) and Acetone (Sigma Aldrich, St. Louis, MO, USA). Solvents were mixed in a ratio of 50/50 wt%, simultaneous heating and stirring at 70°C for 30 mins.

3.2.2 Electrospinning

Electrospinning of fiber mats was carried out using an electrospinning setup as shown in Figure 3.1. The setup consists of three main components such as a syringe with polymer solution, high voltage power supply and grounded collector to collect fibers.

TABLE 3.1
Conditions Used and Fiber Diameter Calculated

Concentration of PVDF (wt%)	DMF:Acetone mass ratio	Tip to collector distance (cm)	Voltage (kV)	Flow rate (mL/hr)	Ave fiber diameter (nm)	Standard deviation (nm)	Rotations per min (rpm)
18	1:1	20	27	5	1239	512	30

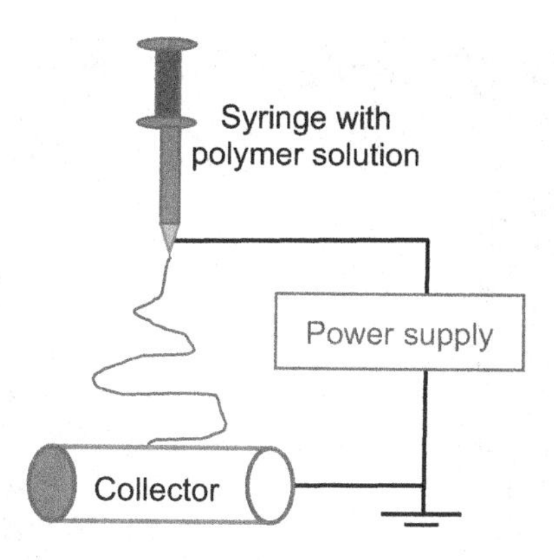

FIGURE 3.1 Schematic of Electrospinning setup [2].

3.2.3 Solid Aerosol Separation Experiment (TSI 8130)

Aerosol filtration experiments were performed using a filter tester named TSI8130 to compare the performances of as-spun and polarized electrospun PVDF nanofiber media. The performance was evaluated in terms of pressure drop and efficiency. Filter index was calculated using the following equation (1.1) [15]:

$$FI = \frac{-\ln\left(\frac{C_{out}}{C_{in}}\right)}{\Delta P} \qquad (1.1)\ [14]$$

Where C_{out} and C_{in} symbolize outlet and inlet stream particle concentrations respectively. The ratio $\frac{C_{out}}{C_{in}}$ is the penetration and ΔP is the pressure drop across the filter media. [15]

A custom-made filter holder having a top plate measuring (14cm × 14cm × 0.3cm) and bottom plate measuring (14cm × 14cm × 2.3cm) was used to hold PVDF fiber samples. The plates had a hole measuring 5cm in diameter and were made using plexiglass as shown in Figure 3.2.

The PVDF fiber mat sample was placed over a polypropylene (PP) mesh (1mm mesh opening, Spectrum Laboratories, Inc.) that supported the fiber mat. The fiber mat and PP mesh were placed covering the 5cm hole and sandwiched between the two plexiglass plates. The assembly was then placed on the base of TSI Filter Tester as shown in Figure 3.3. Impact of mesh was studied by running blank tests without the samples by just keeping the mesh, resulting in zero pressure drop and 100% penetration. The tests showed PP mesh was highly permeable and did not contribute to filter performance when tested with PVDF filter media.

FIGURE 3.2 Assembled filter holder used for placing PVDF fiber mats on PP mesh and between two plexiglass plates [15].

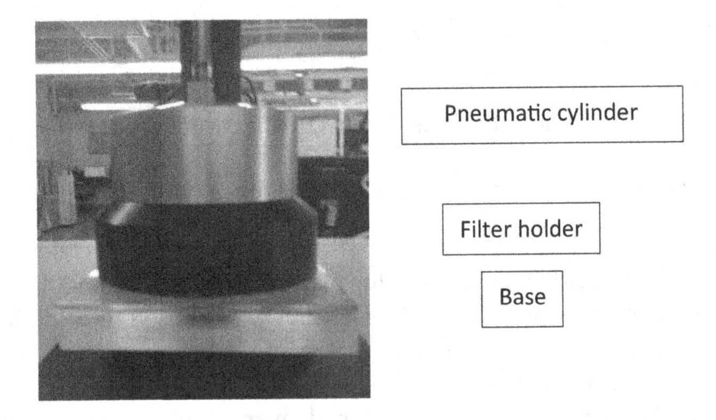

FIGURE 3.3 Filter holder positioned on filter tester as clamped by pneumatic cylinder [15].

3.2.4 Polarization Setup

As-spun media were polarized by placing in a custom-made holder shown in Figure 3.4 (a) and (b). The custom-made holder helped in 10% uniaxial stretching heating up at 150°C and electrically poling of as-spun media sample. [15]

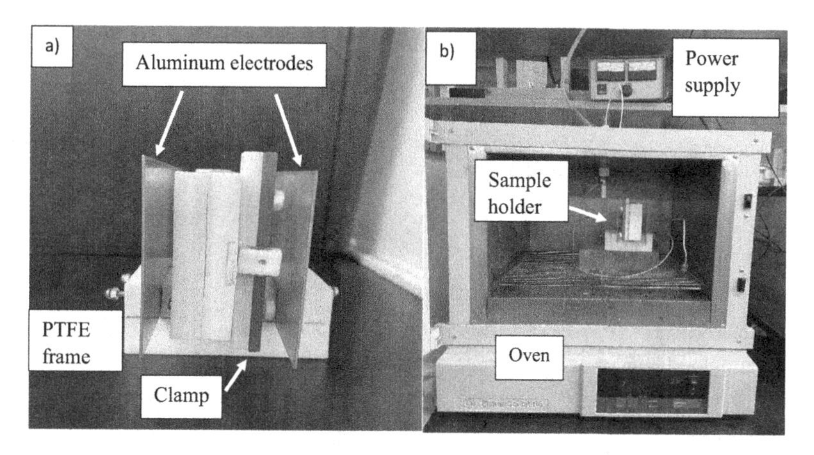

FIGURE 3.4 (a) End view of fabricated sample holder showing the two planar electrodes used to apply the electric field for poling the sample. (b) Photo of sample holder inside of oven and high voltage power supply for charging the electrode above the oven [16].

3.2.5 Characterization

Electric charge of fiber samples was measured using a Faraday bucket. Details regarding working mechanism, measurement techniques, procedures and applications of a Faraday bucket have been explained well in paper [17].

3.3 RESULTS AND DISCUSSIONS

Figure 3.5 shows pore size distributions for as-spun and polarized fiber mats for five different basis weights such as 10, 15, 25, 30 and 45 g/m^2. Pore size distributions were non-symmetric and had longer tails towards larger pore sizes. Increase in basis weights led to smaller pores as well.

Figure 3.6 (a) and (b) shows the mean pore and bubble pore diameter values for as-spun and polarized media for variable basis weights. The plots show that mean and bubble pore diameter continued to decrease with increase in basis weights, but they reached a plateau at 30 g/m^2 and above. In electrospinning, the fibers collected initially form a random mat, but with more fibers being collected, this leads to filling up of larger pores and with time, fewer larger pores get remain exposed leading to increase in thickness as well. Soon a point is reached where collection of more fibers does not cause any reduction of pore size but only leads to enhancement of thickness and strength of mat.

Figure 3.7 (a) and (b) shows the penetration values for two media samples (as-spun and polarized) at five different basis weights such as 10, 15, 25, 30 and 45 g/m^2 and at two different flow rates 20 and 40 lit/min respectively. Polarized media showed lower penetration values which means better efficiencies at all permutations and combinations. Polarization treatments enhance the electrostatic attraction

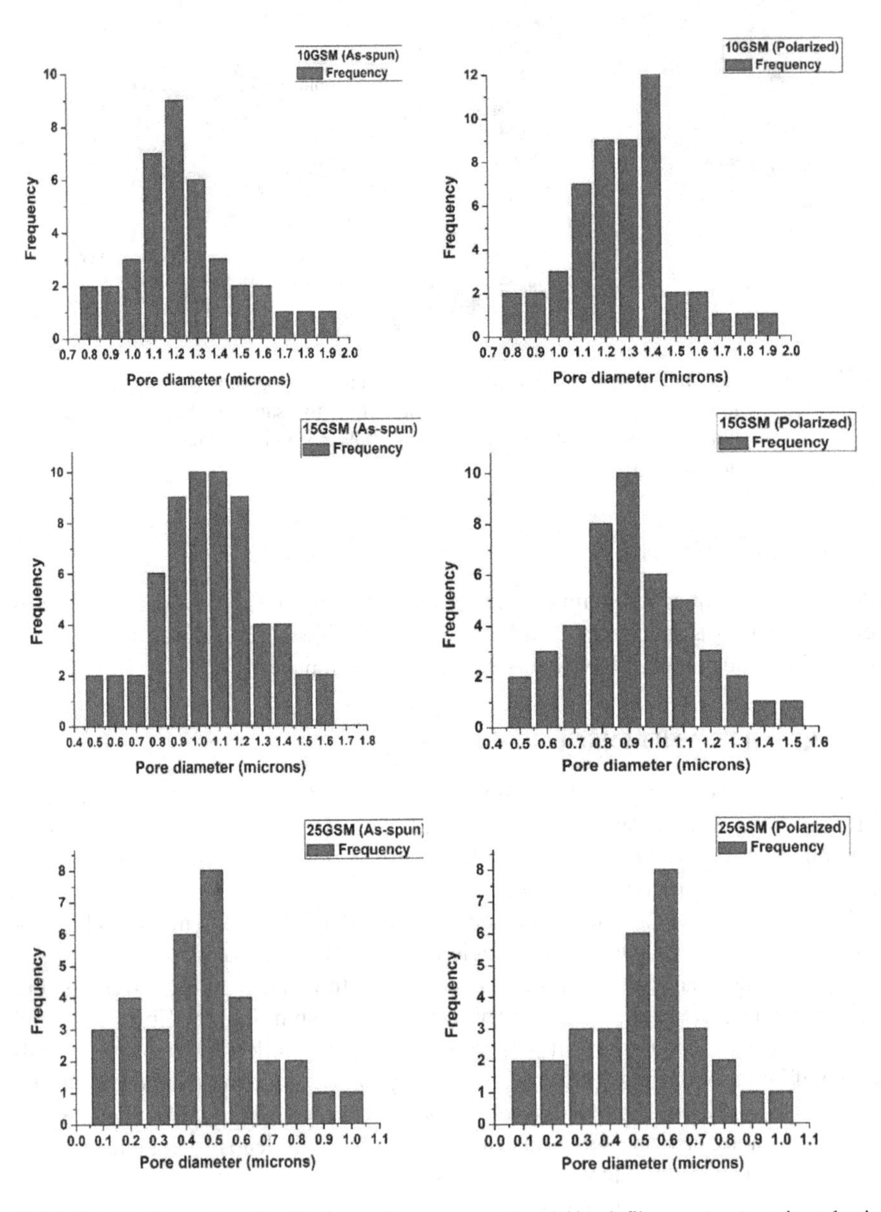

FIGURE 3.5 Pore size distributions of as-spun and polarized fiber mats at various basis weights.

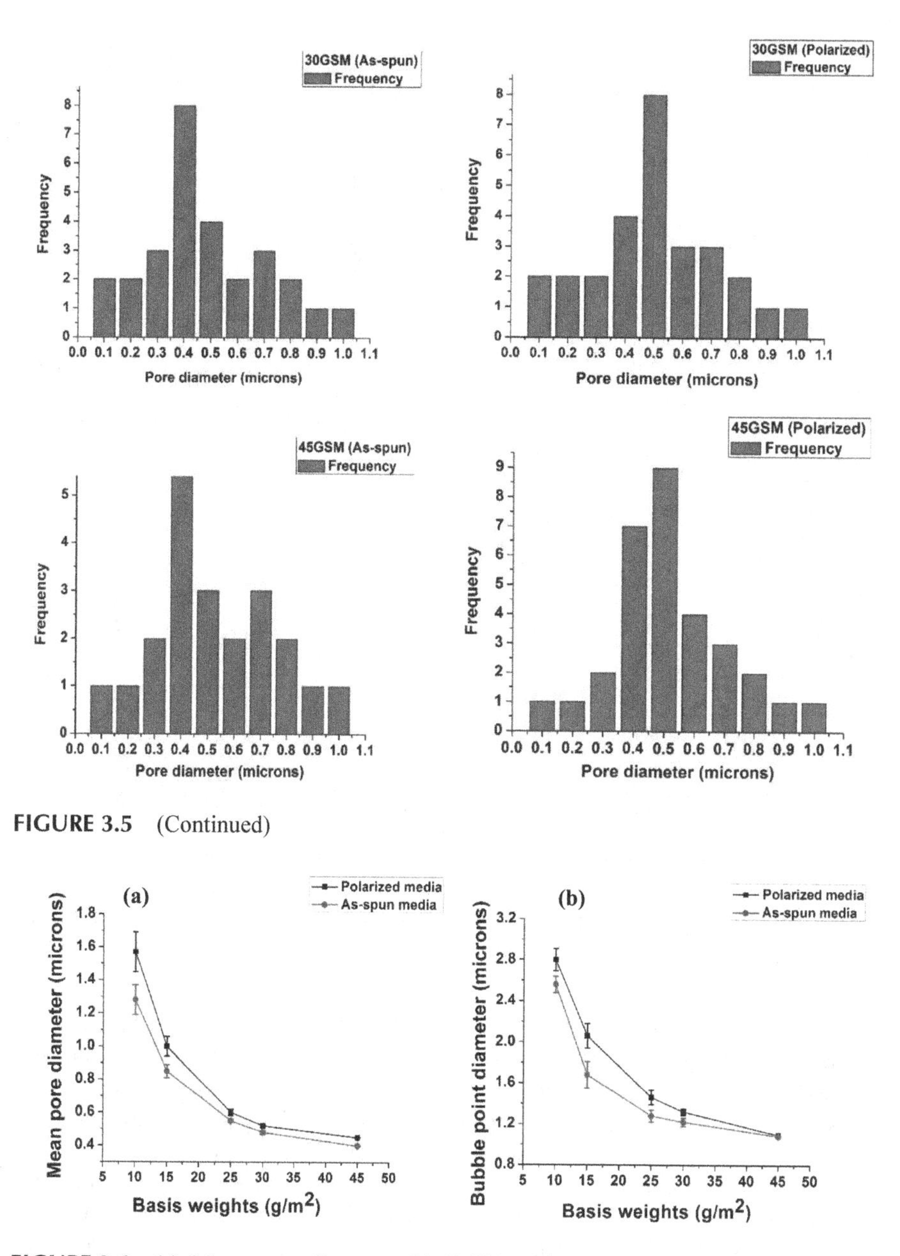

FIGURE 3.5 (Continued)

FIGURE 3.6 (a) Mean pore diameter. (b) Bubble point pore diameter for as-spun and polarized media samples.

mechanism capture and hence, more particles get captured on surface of fibers as compared to as-spun media. As the basis weights increase, the pore sizes decrease and, hence, it leads to more capture than at lower basis weights. Higher flow rate (40 lit/min) may have thrown more particles on sample mats which might lead to more penetration.

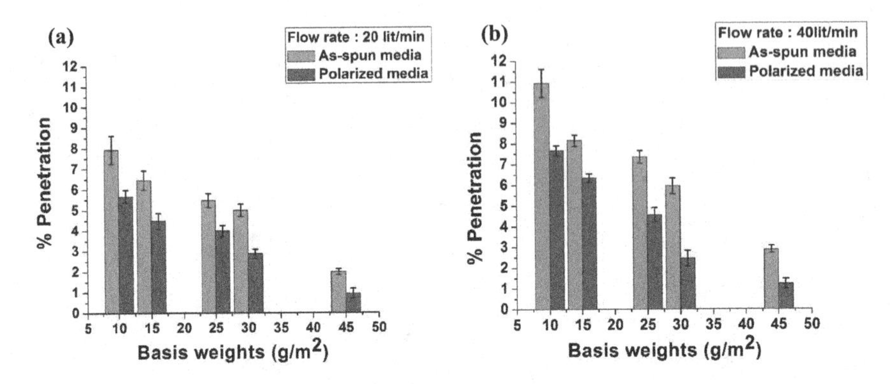

FIGURE 3.7 Penetration values for as-spun and polarized media samples at two flow rates (a) 20 lit/min and (b) 40 lit/min.

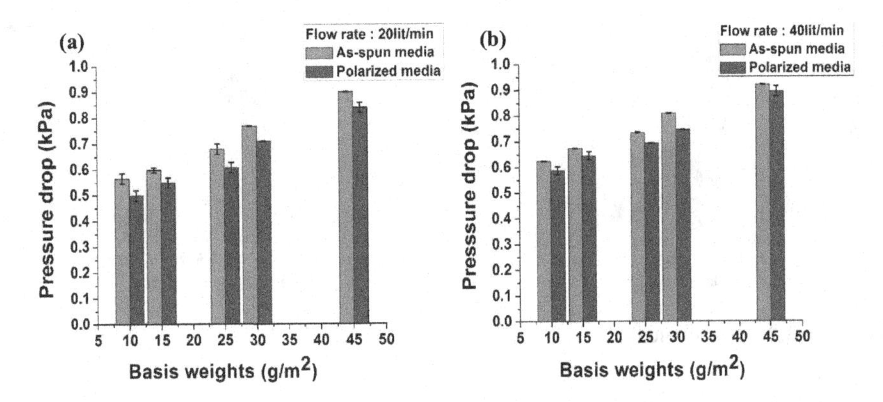

FIGURE 3.8 Pressure values for as-spun and polarized media samples at two flow rates (a) 20 lit/min and (b) 40 lit/min.

Figure 3.8 (a) and (b) shows the pressure drop values for two media samples (as-spun and polarized) at five different basis weights such as 10, 15, 25, 30 and 45 g/m^2 and at two different flow rates such as 20 and 40 lit/min respectively. Similar trends were observed. Polarized fiber media tended to show lower pressure drop at all parameters as compared to as-spun media. This goes to prove that polarized media can be used for a longer time and it will contribute to better cost economics.

Figure 3.9 (a) and (b) shows the penetration values for two media samples (as-spun and polarized) at five basis weights such as 10, 15, 25, 30 and 45 g/m^2 and at two different flow rates such as 20 and 40 lit/min respectively. Filter indices did not change much for both media at all parameters. Filter indices were significantly different for lower basis weights (10 and 20 g/m^2) as compared to higher weights (25, 30 and 45 g/m^2). One reason for this could be that polarization effects had more impact at lower basis weights than higher basis weights.

Figure 3.10 (a) and (b) shows the average surface charge measurements for as-spun PVDF nanofiber samples before and after aerosol testing experiments at different

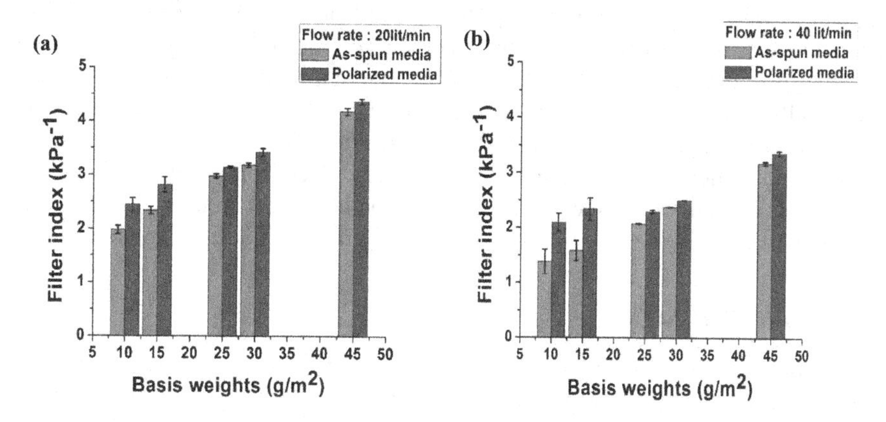

FIGURE 3.9 Filter indices values for as-spun and polarized media samples at two flow rates (a) 20 lit/min and (b) 40 lit/min.

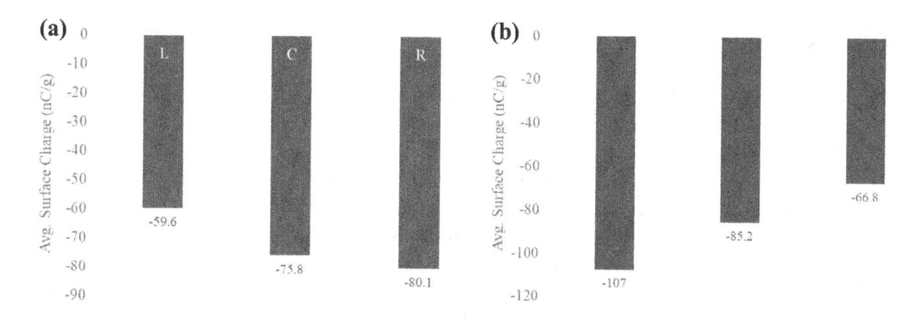

FIGURE 3.10 Surface charge measurements for as-spun PVDF nanofiber samples from three different positions of electrospun mat L (Left), C (Center) and R (Right) (a) pre-salt loading and (b) post-salt loading.

positions of electrospun nanofiber mats. The plots show there was increase in surface charge after salt loading, suggesting capture of salt particles did increase the surface charge. The trend was similar for all the samples cut from all the positions.

3.4 CONCLUSIONS

As-spun and polarized PVDF nanofiber samples were synthesized and tested for salt loading experiments using TSI automated test filter. Polarized filter samples showed greater efficiencies and lower pressure drops than as-spun samples at lower basis weights (10, 15 and 25 g/m^2). The performance values were not very different at higher basis weights. Electrospinning induces electrical and mechanical stresses to PVDF backbone structures which respond to some associated charge. Fiber samples when loaded with salt showed an increase in charge value when measured with a Faraday bucket. This phenomenon can be explained by several assumptions and theories. One of the explanations could be interaction between air-borne NaCl ions with the salt particles on nanofiber matrix. During salt loading, the kinetic energy developed due to collisions between electrons could also impact the charge accumulation on nanofiber matrix. It can be hypothesized that salt loading can generate a new electrically charged nanostructure that can cause positive effects on neutralizing or influencing other particulates passing through the mats [11]. It can affect the already present salt particles present on nanofiber matrices. This explanation is subjected to further experimentation which is not in the scope of present studies. These studies can be pretty useful in some filtration areas involving biomolecular applications.

ACKNOWLEDGMENTS

This work was funded by Coalescence Filtration Fibers Consortium (CFNC) comprising of major filtration companies including Parker Hannifin, Cummins Filtration, Hollingsworth & Vose, Donaldson, Alhstrom. We appreciate all the technical assistance provided by technicians Steve Roberts and William Imes for the building up of a Faraday bucket.

REFERENCES

[1] Henry F.; Ariman T.; Cell Model of Aerosol Collection by Fibrous Filters in an Electrostatic Field. *Journal of Aerosol Science*, 1981, 12, 91–103. DOI: 10.1016/0021-8502(81)90041-0

[2] Lolla D.; Pan L.; Gade H.; Chase G.G.; *Functionalized Polyvinylidene Fluoride Electrospun Fibers and Applications, in Electrospinning Method Used to Create Functional Nanocomposite Films* (Ed. T Tański), Volume 8. London: IntechOpen Limited, 2018. DOI: 10.5772/intechopen.76261

[3] Tomasz T.; Jarka P.; Matysiak W.; eds. *Electrospinning Method Used to Create Functional Nanocomposites Films*. BoD–Books on Demand, 2018. DOI: 10.5772/intechopen.70984

[4] Gade H.; Nikam S.; Chase G.G.; Reneker D.H.; Effect of Electrospinning Conditions on β-phase and Surface Charge Potential of PVDF Fibers. *Polymer*, 2021, 228, 123902. DOI: 10.1016/j.polymer.2021.123902

[5] Kasbe P.; Gade H.; Liu S.; Chase G.G.; Xu W.; Ultrathin Polydopamine-Graphene Oxide Hybrid Coatings on Polymer Filters with Improved Filtration Performance and Functionalities. *ACS Applied BioMaterials*, 2021. DOI: 10.1021/acsabm.1c00367

[6] Liu B.Y.H.; Rubow K.L.; Performance of HEPA and ULPA Filters. *Proceedings, Annual Technical Meeting—Institute of Environmental Sciences*, 1985, 25–8.

[7] Ahn J.H.; Raghavan P.; Sherrington P.C.; Ahn H.J.; Electrospun Polymer Nanofibers: The Booming Cutting Edge Technology. *Reactive and Functional Polymers*, 2012, 72, 915–30. DOI: 10.1016/j.reactfunctpolym.2012.08.018

[8] Yun K.M.; Bao L.; Okuyama K.; Morphology Optimization of Polymer Nanofiber for Applications in Aerosol Particle Filtration. *SEPPUR*, 2010, 75, 340–45. DOI: 10.1016/j.seppur.2010.09.002

[9] Kumar K.R.; Narasimhulu K.; Reddy B.S.K.; Gopal K.R.; Characterization of Aerosol Black Carbon Over a Tropical Semi-arid Region of Anantapur, India. *Atmospheric Research*, 2011, 100, 12–27. DOI: 10.1016/j.atmosres.2010.12.009

[10] Chase G.G.; Reneker D.H.; Lolla D.; Gade H.; Atomic Scale Engineering of Polymer Fibers for Enhanced Filtration Performance. In *8th World Congress on Particle Technology*, 2018 Apr 23. AIChE. https://aiche.confex.com/aiche/wcpt18/meetingapp.cgi/Paper/506343

[11] Kumar B.; ed. *Nanofibers—Synthesis, Properties and Applications*. London: IntechOpen, 2021. DOI: 10.5772/intechopen.92484

[12] Kantanwein R.; Investigation of Surface Charge using Faraday Bucket Measurements: Analysis on Electrospun Polyvinylidene Fluoride Fiber Mats. *Williams Honors College, Honors Research Projects*, 2021, 1280. https://ideaexchange.uakron.edu/honors_research_projects/1280/

[13] Fjeld R.A.; Owens T.M.; The Effect of Particle Charge on Penetration in an Electret Filter. *IEEE Transactions on Industry Applications*, 1988, 24, 725–31.

[14] Lolla D.; Lolla M.; Abutaleb A.; Shin H.U.; Reneker D.H.; Chase G.G.; Fabrication, Polarization of Electrospun Polyvinylidene Fluoride Electret Fibers and Effect on Capturing Nanoscale Solid Aerosols. *Materials*, 2016, 9, 671. DOI: 10.3390/ma9080671

[15] Zhang X.; Yang X.; Chase G.G.; Filtration Performance of Electrospun Acrylonitrile-butadiene Elastic Fiber Mats in Solid Aerosol Filtration. *Separation and Purification Technology*, 2017, 186, 96–105. DOI: 10.1016/j.seppur.2017.06.002

[16] Gade H.; Bokka S.; Chase G.G.; Polarization Treatments of Electrospun PVDF Nanofibers. *Polymer*, 2020, 212, 123152. DOI: 10.1016/j.polymer.2020.123152

[17] Gade H.; Parsa N.; Chase G.G.; Renekar D.H.; Roberts O.S.; Charge Measurement of Electrospun Polyvinylidene Fluoride Fibers using a Custom-made Faraday Bucket. *Review of Scientific Instruments*, 2020, 91, 075107. DOI: 10.1063/1.5142386

4 Experimental Investigation of the Synthetic Polymeric Nanofibers Prepared via Electrospinning

Hossein Ebrahimnezhad-Khaljiri

4.1 INTRODUCTION

Fabricating the materials with nano-scale size not only enhances their properties but also, can introduce new features beyond bulk materials. One of the major groups of nanomaterials is nanofibers, which can be fabricated by various methods. Among them, the electrostatic drawing or electrospinning method is one of the simplest and cheapest methods, which has a proper capability for producing micro and nanofibers from polymers. Given the research, the nanorod and nanotubes offer the fiber length from 100 nm to 10 microns, whereas the electrospun fibers can be kilometers in length. Having caused these features, the electrospinning method attracted much attention during the last decades, so that a tremendous number of experimental studies have been done by researchers [1].

Generally, there are three important sections in the electrospinning systems of polymers, which are the polymer reservoir with the needle, high voltage supplier, and collectors. It should be noted that in some new methods, electrospinning can be performed without the nozzle, which is known as the nozzle-less electrospinning method. There are several parameters, which can influence the final features of electrospun polymeric nanofibers, which can be categorized into two major groups. The first one is processing parameters, which include the distance from the nozzle to the collector, volumetric charge density, relaxation time, solution feed rate, nozzle orifice diameter, ambient conditions, etc. The second group can be related to the features of precursor materials: the kind of solvent, properties of the polymer, additive features and roles, solvent vapor pressure, solution density, etc. [2]. From the experimental viewpoint, the electrospinning of the synthetic polymeric nanofibers can be categorized according to each of mentioned effective factors. Therefore, this chapter aims to review the electrospinning of synthetic polymeric nanofibers according to the type of polymer. Hence, this chapter includes six sections, which are polyester-based nanofibers, vinyl-based

DOI: 10.1201/9781003333814-4

nanofibers, high-performance-based nanofibers, ether-based nanofibers, formaldehyde-based nanofibers, and rubber-based nanofibers.

4.2 POLYESTER-BASED NANOFIBERS

One of the major groups which has been investigated by various scientists is polyester-based electrospun nanofibers. Polyethylene terephthalate (PET), polycaprolactone (PCL), polylactic acid (PLA), and unsaturated polyester belong to this group. These polymers can be used as homopolymers, copolymers, or blending polymers for many applications especially medical applications. One of the important parameters for the electrospinning of nanofibers especially from these polymers can be selecting the proper solvent, which can influence the diameter of electrospun fibers, defect of fibers, and processibility. In an admirable work, Casasola et al. [3] investigated the effect of pure solvent and binary solvent on the morphology and diameter of electrospun PLA nanofibers. In the pure solvent system, they used the acetone, 1,4-dioxane, tetrahydrofuran (THF), dichloromethane (DCM), chloroform (CHL), dimethyl formamide (DMF), and dimethyl acetamide (DMAc). Among them, only acetone solvent could produce the nanofibers. The content of PLA in this solvent was 10% w/v. Therefore, they used the binary solvent system with 10% w/v PLA based on the mixture of acetone and a second solvent. Among these binary systems, acetone/DMF and acetone/DMAc could produce defect-free nanofibers, which can be due to higher electrical conductivity, as compared with other solvents. The thinner nanofibers and narrow diameter distribution were achieved, when the mixing ratio of acetone/DMF was 60 to 40 v/v. By increasing the percentage of acetone, the defect-free nanofibers were more accessible, but the diameter of electrospun fibers showed an increasing trend.

In the interesting work, Qin et al. [4] used self-designed melt electrospinning instead of the conventional method. Also, they used acetyl tributyl citrate (ATBS) as a nontoxic solvent instead of conventional solvents. Using airflow was another effective processing parameter. The reported results showed, by increasing the airflow velocity, the diameter of nanofibers showed a reducing trend. So, it reached 236 nm at the highest air flow velocity. They found that the spinning temperature of PLA with 6 wt.% ATBS was about 240°C. The combination of ATBS and airflow caused the introduction of a new way for producing green PLA nanofibers. Oliveira et al. [5] used the solution blow spinning method for fabricating PLA nanofibers from solvents like CHL, DCM, and dichloroethane (DCE) containing 4, 6, and 8% (w/v) PLA. The nanofibers with the smallest diameter in CHE, DCM, and DCE were obtained when the concentrations of PLA were 6, 8, and 4% w/v, respectively. The size of nanofibers in the mentioned concentrations of solvents were 126, 103, and 151 nm, respectively.

The second polyester-based nanofibers can be electrospun from PET, especially recycled PET. So, it can be said that this polymer can be one of the cheap precursors for electrospinning. For example, the PET bottle can be recycled and used again as electrospun fibers, which was completely studied by Mehdi et al. [6]. Also, these nanofibers can be used as an absorbent for the micro-extraction of heavy metals like chromium (Cr) from water. The recovery of Cr from these nanofibers was about 96.9–99.1% in the study of Sereshti et al. [7]. In another work, Pereao et al. [8]

functionalized PET with diglycolic acid and then fabricated the PET nanofibers for the recovery of rare earth elements.

The third polyester-based polymer for electrospinning nanofibers is PCL. Jeun et al. [9] investigated the morphology of PCL nanofibers. For doing this, they prepared 8–16% (w/v) PCL solutions by using methylene chloride, chloroform, DCE, DMF, n-hexane, and methanol solvents. The obtained results showed that by increasing the concentration of the solution, the morphology changed from beaded fibers to uniform fibers. In the interesting work, Van der Schueren et al. [10] used an alternative solvent instead of a conventional solvent like chloroform. They found that using the formic acid/acetic acid as a mixing solvent caused the diameter of electrospun PCL fibers to be 10 times smaller than electrospun nanofibers from chloroform solution. This mixing system provided the proper situation for fabricating nanofibers under steady-state conditions and nano-scale fibers. So, it can be a better choice than other conventional solvents for electrospinning PCL nanofibers.

Sivan et al. [11] investigated the effect of AC high voltage signal shape and frequencies on the production and spinnability of PCL nanofibers. They concluded that the triangle and sin waveform (except at 50 and 30 Hz) can produce smooth fibers without beads and spindles. The square waveform can produce trimodal fibers morphology consisting of the spindle, beads, and helical fibers.

The unsaturated polyesters have a thermoset nature. Hence, these polymers need crosslinking reactions during the electrospinning process. So, it can be said that controlling these reactions during the process is so difficult and sometimes it can be uncontrollable. However, some researchers succeed to fabricate the nanofibers from these polymers by electrospinning methods. For example, Chiaradia et al. [12] used the in-situ crosslinking method for fabricating electrospun nanofibers from poly(globalide) (PGI) as unsaturated aliphatic polyester (seen Figure 4.1). The crosslinking agents were hexamethylene bis-triazolinediones and methylene diphenyl bis-triazolinediones. In the other work, Dia et al. [13] used the thermal crosslinking method for electrospinning unsaturated polyester macro-monomers (UPM). This macro-monomer was prepared by reactions between poly(2-methyl-1,3 propylene adipate) diol terminated, isophorone diisocyanate, and 2-hydroxyethyl methacrylate in two steps. Also, for better processability, the poly(3-hydroxybutyrate-co-3-hydroxyl valerate) (PHBV) was used. Oliveira et al. [14] crosslinked PGI electrospun fibers through UV-triggered thiol-ene method. Actually, adding a dithiol cross-linker and photo-initiator to the PGI solution created intra-cross-liking during the spinning process.

4.3 VINYL-BASED NANOFIBERS

The second group of polymers, which were electrospun by various science groups for different applications, is vinyl-based polymers. Propylene (PP), polyethylene (PE), polyacrylonitrile (PAN), poly methyl methacrylate (PMMA), polystyrene, polyvinyl alcohol (PVA), and polyvinyl acetate (PVAc) are examples of this group, from which the electrospinning nanofibers have been investigated up to now. Among them, PP is one of the polymers which was electrospun by various methods. For example, Fang et al. [15] fabricated PP nanofibers by the needless melt-electrospinning method. The finest fibers in this work were about 400 nm, which were obtained by using 3% dodecyl trimethyl ammonium bromide. Maeda et al. [16] fabricated the semi-crystalline PP nanofibers

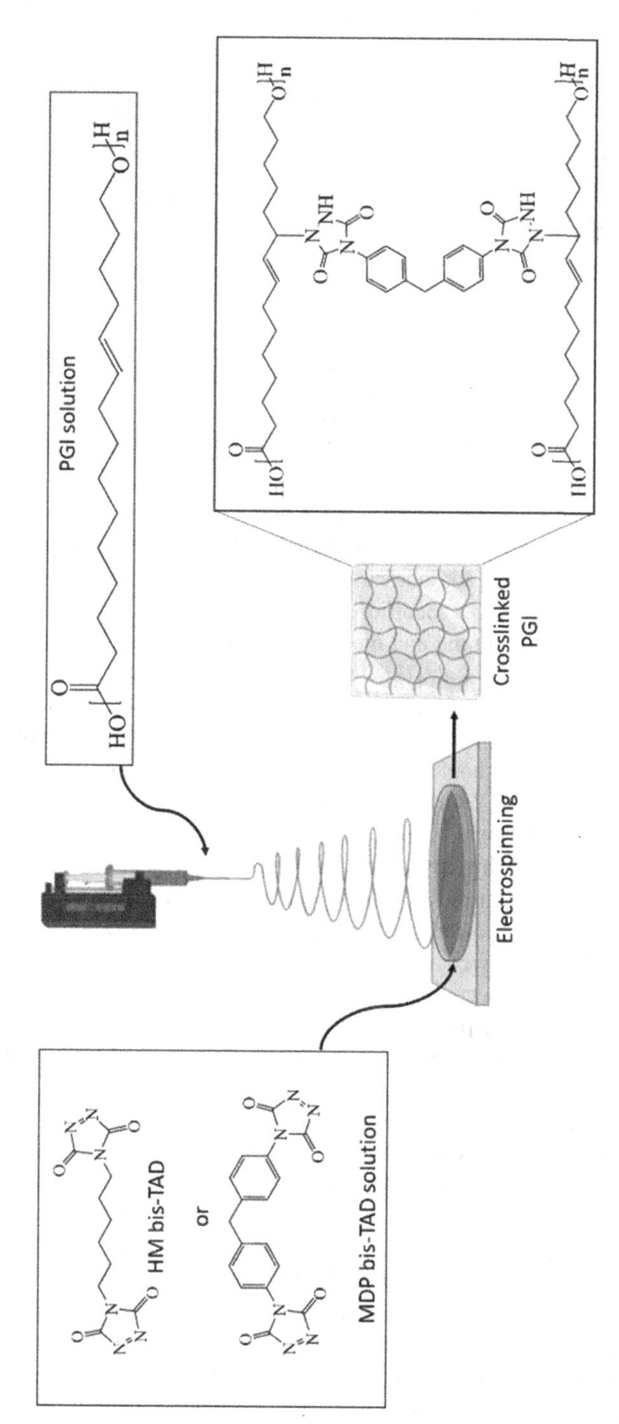

FIGURE 4.1 In-situ crosslinking method for fabricating electrospun PGI nanofibers.

Source: Reprinted from [12] as open access source of MDPI.

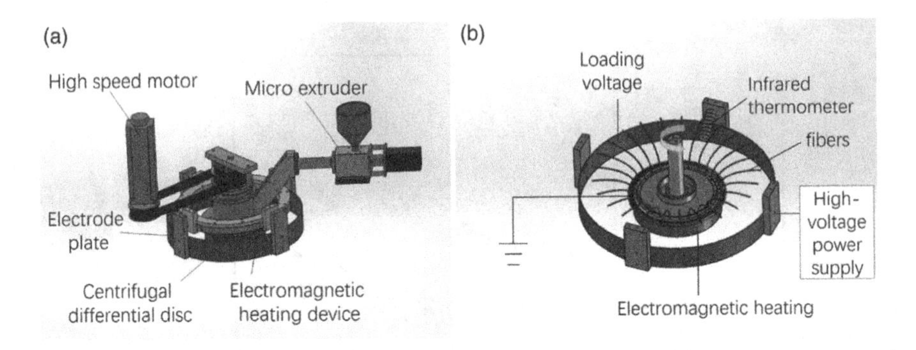

FIGURE 4.2 The schematic form of centrifugal electrospinning, a) spinning section, b) performed centrifugal electrospinning.

Source: Reprinted from [17] with the permission of John Wiley and Sons (License number: 5424710927380).

by using methylcyclohexane as the solvent, which had a 230 nm diameter. They concluded that by selecting the lower voltage and higher viscosity, the non-crystalline PP nanofibers can be produced. But, for fabricating the semi-crystalline PP nanofibers, the moderate gelation speed and lower specific viscosity of the solution should be considered. Combining the electrospinning and melt centrifugal spinning for fabricating the PP nanofibers was done by Liu et al. [17]. The schematic diagram of this method can be seen in Figure 4.2. The advantages of this method were high efficiency, solvent-free, small diameter, and multiple jets for forming nanofibers. The diameter of electrospun PP nanofibers in this method was about 79 nm. This method can be introduced as an eco-friendly method for fabricating electrospun nanofibers.

Characterizing the mat fibers, which have been fabricated by electrospinning PMMA nanofibers, was done by Koysuren et al. [18]. These researchers investigated the effect of solution concentration, solvent type, applied voltage, etc. parameters on the shape and diameter of nanofibers. By enhancing the concentration of the solution, the diameter of fibers showed an increasing trend. The obtained fibers from the dimethyl formamide solution were thicker than the achieved fibers from the acetone solution. Changing the electric potential had a negligible effect on the thickness of fibers, whereas increasing the distance between electrodes caused a reduction in the thickness of PMMA fibers.

One of the parameters which may influence on the properties of electrospun nanofibers can be needle diameter. Macossay et al. [19] investigated the effect of this parameter on the thermal features of PMMA nanofibers. They found that there was no correlation between the used needles with the various diameters, but using the needle with the smaller diameter showed the broader range of diameter of PMMA nanofibers. Comparing the PMMA nanofibers and pure PMMA showed that the thermal stability increased by electrospinning PMMA, but the glass transition temperature of that reduced, which can be due to absorbing water during the spinning process.

The third vinyl-based polymer, which is intensely popular among researchers, is polyvinyl cyanide with the trade name polyacrylonitrile. Although this polymer can be electrospun for using many applications, the nanofibers from that can be converted to oxidized PAN or carbon fibers by heat treating in high temperatures.

Therefore, this polymer is popular, because it is one of the important sources for fabricating carbon fibers, especially electrospun nanofibers. Hence, the features of this polymer and electrospinning conditions can influence the final properties of carbon nanofibers. Gathering the electrospun PAN nanofibers and twisting them into the yarn is one of the methods which can produce fibers with high mechanical features. Yan et al. [20] investigated the effect of twisting angles on the mechanical properties of PAN fibers. The summary results from this work can be seen in Figure 4.3. As can be observed, the fibers with a twisting angle of 35° had the highest mechanical properties compared to the others.

In an interesting work, Wu et al. [21] optimized the stabilization conditions of the electrospun PAN nanofibers. The stabilization of PAN fibers can enhance the conductivity of those for electrical applications. The results showed that the PAN nanofibers had the highest electrical conductivity, when they stabilized in an air atmosphere at the temperature of 280 °C for 2 hours. Aykut et al. [22] studied the effect of various surfactants on the microstructure and features of PAN and carbonized PAN nanofibers. The used surfactants were Triton X-100, sodium dodecyl sulfate (SDS), and hexadecyl trimethyl ammonium bromide, which were nonionic, anionic, and cationic surfactants, respectively. The uniform electrospun nanofibers were obtained when the anionic and cationic surfactants were used. In contrast, the morphology of a bead-on-a-string was seen when these surfactants were used. Also, the cationic surfactant enhanced the defect structures. After carbonization, it was found adding surfactant reduced the disordering in the structures, which was more obvious with the anionic surfactant sample. The latter defects in nanofibers with the anionic surfactant could create the mid-band gap for enhancing electron emission. This means that these nanofibers are proper for electrical applications like new battery generations.

Jia and Qin [23] investigated the effect of using surfactants on the electrospinning PVA nanofibers. They used Gemini quaternary, rosin acid sodium, heterogeneous polyoxyethylene polyoxypropylene ether, and lauryl betaine surfactants, which were cationic, anionic, nonionic, and amphoteric surfactants, respectively. Adding less than 1% surfactant intensely reduced the surface tension of the spinning solution. By increasing the content of anionic and cationic surfactants, the viscosity and electric conductivity of the PVA solution were increased. The fiber diameter was reduced from 405 to 100 nm by adding 1% (v/v) non-ionic surfactant. The heat of fusions and crystallinity of nanofibers were significantly increased by the increment of surfactant content. Using PVA nanofibers for producing carbon nanofibers was done by Fatema et al. [24]. They added the nickel (II) acetate tetrahydrate to the PVA solution during the electrospinning process. After carbonizing these nanofibers, the carbon yield of those was about 34%. This can be due to the accelerating formation of graphite structures at a lower carbonization temperature of 1200 °C.

Optimizing the conditions for electrospinning the PVAc nanofibers was done by Park et al. [25]. They used ethanol as a solvent for the preparation of the PVAc solution. By increasing the spinning solution, the fiber diameters increased. Beads were formed at the concentration of 10 wt.%, whereas distinct fibers were not seen when the concentration was higher than 25 wt.%. The intensity of the electric field became weak when the collector distance was more than 10 cm. Applying 15 kV voltage, using a spinning flow rate of 100 μL/min, being a collector distance of about 10 cm, and using the 15 wt.% PVAc solutions were the optimized conditions, which caused

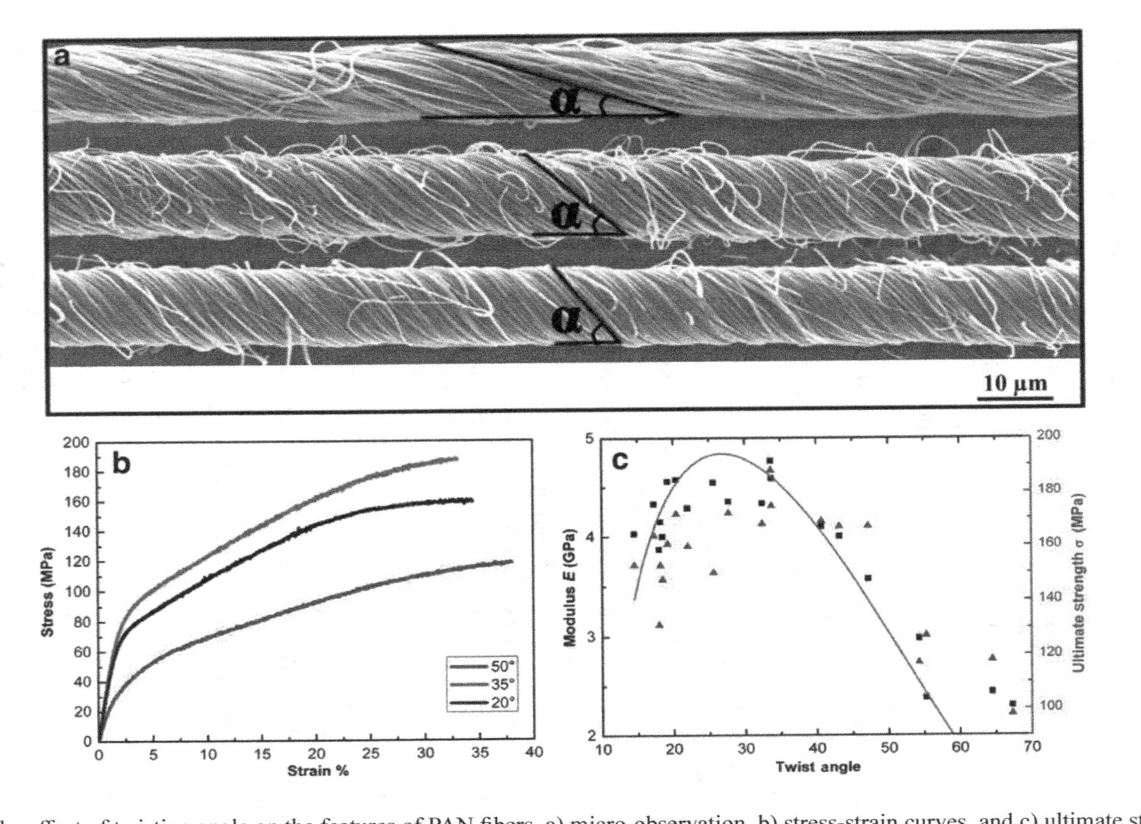

FIGURE 4.3 The effect of twisting angle on the features of PAN fibers, a) micro-observation, b) stress-strain curves, and c) ultimate strength (▲) and elastic modulus (■).

Source: Reprinted from [20] with the permission of Elsevier (License number: 5424711257921).

to produce the nanofibers with 700 nm diameter. In a similar work, Veerabhadraiah et al. [26] used dimethyl formamide instead of ethanol as a solvent. They used 19 kV voltage, 0.5 mL/h flow rate, and an 8 cm distance between the spinneret to the collector to fabricate the nanofibers with a diameter of 24.83 nm.

4.4 HIGH-PERFORMANCE NANOFIBERS

The third group of polymers, which are used for fabricating the nanofibers, can be other high-performance fibers. This group includes polyether ether ketone (PEEK), polysulfone (PSU), polyimide (PI), polyamide (PA), polyurethane (PU), and polycarbonate (PC) polymers. It should be noted that, in most cases, the thermoplastic structure of these polymers (due to their natures) is used for electrospinning the nanofibers, but it can be predicted that the thermoset structures can be electrospun by in-situ crosslinking methods.

One of the most favorable polymers for electrospinning the nanofibers is the polyamide family (with the trade name of nylon), especially polyamide 6.6. For example, Wu et al. [27] characterized the bundles of nanofibers, which were fabricated by electrospinning the PA 66. They first solved the PA 66 in the formic acid with the ratio of 15 wt.% at the temperature of 70 °C for 1 hour. The electrospinning of this solution was performed by applying a 25 kV voltage. The summary data from this work can be seen in Figure 4.4.

Heikkilä et al. [28] electrospun PA with the different chain compositions for filtration applications. The used polymers were PA 612, PA 614, PA 1012, PA 1014, and PA 66. The results showed that the solubility in polar formic acid and the electrospinning ability of PA were improved by changing the chain length. So, PA 66 showed the best results, followed by PA 612 and PA 6. Bazbouz and Stylios [29] optimized the conditions for electrospinning Nylon 6. The investigated parameters were applied voltage (12, 15, and 18 kV), solution concentration (15, 20, and 25 wt.%), and spinning distance

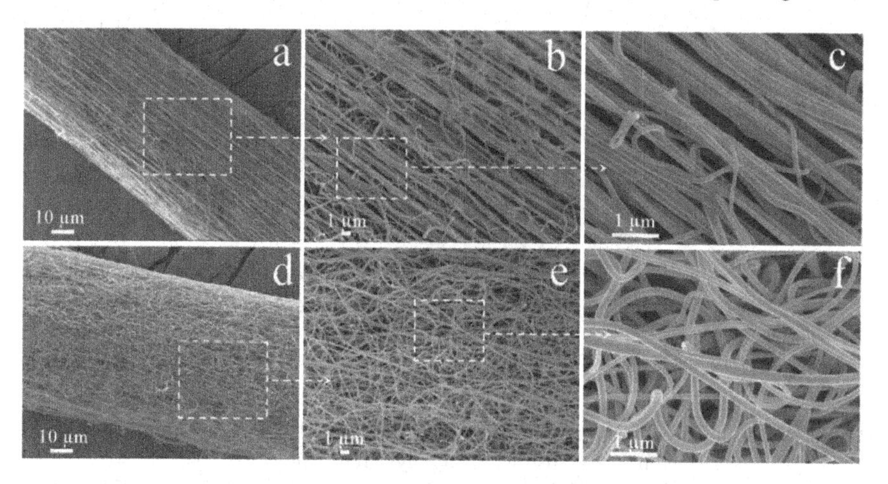

FIGURE 4.4 SEM observation from electrospun PA nanofibers, a-c) PA 66–0, d-F) PA 66–300.

Source: Reprinted from [27] with the permission of Elsevier (License number: 5424711481733).

(5, 8, and 11 cm). The uniform Nylon 6 fibers were obtained by selecting the 20 wt.% solution concentration, 15 kV applied voltage, and spinning distance of 8 cm. Nirmala et al. [30] selected various solvents for electrospinning the PA 6, which were the formic acid and mixtures of formic acid/acetic acid, formic acid/DCM, and formic acid/chlorophenol. As per this work, it was found that formic acid can be a suitable solvent, due to obtaining high aspect ratio nanofibers, in comparison with the mixture solvents.

One of the susceptible polymers for electrospinning the nanofibers can be polystyrene-based. For achieving various morphologies in the electrospun nanofibers, this polymer can be used. For example, Rajak et al. [31] studied the various morphologies of nanofibers by electrospinning the waste polystyrene for aerosol filtration. The used solvents in this study were dimethylformamide and d-limonene. They could produce the nanofibers with the morphologies of smooth, wrinkled, and beaded structures, which had a diameter in the range of 314 to 350 nm only by changing the solvent ratio. The summary of fabricated nanofibers can be seen in Figure 4.5. Uyar

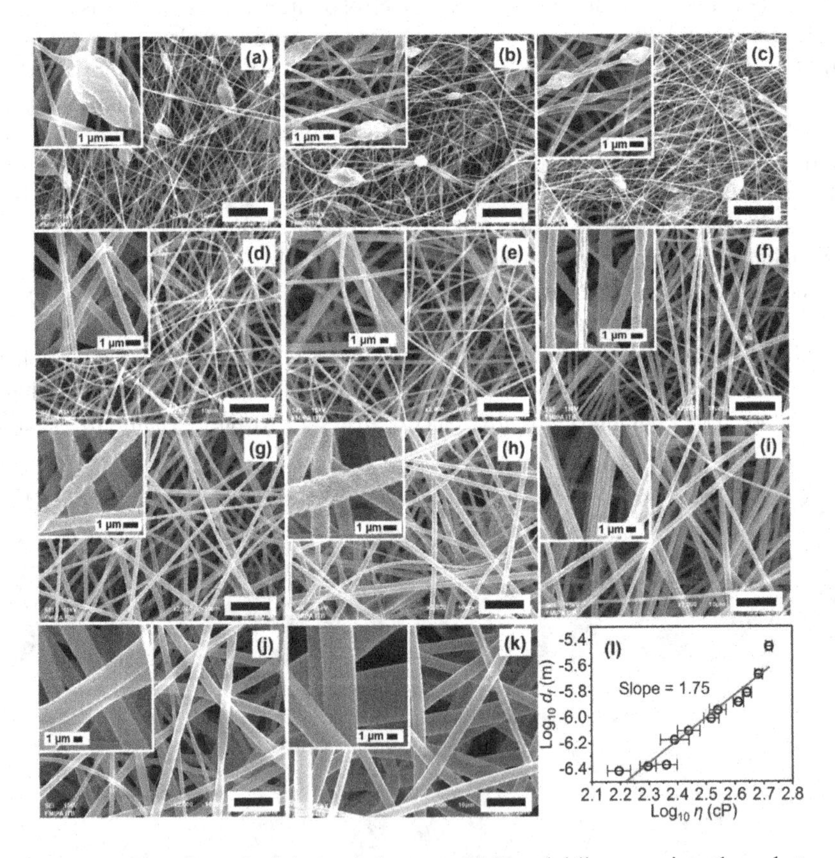

FIGURE 4.5 The effect of mixing ratio between DMF and d-limonene into the polystyrene solution on the microstructure of polystyrene nanofibers: a) 75:00, b) 62.5:12.5, c) 60:15, d) 56.25:18.75, e) 50:25, f) 37.5:37.5, g) 25:50, h) 18.75:56.25, i) 15:60, j) 12.5:62.5, k) 00:75, and l) the relation between fiber diameter and solution viscosity.

Source: Reprinted from [31] with the permission of IOP Publishing (License number: 1288177–1).

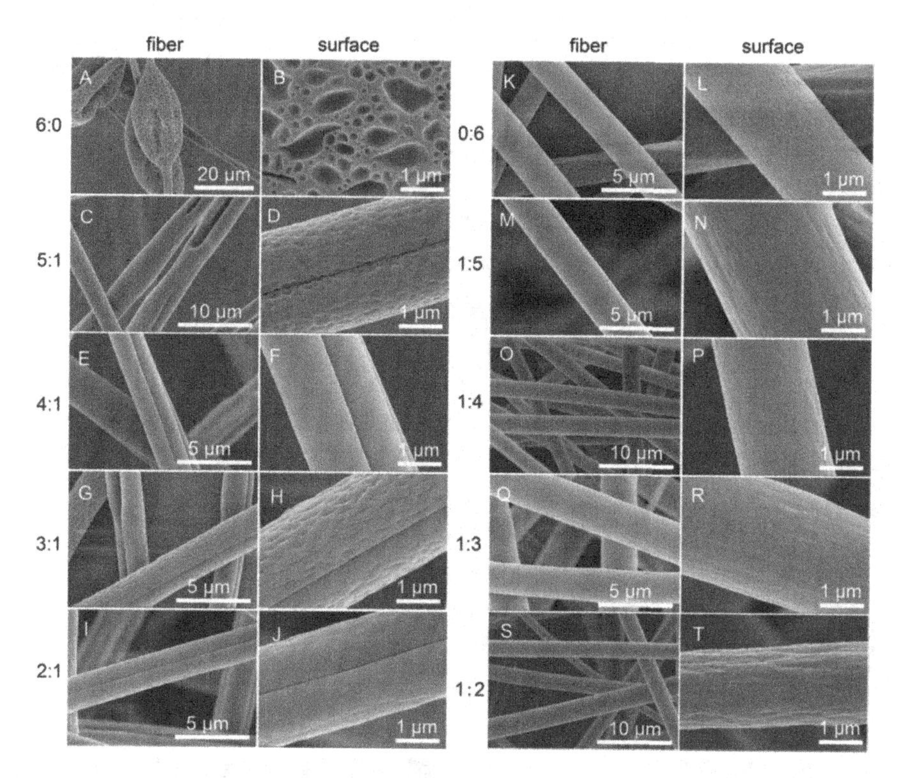

FIGURE 4.6 The effect of various THF/DMF ratios on the morphology of polystyrene nanofibers: (A, B) 6:0, (C, D) 5:1, (E, F) 4:1, (G, H) 3:1, (I, J) 2:1, (K, L) 0:6, (M, N) 1:5, (O, P) 1:4, (Q, R) 1:3, and (S, T) 1:2, v/v.

Source: Reprinted from [33] as open access source of Springer.

and Besenbacher [32] investigated the effect of DMF conductivity on the morphology of polystyrene nanofibers. They confirmed that by using the solvent with the higher conductivity, the bead-free fibers from the solution with lower polymer concentration can be electrospun. It seems that the type of solvents can affect the morphology of polystyrene nanofibers. With this viewpoint, Liu et al. [33] tailored the groove structure of nanofibers by controlling the tetrahydrofuran and DMF ratios, which can be seen in Figure 4.6.

The third polymer in this group, which has the proper capability for electrospinning nanofibers to use various applications is PC. The study about electrospinning this polymer was done by many researchers. For example, Baby et al. [34] fabricated bead-less PC nanofibers. They first dissolved this polymer into the DCM and mixture solvent of DCM/N, N dimethyl formamide (mixing ratio 1:1). Figure 4.7 shows the electrospun nanofibers. The ultrafine, uniform, and bead-less PC fibers with an average diameter of 90 nm were achieved even at 14% w/v concentration.

Da Yong et al. [35] investigated the effect of various surfactants on the morphology of electrospun nanofibers from PC/chloroform solution. The SDS, betaine hydrochloride (BHC), Tween 80, cetane trimethyl ammonium bromide (CTAB),

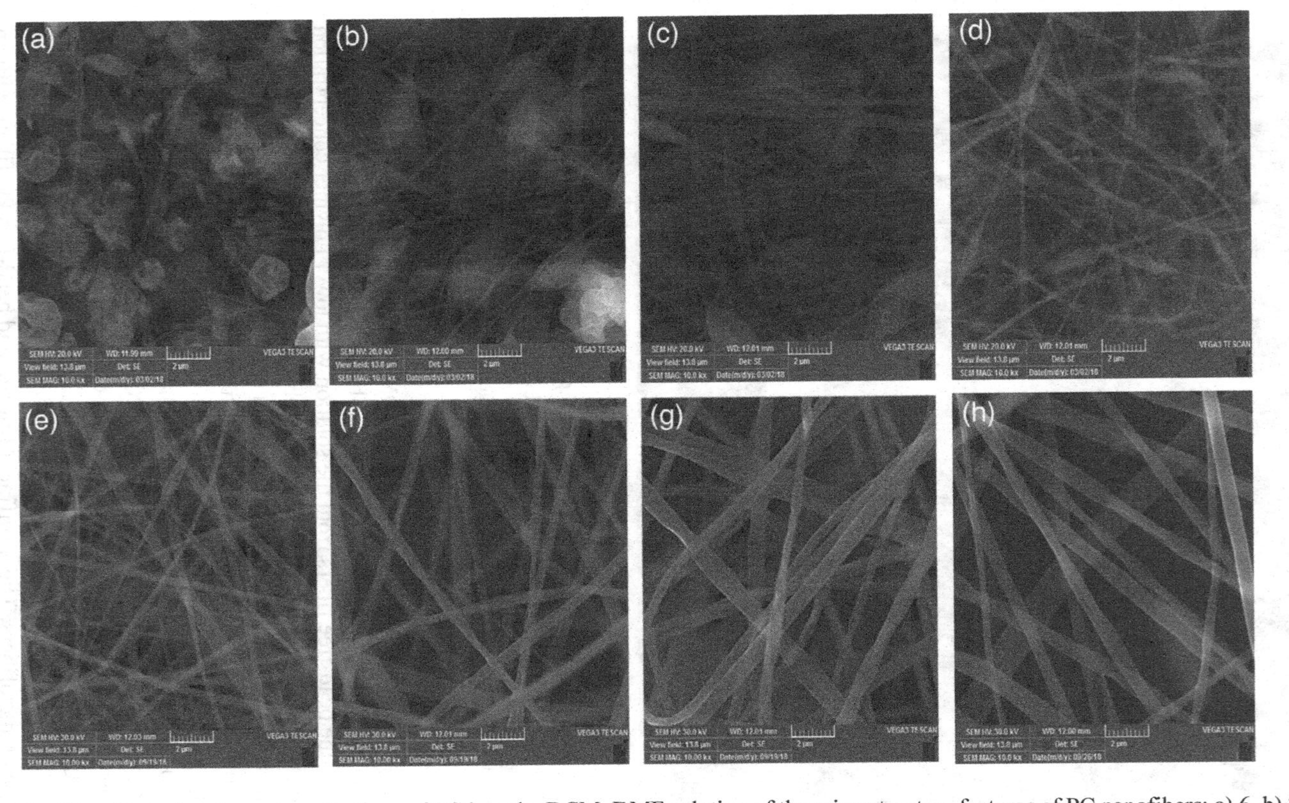

FIGURE 4.7 The effect of various concentrations of PC into the DCM–DMF solution of the microstructure features of PC nanofibers; a) 6, b) 8, c) 10, d) 12, e) 14, f) 16, g) 18, and h) 20 w/v %.

Source: Reprinted from [34] with the permission of John Wiley and Sons (License number: 5424721468834).

and dodecyl trimethyl ammonium bromide (DTAB) were used, which were anionic, zwitterionic, non-ionic, cationic, and cationic surfactants, respectively. According to the results, the CTAB and DTAB as cationic surfactants could form smooth and uniform nanofibers, due to reducing the viscosity of PC solution during the electrospinning process. Moon et al. [36] studied the morphology of PC nanofibers by changing different parameters such as solvent composition, flow rate, and applied voltage. To obtain the aligned PC nanofibers with the best morphology, the fabrication parameters like the polymer concentration, solvent ratio, flow rate, velocity, and applied voltage were optimized, which were 22%, 50:50 (THF:DMF), 0.5 mL/m, 7.3 m/s, and 14 kV, respectively.

Polyimide is introduced as high-performance polymers with the macromolecular repeating unit containing the imide group. This polymer is classified into aliphatic, semi-aromatic, and aromatic materials. Due to having superior thermal stability and mechanical properties, this polymer can be a good choice for fabricating nanofibers by the electrospinning method. Lasprilla-Botero et al. [37] investigated the effect of electrospinning parameters and type of solvent on the morphology and diameter of electrospun PI nanofibers. The summary of obtained data from this work has been illustrated in Figure 4.8. In a similar work, Fukushima et al. [38] assessed the effect of applied voltage and distance between nozzle and collector on electrospun PI nanofibers, which can be seen in Figure 4.9. In an interesting work, Chenge et al. [39] characterized the mechanical features of electrospun 6F-PI nanofibers. According to this work, the ultimate tensile strength, modulus, and elongation at the break of 6F-PI nanofibers were 308 MPa, 2.08 GPa, and 202%, respectively.

Among the various polyurethane families, the thermoplastic PU (TPU) has the proper capability for electrospinning the nanofibers. For this reason, some researchers tried to optimize the fabrication process of these nanofibers. For example, Li et al. [40] investigated the effect of solvent type on the electrospinning of TPU nanofibers. They used the various mixing ratios between TCM/TFE, and compared them with the DMF/THF (as a conventional solvent) to characterize the features of TPU nanofibers. The summary results from this work have been illustrated in Figure 4.10. This polymer can create self-assembled honeycomb structures during the electrospinning process. This unique feature was introduced by Thandavamoorthy et al. [41], which can be seen in Figure 4.11. Similar work was reported by Hu et al. [42], which shows the 3D-assembling of the PU nanofibers. This 3D structure can provide a new platform for fabricating ultrasensitive sensors, ultra-filters, etc.

PEEK is another candidate, which can be used for producing electrospun nanofibers. According to the performed research, the sulfonated PEEK (SPEEK) is generally used for fabricating the nanofibers, due to having the ability for proton changing in electrical applications. One of the interesting works about electrospinning the SPEEK was performed by Boaretti et al. [43]. These researchers investigated the effect of solvent, solution concentrations, and other parameters on the electrospinning ability of this polymer. The reported results from this work can be seen in Figure 4.12.

The last polymer, which is introduced in this group for electrospinning nanofibers, is polysulfone. Yongyi et al. [44] electrospun this polymer by using the gas/jet electrospinning method. The diameter of nanofibers under the optimized conditions (applied voltage of 15 kV, distance between collector and spinneret~19 cm, and solution of 10%) was about 150 nm.

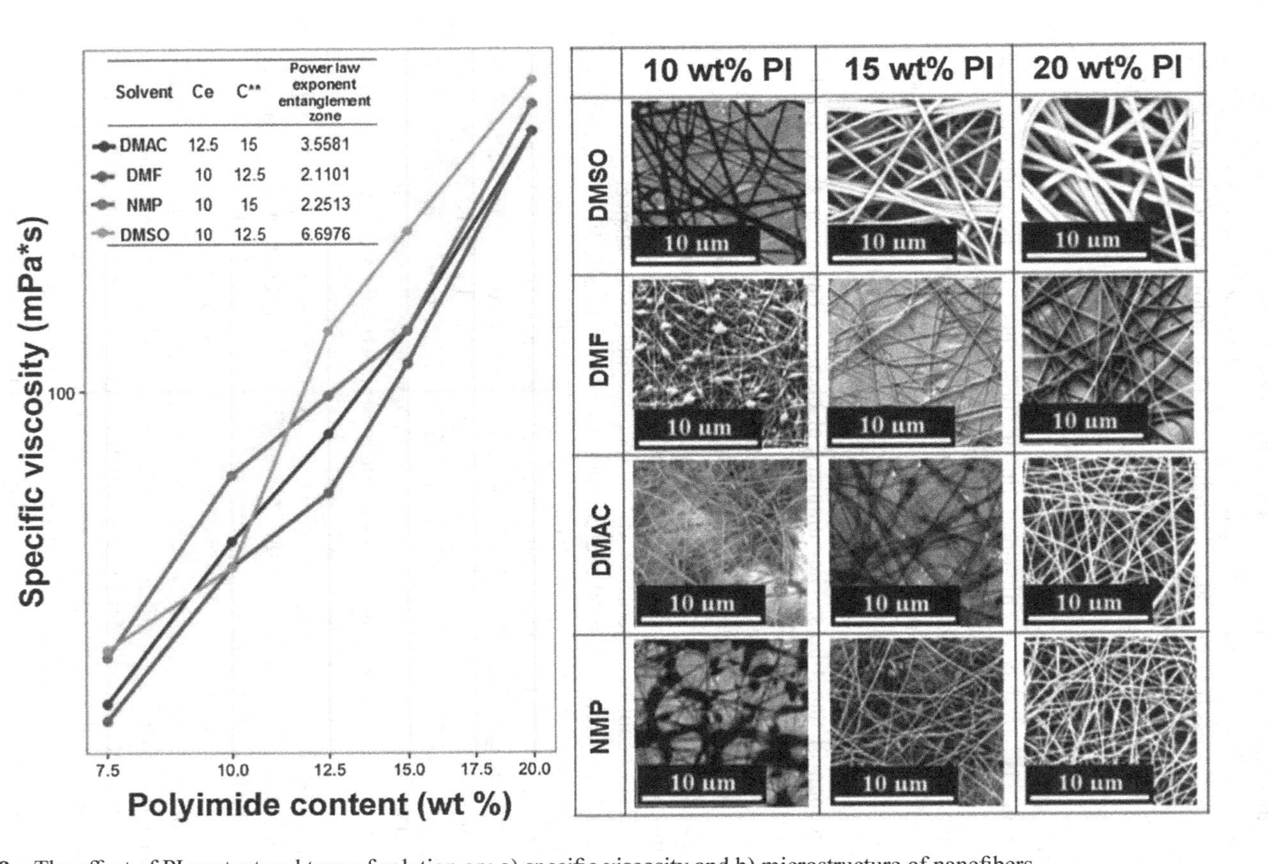

FIGURE 4.8 The effect of PI content and type of solution on: a) specific viscosity and b) microstructure of nanofibers.

Source: Reprinted from [37] with the permission of Elsevier (License number: 5424731242630).

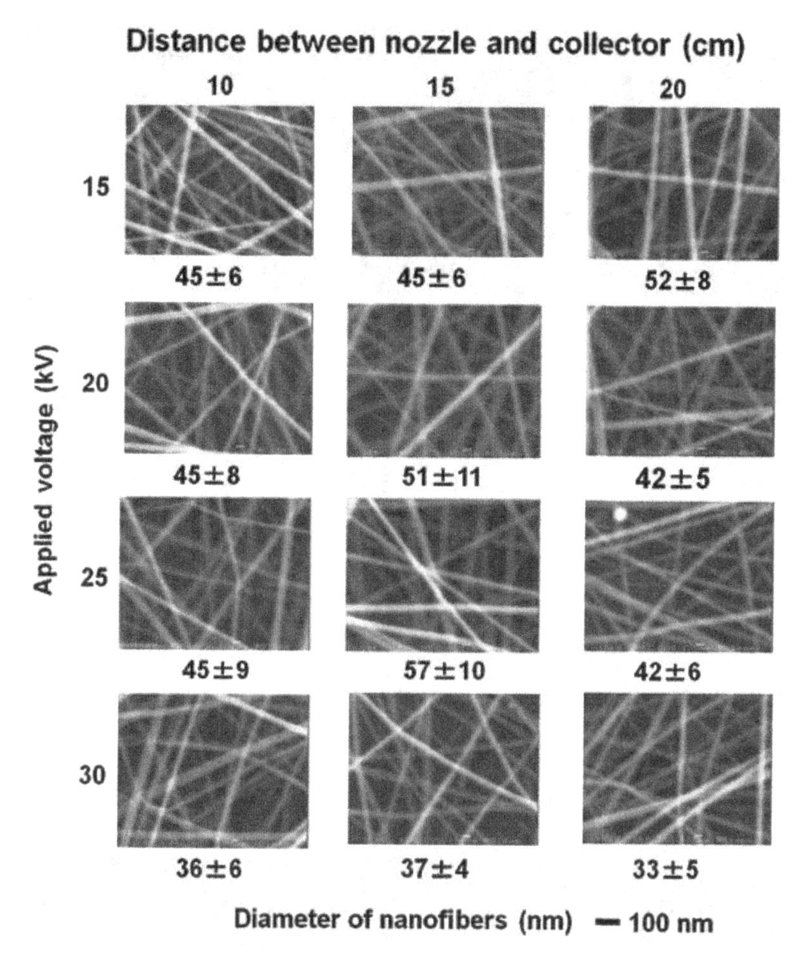

FIGURE 4.9 The effect of distance between nozzle/collector, and applied voltage on the morphology of electrospun PI nanofibers.

Source: Reprinted from [38] with the permission of Springer Nature (License number: 5424731470293).

4.5 ETHER-BASED NANOFIBERS

As one of the interesting polymeric groups, ether-based polymers have the capability for use in electrospinning technologies. In this group, the polyethylene oxide (PEO) (polyethylene glycol) and epoxy families were electrospun for fabricating nanofibers. The most interesting works about polymers are introduced in the following. Song et al. [45] investigated the effect of solvent type on the structure and properties of PEO nanofibers. The used solvents were the various mixing ratios between ethanol and deionized water. According to the results, it was found that the mixed solvent system had no noteworthy effect on the properties of PEO nanofibers, including surface morphology, crystallinity degree, molecular chain orientation,

FIGURE 4.10 The effect of various mixed solutions on the a) morphology of TPU, b) TPU fibers diameter, and c) viscosity of TPU solution, during the electrospinning process.

Source: Reprinted from [40] as open access source of MDPI.

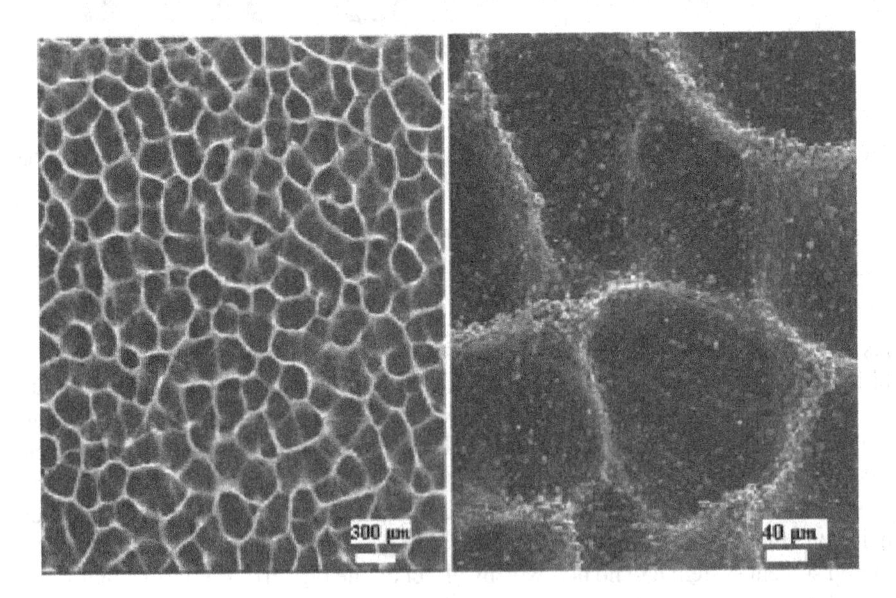

FIGURE 4.11 The microstructure of self-assembled PU electrospun nanofiber.

Source: Reprinted from [41] with the permission of John Wiley and Sons (License number: 5424740153905).

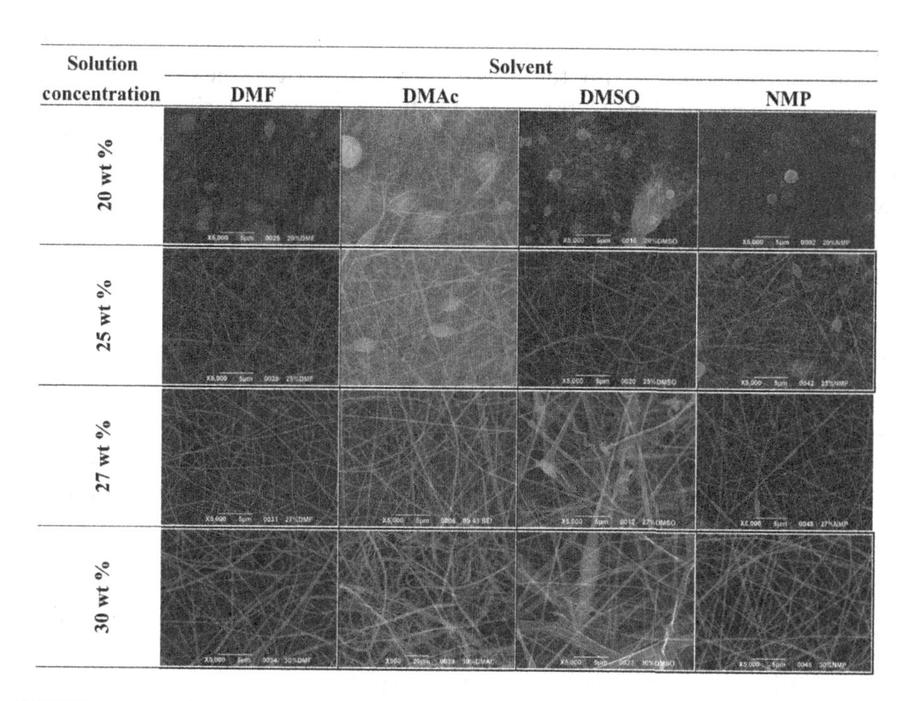

FIGURE 4.12 The effect of SPEEK concentration and various solutions on the morphology of electrospun SPEEK nanofibers.

Source: Adapted from [43] as open access source of MDPI.

diameter distribution, fiber diameters, spinnability, and productivity. The wrinkly morphology was seen when the concentration of ethanol was increased in the mixed solvent. The increment of the evaporation rate of ethanol in the solvent was the reason for this phenomenon. In another work, Tripatanasuwan et al. [46] assessed the effect of evaporation and solidification of PEO from its solution during electrospinning nanofibers. By enhancing the humidity from 5.1 to 48.7%, the diameter of nanofibers was reduced from 253 to 144 nm. In humidity more than 50%, the diameter of fibers was reduced, but the beads were formed on these fibers.

The epoxy resins have a thermoset nature and need the curing process to be solid state. Therefore, performing the curing process is an important challenge for this polymer. For this reason, limited research has introduced new ways for solving this issue. The simplest and most accessible way can be the core/sheath method. In this method, the epoxy, its hardener, and the proper solvent are mixed, which is used as the core. The second polymer like PVA was selected as a shell, which is mixed with the proper solvent. By using the coaxial spinning method, the nanofibers with the epoxy core and polymeric shell are formed. The epoxy resin is cured into the shell. Finally, by removing the shell by using a solvent or thermal degradation methods, epoxy nanofibers can obtain. Wang et al. [47] used polyvinyl pyrrolidone (PVP) as the shell for fabricating epoxy nanofibers. After curing the epoxy core, the PVP shell was removed by using ethanol. The second method is using the semi-curing method. In this way, the epoxy and hardener are mixed, but the mixing ratio is different because the aim is semi-curing and is not complete curing. Then, the mixture is electrospun to fabricate the nanofibers. The residue curing can be done by the catalyst or thermal process during the spinning [48]. The third method is creating soft and hard sections in the epoxy polymer. This method was previously used for the fabrication of shape-memory polymers. By this method, Li et al. [49] first synthesized the thermoplastic elastomer-epoxy by expanding the bisphenol structure. Then, they electrospun this polymer to fabricate the epoxy nanofibers.

4.6 FORMALDEHYDE-BASED NANOFIBERS

The next important type of polymers which can be electrospun is formaldehyde-based polymers or resins. This group includes melamine formaldehyde (MF), phenol formaldehyde (PF), and resorcinol formaldehyde (RF) resins. As can be expected, this group of polymers is the thermoset polymers, which can be used in different applications. For example, Wang et al. [50] electrospun the MF for use as membranes for lithium-ion batteries for increasing the cycling performance of those at elevated temperatures. The other application for MF nanofibers can be fabricating carbon nanofibers from those, which was studied by Ma et al. [51]. By researching about electrospinning PF resin, it was realized that this resin is only electrospun for fabricating carbon nanofibers for various applications like dye absorption and super-capacitor electrodes [52].

One of the interesting subjects about thermoset polymers is using PVA beside them. For example, Zu-Sheng et al. [53] electrospun MF fibers by helping PVA. From this work, it was found that by increasing the PVA content in the solution of MF, the average diameter of fibers became larger. Badhe et al. [54] mixed RF solution with PVA solution (mixing ratio~1:1) to fabricate the nanofibers by core/sheath method.

4.7 RUBBER-BASED NANOFIBERS

Rubbers are one of the interesting polymers which can be used for electrospinning nanofibers. For example, Kim et al. [55] electrospun poly benzimidazole for producing nanofibers. For increasing the tensile strength of that, they surface-treated these fibers with sulfuric acid. The diameter of these nanofibers was around 300 nm. Also, because of aligning molecules in fibers, the obtained data showed proper tensile strength. In admirable work, Nie et al. [56] electrospun smooth trans-poly isoprene nanofibers by modifying parameters like solution concentration, solvent, environment, applied voltage, and distance to the collector. The novelty of this work was the chemical crosslinking of this rubber by vulcanization at high temperatures in a short time. Photo-crosslinking the styrene-butadiene electrospun nanofibers was done by Vitale et al. [57] by using a thiol crosslinking agent. These nanofibers can be used for a wide range of applications like stretchable electronics, bio-filtrations, and especially chemical and biological sensing. One of the important rubbers in electrospinning technology can be dimethyl siloxane rubber. This rubber is not directly used for electrospinning, and generally is used as a second polymer in the electrospinning process. This polymer is used in the form of grafting, blending, blocking, compositing, or shelling for other electrospun polymers. Also, as a new application, the electrospun nanofibers of this polymer are heated at high temperatures to fabricate ceramic electrospun nanofibers [58].

4.8 FUTURE PERSPECTIVE

Up to now, many efforts have been done to electrospun synthetic polymers for fabricating nanofibers. As reviewed in this chapter, it can be found that most of the polymers which are used for electrospinning have thermoplastic nature. It seems that thermoset polymers have important issues during the spinning process. Therefore, it can be predicted that developing in-situ crosslinking methods is one of the most important and interesting subjects for researchers in the future. Developing a needless spinning process, combining systems like electrospinning- centrifugal methods, solvent-free electrospinning, melt electrospinning, and introducing eco-friendly solvents are the subjects which will be studied or developed by researchers in the near future.

4.9 CONCLUSION

Electrospinning is one of the progressive methods which has the proper capability for fabricating continuous nanofibers from synthetic polymers. Synthetic polymeric nanofibers can be categorized according to the type of used polymers. In the polyester-based category, unsaturated polyester, polyethylene terephthalate, polycaprolactone, and polylactic acid are the polymers which were electrospun by various researchers. The second major group of polymers for electrospinning can be vinyl-based polymers, which include poly methyl methacrylate, polyacrylonitrile, polystyrene, polyvinyl alcohol, polyvinyl acetate, and polypropylene. The third group of polymers can be polyether ether ketone, polysulfide, polyimide, polyamide, polyurethane, and polycarbonate. Polyethylene oxide and epoxy families can be categorized as ether-based nanofibers. The formaldehyde-based polymers (melamine formaldehyde, phenol, formaldehyde, and resorcinol formaldehyde) can resist in high temperatures. Therefore, some researchers focused to fabricate the nanofibers by the electrospinning method from this family. Some researchers tried to spin nanofibers from rubbers, due to their flexibility features. Among them, dimethyl siloxane has achieved the highest attention.

REFERENCES

[1] Brown, T. D., P. D. Dalton, and D. W. Hutmacher. 2016. "Melt Electrospinning Today: An Opportune Time for an Emerging Polymer Process." Progress in Polymer Science 56: 116–66. https://doi.org/10.1016/j.progpolymsci.2016.01.001.

[2] Valizadeh, A., and S. M. Farkhani. 2014. "Electrospinning and Electrospun Nanofibres." IET Nanobiotechnology 8 (2): 83–92. https://doi.org/10.1049/iet-nbt.2012.0040.

[3] Casasola, R., N. L. Thomas, A. Trybala, and S. Georgiadou. 2014. "Electrospun Poly Lactic Acid (PLA) Fibres: Effect of Different Solvent Systems on Fibre Morphology and Diameter." Polymer 55 (18): 4728–37. https://doi.org/10.1016/j.polymer.2014.06.032.

[4] Qin, Y., L. Cheng, Y. Zhang, X. Chen, X. Wang, X. He, W. Yang, Y. An, and H. Li. 2018. "Efficient Preparation of Poly(Lactic Acid) Nanofibers by Melt Differential Electrospinning with Addition of Acetyl Tributyl Citrate." Journal of Applied Polymer Science 135 (31): 42–5. https://doi.org/10.1002/app.46554.

[5] Oliveira, J., G. S. Brichi, J. M. Marconcini, L. H. C. Mattoso, G. M. Glenn, and E. S. Medeiros. 2014. "Effect of Solvent on the Physical and Morphological Properties of Poly (Lactic Acid) Nanofibers Obtained by Solution Blow Spinning." Journal of Engineered Fibers and Fabrics 9 (4): 117–25. https://doi.org/10.1177/155892501400900414.

[6] Mehdi, M., F. K. Mahar, U. A. Qureshi, M. Khatri, Z. Khatri, F. Ahmed, and I. S. Kim. 2018. "Preparation of Colored Recycled Polyethylene Terephthalate Nanofibers from Waste Bottles: Physicochemical Studies." Advances in Polymer Technology 37 (8): 2820–27. https://doi.org/10.1002/adv.21954.

[7] Sereshti, H., F. Amini, and H. Najarzadekan. 2015. "Electrospun Polyethylene Terephthalate (PET) Nanofibers as a New Adsorbent for Micro-Solid Phase Extraction of Chromium(VI) in EnVIronmental Water Samples." RSC Advances 5 (108): 89195–203. https://doi.org/10.1039/c5ra14788c.

[8] Pereao, O., K. Laatikainen, C. Bode-Aluko, O. Fatoba, E. Omoniyi, Y. Kochnev, A. N. Nechaev, P. Apel, and L. Petrik. 2021. "Synthesis and Characterisation of Diglycolic Acid Functionalised Polyethylene Terephthalate Nanofibers for Rare Earth Elements Recovery." Journal of Environmental Chemical Engineering 9 (5): 105902. https://doi.org/10.1016/j.jece.2021.105902.

[9] Jeun, J. P., Y. M. Lim, and Y. C. Nho. 2005. "Study on Morphology of Electrospun Poly (Caprolactone) Nanofiber." Journal of Industrial and Engineering Chemistry 11 (4): 573–78.

[10] Schueren, L. V. D., B. D. Schoenmaker, Ö. I. Kalaoglu, and K. D. Clerck. 2011. "An Alternative Solvent System for the Steady State Electrospinning of Polycaprolactone." European Polymer Journal 47 (6): 1256–63. https://doi.org/10.1016/j.eurpolymj.2011.02.025.

[11] Sivan, M., D. Madheswaran, J. Valtera, E. K. Kostakova, and D. Lukas. 2022. "Alternating Current Electrospinning: The Impacts of Various High-Voltage Signal Shapes and Frequencies on the Spinnability and Productivity of Polycaprolactone Nanofibers." Materials and Design 213: 110308. https://doi.org/10.1016/j.matdes.2021.110308.

[12] Chiaradia, V., S. B. Hanay, S. D. Kimmins, D. D. Oliveira, P. H. H. Araújo, C. Sayer, and A. Heise. 2019. Tad, "Crosslinking of Electrospun Fibres from Unsaturated Polyesters Bis-triazolinedione." Polymers 11: 1808. https://doi.org/10.3390/polym11111808

[13] Dai, T.-H., H. Yu, K. Zhang, M.-F. Zhu, Y.-M. Chen, and H.-J. Adler. 2008. "Fabricating Novel Thermal Crosslinked Ultrafine Fibers via Electrospinning." Journal of Applied Polymer Science 107: 2142–49. https://doi.org/10.1002/app.26655

[14] Oliveira, F. C. S. D., D. Olvera, M. J. Sawkins, S. A. Cryan, S. D. Kimmins, T. E. D. Silva, D. J. Kelly, G. P. Duffy, C. Kearney, and A. Heise. 2017. "Direct UV-Triggered Thiol-Ene Cross-Linking of Electrospun Polyester Fibers from Unsaturated Poly(Macrolactone)s and Their Drug Loading by Solvent Swelling." Biomacromolecules 18 (12): 4292–98. https://doi.org/10.1021/acs.biomac.7b01335.

[15] Fang, J., L. Zhang, D. Sutton, X. Wang, and T. Lin. 2012. "Needleless Melt-Electrospinning of Polypropylene Nanofibres." Journal of Nanomaterials 2012: 382639. https://doi.org/10.1155/2012/382639.

[16] Maeda, T., K. Takaesu, and A. Hotta. 2016. "Syndiotactic Polypropylene Nanofibers Obtained from Solution Electrospinning Process at Ambient Temperature." Journal of Applied Polymer Science 133 (13): 43238. https://doi.org/10.1002/app.43238.

[17] Liu, Y. J., J. Tan, S. Y. Yu, M. Yousefzadeh, T. Lyu, Z. W. Jiao, H. Li, and S. Ramakrishna. 2020. "High-Efficiency Preparation of Polypropylene Nanofiber by Melt Differential Centrifugal Electrospinning." Journal of Applied Polymer Science 137 (3): 48299. https://doi.org/10.1002/app.48299.

[18] Koysuren, O., and H. N. Koysuren. 2016. "Characterization of Poly(Methyl Methacrylate) Nanofiber Mats by Electrospinning Process." Journal of Macromolecular Science, Part A: Pure and Applied Chemistry 53 (11): 691–98. https://doi.org/10.1080/10601325.2016.1224627.
[19] Macossay, J., A. Marruffo, R. Rincon, T. Eubanks, and A. Kuang. 2007. "Effect of Needle Diameter on Nanofiber Diameter and Thermal Properties of Electrospun poly(methyl methacrylate)." Polymers for Advanced Technologies 18: 180–3. https://doi.org/10.1002/pat.844.
[20] Yan, H., L. Liu, and Z. Zhang. 2011. "Continually Fabricating Staple Yarns with Aligned Electrospun Polyacrylonitrile Nanofibers." Materials Letters 65 (15–16): 2419–21. https://doi.org/10.1016/j.matlet.2011.04.091.
[21] Wu, M., Q. Wang, K. Li, Y. Wu, and H. Liu. 2012. "Optimization of Stabilization Conditions for Electrospun Polyacrylonitrile Nanofibers." Polymer Degradation and Stability 97 (8): 1511–19. https://doi.org/10.1016/j.polymdegradstab.2012.05.001.
[22] Aykut, Y., B. Pourdeyhimi, and S. A. Khan. 2013. "Effects of Surfactants on the Microstructures of Electrospun Polyacrylonitrile Nanofibers and Their Carbonized Analogs." Journal of Applied Polymer Science 130 (5): 3726–35. https://doi.org/10.1002/app.39637.
[23] Jia, L., and X. H. Qin. 2013. "The Effect of Different Surfactants on the Electrospinning Poly(Vinyl Alcohol) (PVA) Nanofibers." Journal of Thermal Analysis and Calorimetry 112 (2): 595–605. https://doi.org/10.1007/s10973-012-2607-9.
[24] Fatema, U. K., A. J. Uddin, K. Uemura, and Y. Gotoh. 2011. "Fabrication of Carbon Fibers from Electrospun Poly(Vinyl Alcohol) Nanofibers." Textile Research Journal 81 (7): 659–72. https://doi.org/10.1177/0040517510385175.
[25] Park, J. Y., I. H. Lee, and G. N. Bea. 2008. "Optimization of the Electrospinning Conditions for Preparation of Nanofibers from Polyvinylacetate (PVAc) in Ethanol Solvent." Journal of Industrial and Engineering Chemistry 14 (6): 707–13. https://doi.org/10.1016/j.jiec.2008.03.006.
[26] Veerabhadraiah, A., S. Ramakrishna, G. Angadi, M. Venkatram, V. K. Ananthapadmanabha, N. M. H. NarayanaRao, and K. Munishamaiah. 2017. "Development of Polyvinyl Acetate Thin Films by Electrospinning for Sensor Applications." Applied Nanoscience (Switzerland) 7 (7): 355–63. https://doi.org/10.1007/s13204-017-0576-9.
[27] Wu, S., B. Wang, G. Zheng, S. Liu, K. Dai, C. Liu, and C. Shen. 2014. "Preparation and Characterization of Macroscopically Electrospun Polyamide 66 Nanofiber Bundles." Materials Letters 124: 77–80. https://doi.org/10.1016/j.matlet.2014.03.048.
[28] Heikkilä, P., A. Taipale, M. Lehtimäki, and A. Harlin. 2008. "Electrospinning of Polyamides With Different Chain Compositions for Filtration Application." Polymer Engineering and Science 48 (6): 1168–76. https://doi.org/10.1002/pen.21070.
[29] Bazbouz, M. B., and G. K. Stylios. 2008. "Alignment and Optimization of Nylon 6 Nanofibers by Electrospinning." Journal of Applied Polymer Science 107 (5): 3023–32. https://doi.org/10.1002/app.27407.
[30] Nirmala, R., H. R. Panth, C. Yi, K. T. Nam, S. J. Park, H. Y. Kim, and R. Navamathavan. 2010. "Effect of Solvents on High Aspect Ratio Polyamide-6 Nanofibers via Electrospinning." Macromolecular Research 18 (8): 759–65. https://doi.org/10.1007/s13233-010-0808-2.
[31] Rajak, A., D. A. Hapidin, F. Iskandar, M. M. Munir, and K. Khairurrijal. 2019. "Controlled Morphology of Electrospun Nanofibers from Waste Expanded Polystyrene for Aerosol Filtration." Nanotechnology 30 (42): 425602. https://doi.org/10.1088/1361-6528/ab2e3b
[32] Uyar, T., and F. Besenbacher. 2008. "Electrospinning of Uniform Polystyrene Fibers: The Effect of Solvent Conductivity." Polymer 49 (24): 5336–43. https://doi.org/10.1016/j.polymer.2008.09.025.
[33] Liu, W., C. Huang, and X. Jin. 2014. "Tailoring the Grooved Texture of Electrospun Polystyrene Nanofibers by Controlling the Solvent System and Relative Humidity." Nanoscale Research Letters 9: 350. https://doi.org/10.1186/1556-276X-9-350.

[34] Baby, T., T. Jose E., P. C. Thomas, and J. T. Mathew. 2019. "A Cost Effective and Facile Approach to Prepare Beadless Polycarbonate Nanofibers with Ultrafine Fiber Morphology." Polymer Engineering and Science 59 (9): 1799–809. https://doi.org/10.1002/pen.25180.

[35] Yang, D. Y., Y. Wang, D. Z. Zhang, Y. Y. Liu, and X. Y. Jiang. 2009. "Control of the Morphology of Micro/Nano-Structures of Polycarbonate via Electrospinning." Chinese Science Bulletin 54 (17): 2911–17. https://doi.org/10.1007/s11434-009-0241-0.

[36] Moon, S. C., and R. J. Farris. 2008. "The Morphology, Mechanical Properties, and Flammability of Aligned Electrospun Polycarbonate (PC) Nanofibers." Polymer Engineering and Science 48 (9): 1848–54. https://doi.org/10.1002/pen.21158.

[37] Lasprilla-Botero, J., M. Álvarez-Láinez, and J. M. Lagaron. 2018. "The Influence of Electrospinning Parameters and Solvent Selection on the Morphology and Diameter of Polyimide Nanofibers." Materials Today Communications 14: 1–9. https://doi.org/10.1016/j.mtcomm.2017.12.003.

[38] Fukushima, S., Y. Karube, and H. Kawakami. 2010. "Preparation of Ultrafine Uniform Electrospun Polyimide Nanofiber." Polymer Journal 42 (6): 514–18. https://doi.org/10.1038/pj.2010.33.

[39] Cheng, C., J. Chen, F. Chen, P. Hu, X.-F. Wu, D. H. Reneker, and H. Hou. 2010. "High-Strength and High-Toughness Polyimide Nanofibers: Synthesis and Characterization." Journal of Applied Polymer Science 116: 1581–86. https://doi.org/10.1002/app.31523.

[40] Li, B., Y. Liu, S. Wei, Y. Huang, S. Yang, Y. Xue, H. Xuan, and H. Yuan. 2020. "A Solvent System Involved Fabricating Electrospun Polyurethane Nanofibers for Biomedical Applications." Polymers 12 (12): 3038. https://doi.org/10.3390/polym12123038.

[41] Thandavamoorthy, S., N. Gopinath, and S. S. Ramkumar. 2006. "Self-Assembled Honeycomb Polyurethane Nanofibers." Journal of Applied Polymer Science 101 (5): 3121–24. https://doi.org/10.1002/app.24333.

[42] Hu, J., X. Wang, B. Ding, Ji. Lin, J. Yu, and G. Sun. 2011. "One-Step Electro-Spinning/Netting Technique for Controllably Preparing Polyurethane Nano-Fiber/Net." Macromolecular Rapid Communications 32 (21): 1729–34. https://doi.org/10.1002/marc.201100343.

[43] Boaretti, C., M. Roso, A. Lorenzetti, and M. Modesti. 2015. "Synthesis and Process Optimization of Electrospun PEEK-Sulfonated Nanofibers by Response Surface Methodology." Materials 8 (7): 4096–117. https://doi.org/10.3390/ma8074096.

[44] Yongyi, Y., Z. Puxin, Y. Hai, N. Anjian, G. Xushan, and W. Dacheng. 2006. "Polysulfone Nanofibers Prepared by Electrospinning and Gas/Jet- Electrospinning." Frontiers of Chemistry in China 1 (3): 334–39. https://doi.org/10.1007/s11458-006-0041-4.

[45] Song, Z., S. W. Chiang, X. Chu, H. Du, J. Li, L. Gan, C. Xu, et al. 2018. "Effects of Solvent on Structures and Properties of Electrospun Poly(Ethylene Oxide) Nanofibers." Journal of Applied Polymer Science 135 (5): 45787. https://doi.org/10.1002/app.45787.

[46] Tripatanasuwan, S., Z. Zhong, and D. H. Reneker. 2007. "Effect of Evaporation and Solidification of the Charged Jet in Electrospinning of Poly(Ethylene Oxide) Aqueous Solution." Polymer 48 (19): 5742–46. https://doi.org/10.1016/j.polymer.2007.07.045.

[47] Wang, X., W. J. Zhang, D. G. Yu, X. Y. Li, and H. Yang. 2013. "Epoxy Resin Nanofibers Prepared Using Electrospun Core/Sheath Nanofibers as Templates." Macromolecular Materials and Engineering 298 (6): 664–69. https://doi.org/10.1002/mame.201200174.

[48] Shneider, M., X. M. Sui, I. Greenfeld, and H. D. Wagner. 2021. "Electrospinning of Epoxy Fibers." Polymer 235: 124307. https://doi.org/10.1016/j.polymer.2021.124307.

[49] Li, W., Y. Ding, M. Tebyetekerwa, Y. Xie, L. Wang, H. Li, R. Hu, Z. Wang, A. Qin, and Ben Z. Tang. 2019. "Fluorescent Aggregation-Induced Emission (AIE)-Based Thermosetting Electrospun Nanofibers: Fabrication, Properties and Applications." Materials Chemistry Frontiers 3 (11): 2491–98. https://doi.org/10.1039/c9qm00342h.

[50] Wang, Q., Y. Yu, J. Ma, N. Zhang, J. Zhang, Z. Liu, and G. Cui. 2016. "Electrospun Melamine Resin-Based Multifunctional Nonwoven Membrane for Lithium Ion Batteries

at the Elevated Temperatures." Journal of Power Sources 327: 196–203. https://doi.org/10.1016/j.jpowsour.2016.07.063.

[51] Ma, C., Y. Song, J. Shi, D. Zhang, Q. Guo, and L. Liu. 2013. "Preparation and Electrochemical Performance of Heteroatom-Enriched Electrospun Carbon Nanofibers from Melamine Formaldehyde Resin." Journal of Colloid and Interface Science 395: 217–23. https://doi.org/10.1016/j.jcis.2013.01.009.

[52] Ma, C., Y. Song, J. Shi, D. Zhang, X. Zhai, M. Zhong, Q. Guo, and L. Liu. 2013. "Preparation and One-Step Activation of Microporous Carbon Nanofibers for Use as Supercapacitor Electrodes." Carbon 51: 290–300. https://doi.org/10.1016/j.carbon.2012.08.056.

[53] Zu-Sheng, H., T. Ling-Hua, C. Xiao-Miao, J. Fa-Yin, Y. San-Jiu, and X. Fu-Ming. 2011. "Preparation of Melamine Microfibers by Reaction Electrospinning." Materials Letters 65 (7): 1079–81. https://doi.org/10.1016/j.matlet.2011.01.010.

[54] Badhe, Y., and K. Balasubramanian. 2015. "Nanoencapsulated Core and Shell Electrospun Fibers of Resorcinol Formaldehyde." Industrial and Engineering Chemistry Research 54 (31): 7614–22. https://doi.org/10.1021/acs.iecr.5b00929.

[55] Kim, J. S., and D. H. Reneker. 1999. "Polybenzimidazole Nanofiber Produced by Electrospinning." Polymer Engineering and Science 39 (5): 849–54. https://doi.org/10.1002/pen.11473.

[56] Nie, H.-R., C. Wang, and A.-H. He. 2016. "Fabrication and Chemical Crosslinking of Electrospun Trans-Polyisoprene Nanofiber Nonwoven." Chinese Journal of Polymer Science (English Edition) 34 (6): 697–708. https://doi.org/10.1007/s10118-016-1796-2.

[57] Vitale, A., G. Massaglia, A. Chiodoni, R. Bongiovanni, C. F. Pirri, and M. Quaglio. 2019. "Tuning Porosity and Functionality of Electrospun Rubber Nanofiber Mats by Photo-Crosslinking." ACS Applied Materials and Interfaces 11 (27): 24544–51. https://doi.org/10.1021/acsami.9b04599.

[58] Al-Ajrah, S., K. Lafdi, Y. Liu, and P. L. Coustumer. 2018. "Fabrication of Ceramic Nanofibers Using Polydimethylsiloxane and Polyacrylonitrile Polymer Blends." Journal of Applied Polymer Science 135 (10): 45967. https://doi.org/10.1002/app.45967.

5 Influence of Process Parameters on the Performance of the Hybrid Nanofibers Prepared via Electrospinning

Lin Feng Ng, Mohd Yazid Yahya, Senthilkumar Krishnasamy and Zaleha Mustafa

5.1 INTRODUCTION

Many sectors have started to utilize fiber-reinforced composites for a wide variety of applications, ranging from transportation to biomedical sectors. It is well known that delamination or fiber-matrix debonding has been a common challenging issue in composite materials since these materials are formed by combining multiple solid phases. Innumerable literature studies have noticed delamination or fiber-matrix debonding in composite material [1]–[4]. Delamination can be very dangerous as it can sharply deteriorate the mechanical performance of the composite materials while remaining invisible. On this note, incorporating nanofibers in the composite materials is considered one of the viable options to mitigate or resolve the delamination due to their nanoscale geometry. In recent years, nanofibers have attracted the interest of researchers and engineers owing to their attractive attributes, such as high aspect ratio and large specific surface area. These attributes interact with the polymer matrix in a significant way to improve mechanical performance. Generally, synthetic fibers such as carbon, glass and aramid as reinforcing materials of composites are in the form of microfibers due to their micro-scale geometry. However, the mechanical performance of microfiber-reinforced composites is weaker than that of nanofiber-reinforced composites. This leads to continuous growth in the interest in using nanofibers as the reinforcement to develop nanocomposites. Furthermore, nanofibers possess an excellent surface area-to-volume ratio, which is the driving factor contributing to the superior performance of such fibers. Typically, nanomaterials can be classified based on their geometries, including particle, layered and

DOI: 10.1201/9781003333814-5

fibrous nanomaterials. Nanofibers, nanotubes, nanowires, clays, fullerenes and metallic nanoparticles are the most commonly seen nanomaterials.

As of today, there are several well-established methods used to produce nanofibers. However, electrospinning is currently the most efficient and widely used method to synthesize nanofiber. It is a unique technique which is capable of producing ultrafine fibers through electrostatic forces. Undeniably, the morphological properties and diameter of nanofibers produced from electrospinning can be affected by the process parameters. Besides the process parameters, environmental factors and the type of polymer solution might also leave a profound impact on the properties of the nanofibers. The growth of electrospinning technology has increased its versatility and improved the quality of nanofibers. Electrospun nanofibers can be produced from about 100 different polymers, including synthetic and natural polymers [5]. However, the understanding of the electrospinning technology is still regarded as inadequate despite its widespread use, and the quality of its product and its throughput still have room for improvement.

The formation of beads on the fiber surface is one of the key challenges of electrospinning. Although nanoscale ultrafine fibers exhibit a high specific area, the presence of beads on the fibers could significantly reduce the specific area. Therefore, the beads on the electrospun fibers are often regarded as defects and should be eliminated entirely. In order to control and purge beads during the electrospinning process, identifying the driving factors toward the formation of beads is of utmost importance. To date, several research studies have been performed to identify the factors triggering bead formation and altering the fiber diameter, and it was found that the process parameters have a close relationship with bead formation and fiber diameter. Thus, controlling the process parameters during electrospinning is essential in limiting bead formation and reducing the fiber diameter. However, it is worth noting that beads in micro-size are beneficial for drug-loading purposes [6].

5.2 INTRODUCTION TO NANOCOMPOSITES

Nanocomposites can be defined as materials formed by combining two or more phases in which at least one of the phases has one dimension in the nanoscale. The nanoscale constituent in nanocomposites can be one-, two or three-dimensional. Apart from the material selection, the functional properties of the nanocomposites are influenced by the fiber size. Small fiber size offers nanocomposites with greater mechanical properties. Nanocomposites can have different functional properties when compared to microcomposites with the same fiber composition because their surface properties and topography are varied. The surface properties and topography determine the level of fiber-matrix interaction, which has a decisive effect on the functional properties of the composites. When microfibers are replaced with nanofibers, the physical and mechanical properties of composites may alter. Variations in the physical and mechanical properties of composites are attributed to changes in the total surface area for a given volume. The transition from micro to nano scales increases the specific surface area, thus enhancing the fiber-matrix interaction. For nanocomposites with fixed nanofiber content, a reduction in the nanofiber diameter can undoubtedly improve the fiber aspect ratio (length/diameter).

Indeed, the types of nanomaterials can influence the mechanical properties of nanocomposites. Therefore, it is vital to select the most appropriate nanomaterials having strong compatibility with the polymer matrices to guarantee the optimum mechanical properties of the materials. Aside from compatibility, fiber dispersion is another criterion that should not be neglected during nanocomposite production. The nanocomposites are not at their optimum condition if the nanofibers are not dispersed well in the polymer matrix. Instead of improving the mechanical properties, poor nanofiber dispersion could degrade the mechanical properties of the materials [7].

5.3 HYBRID NANOFIBERS

Electrospun nanofibers have found applications primarily in biomedical sectors due to their large specific surface area, good biocompatibility and high porosity. In order to further improve the mechanical properties and biological activity, combining inorganic substances and organic polymers can be regarded as one of the feasible options. Examples of inorganic nanoparticles include carbon nanotubes, halloysite, graphene, nano-hydroxyapatite, etc. Hybrid nanofibers are promising as they manifest enhanced performance in various aspects, such as mechanical properties, cell adhesion, migration, proliferation and differentiation [8]. Although the addition of nanoparticles can improve the performance of electrospun nanofibers, precautions should be given to ensure uniform distribution and avoid the aggregation of the nanoparticles.

There are several methods that can be used to prepare hybrid electrospun nanofibers. Directly mixing nanoparticles with the polymer solution before electrospinning is considered the simplest way to produce hybrid electrospun nanofibers. Stirring the mixture for a certain period of time is required to ensure the uniform distribution of the nanoparticles in the polymer. However, the distribution of the nanoparticles is highly dependent on the viscosity of the polymer solution. Sol-gel technology is another well-established approach to producing hybrid electrospun nanofibers. In this method, the nanoparticles are firstly dispersed into a solvent. The hydrolysis and condensation of the inorganic nanoparticles result in gel formation. The gel is then mixed with the polymer solution, and subsequently, the mixture is subjected to electrospinning to form hybrid electrospun nanofibers. Apart from the aforementioned methods, organic/inorganic hybrid electrospun nanofibers can also be produced through the *in situ* reactions within the polymer solution. The particle is formed *in situ* within the polymer solution, which is followed by the electrospinning process. In this case, nanofibers doped with inorganic metal oxide can be produced with subsequent *in situ* growth of the inorganic nanoparticles [8]. In addition to creating hybrid nanofibers with evenly scattered nanoparticles, this technique is particularly effective in preventing particle aggregation. Nevertheless, it is worthy of note that this technique is only appropriate for hybrid nanofibers doped with metal or metal oxide as nanoparticles. As an alternative, using the *in situ* reduction procedure and electrospinning technology is another way to produce uniformly distributed organic/inorganic hybrid nanofibers. This approach involves electrospinning of the solution consisting of polymer and metal salts, followed by chemical or physical treatment to yield the nanoparticles-containing nanofibers.

5.4 ELECTROSPINNING

Currently, several established and advanced methods are available to produce nanofibers: drawing processing, template-assisted synthesis, solvent casting, electrospinning, self-assembly and phase separation. Among these synthesis techniques, electrospinning is the most popular and widely studied technique to yield high-performance nanofibers. Even though the primary function of these techniques is to synthesize nanofibers, different production methods could produce nanofibers with distinct properties.

The electrospinning technique was initially noticed by Rayleigh in 1897, and Zenely then conducted a detailed investigation of it in regard to electrospraying in 1914. Finally, Formhals patented electrospinning in 1934, which outlined an experimental setup for generating polymer filaments. It is a powerful technique that can synthesize nanofibers with a diameter ranging from 40 nm to 2 µm [9]. The electrospinning process has been the most popular technique to produce nanofibers, not only because of its versatility in spinning a wide range of polymer solutions but also its ability to consistently produce nanoscale fiber threads, which is difficult to be realized by standard mechanical spinning techniques [10],[11]. In addition, the electrospun nanofibers exhibit a large number of inter/intra fibrous pores. Simply put, nanofibers from a wide range of polymer solutions can be produced with the help of relatively simple and advanced electrospinning. Figure 5.1 illustrates the electrospinning apparatus used to yield nanofibers from polymer solutions. Either natural or synthetic polymer solutions can be used to synthesize nanofiber threads. When referring to Figure 5.1, the electrospinning apparatus consists of three main components: a high-voltage power supply, a spinneret connected to a pump and a metal collecting plate placed horizontally or vertically. Electrospinning is not a complicated synthesis method as it is conducted at ambient temperature and atmospheric conditions. This method requires a high DC voltage to initiate the electrospinning by overcoming the surface tension of the polymer droplet. Thus, the applied voltage must be higher than the threshold voltage in electrospinning. However, this threshold voltage is dependent on the polymer solutions. Before electrospinning, most polymers are firstly dissolved in certain solvents to form a polymer solution. The aqueous polymer is then moved to the capillary tube for electrospinning. After that, an electric charge is applied to the polymer droplet at the tip of the capillary tube, which is kept in place by its surface tension. This electric charge is necessary to generate electrospinning. The high voltage induces a strong mutual electrical repulsive force between polymer chains to counterbalance the surface tension. The viscoelastic force and the surface tension of the polymer solution at the tip of the capillary tube will be overcome by the electrostatic force once the strength of the applied electric field exceeds the threshold value. This allows the surface of the polymer solution to deform into a cone shape, which is commonly known as the Taylor cone [12]. This

a)

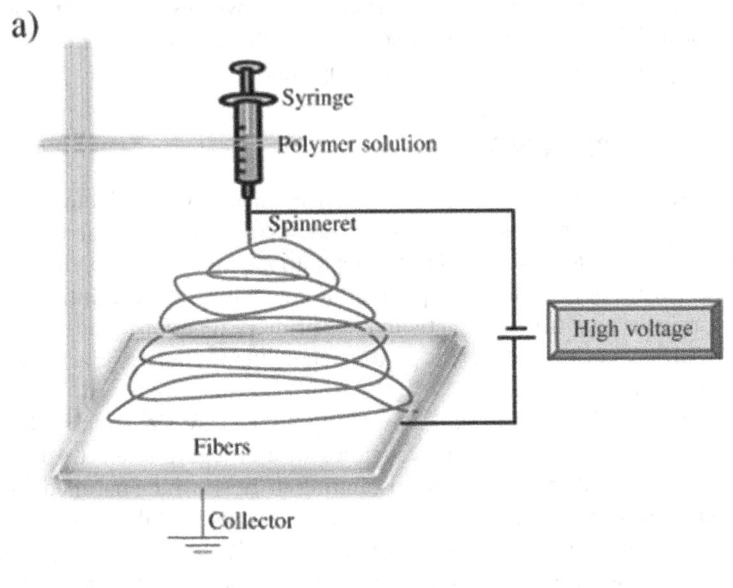

b)

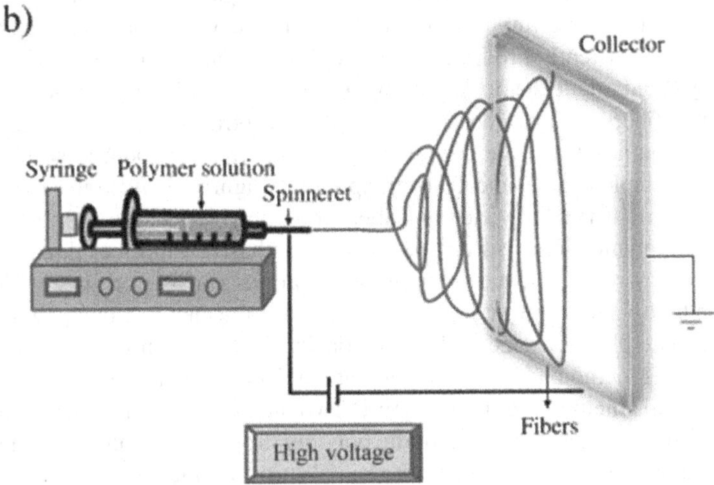

FIGURE 5.1 Illustration of the electrospinning apparatus (a) vertical position and (b) horizontal position.

Source: (Reproduced with the permission obtained from [13]).

Taylor cone then travels from the spinneret to the metal collecting plate. Meanwhile, the solvent molecules in the Taylor cone evaporate rapidly in the pathway, leaving nanofibers lying on the metal collecting plate.

The nanofibers produced from the electrospinning process provide several merits, such as high ductility to form a wide range of sizes and shapes, superior surface-to-volume ratio, controllable porosity, and freedom of tailoring the functionality and properties [13]. Although electrospinning is promising, the production rate has been an obstacle to this technique. In order to resolve the shortcoming of the electrospinning process, various research groups have started studying a two-layer electrospinning system with the aim of improving the throughput of nanofibers. Instead of a single jet, multiple jets can be employed to scale up the quantity of the nanofibers produced from electrospinning. When using multiple jets to produce nanofibers, increasing the number of pores and length of the porous hollow tube is one of the ways to improve the production rate. Apart from modifying the electrospinning apparatus, increasing the voltage during the electrospinning can also increase the amount of polymer solution drawn from the tip of the syringe, thus accelerating the electrospinning jet. However, it should be noted that the increase in the voltage will eventually alter the fiber geometry.

The past few decades have witnessed the dramatic growth of electrospinning technology to produce various nanostructures for many applications. As one of the advanced nanotechnologies, electrospinning has shown its potential to efficiently produce nanofibers and microfibers with superior surface area-to-volume ratio. Today, it can be seen that electrospun nanofibers with well-controlled diameters, compositions, and morphologies have been produced from a range of materials, including natural polymers, synthetic polymers, ceramics, and metals, or a combination of them [14]–[17].

5.5 INFLUENCE OF PROCESS PARAMETERS ON ELECTROSPINNING

Electrospinning with a simple setup can be easily operated to produce electrospun micro or nanofibers. Even though the electrospinning setup is simple, it is possible to modify the process parameters to produce nanofibers with different physical and mechanical properties. Indeed, the nanofiber properties and the electrospinning process are profoundly influenced by several process parameters, including the intensity of the electric field, the distance between the spinneret and the metal collecting plate, the solution flow rate and the needle diameter. The process parameters of electrospinning, which affect the electrospun nanofiber morphology and diameter, are summarized in Figure 5.2. It should be emphasized that the performance of electrospun nanofibers is closely linked to fiber density. This is the primary rationale for reducing the diameter of nanofibers as much as possible. It is essential to investigate the impact of process parameters on the electrospinning and nanofiber qualities in order to better understand the electrospinning process and continually improve the quality and properties of nanofibers.

5.5.1 EFFECT OF VOLTAGE

When the polymer solution is charged, the polymer droplet at the tip of the syringe deforms into a Taylor cone and forms nanofibers with a small diameter, travelling from the syringe to the metal collecting plate. Adjusting the voltage will alter the intensity of the electric charge, changing the amount of electric charge applied to the polymer solution. As mentioned in the previous section, the electric charge is required to overpower the viscoelastic force and surface tension of the polymer droplet. Therefore, the applied voltage must be higher than the threshold value to ensure the electric charge is high enough to initiate the electrospinning. However, it should

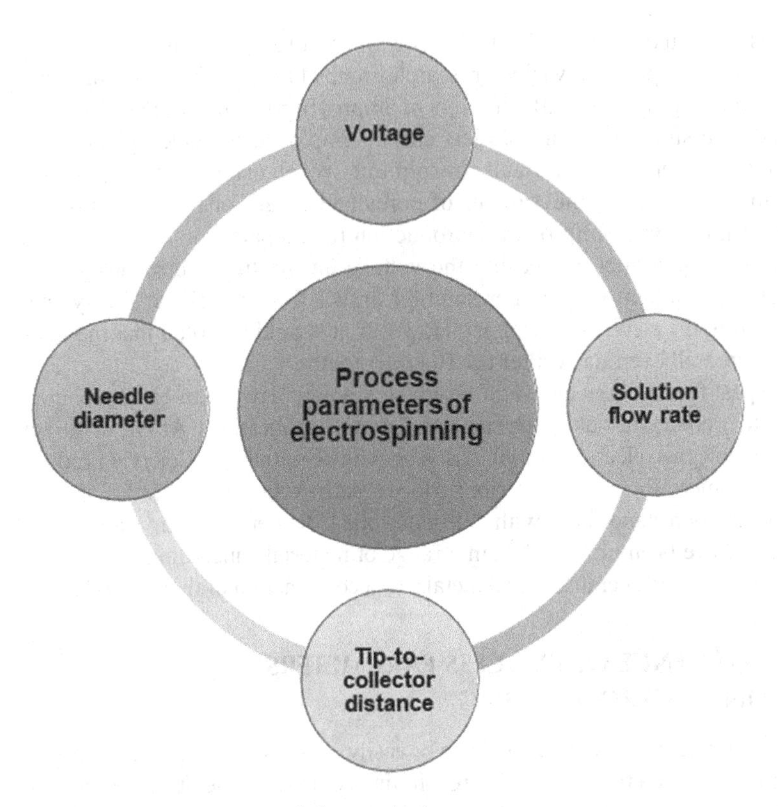

FIGURE 5.2 Process parameters of electrospinning.

be noted that the threshold value depends on the polymer types. Since the voltage determines the intensity of electric charge, the amount of polymer solution extracted from the syringe and the diameter of the nanofibers are both affected by augmenting the applied voltage. It has been shown that voltage has the most significant impact on the behaviors of electrospun fibers. In particular, the voltage determines the geometric shape of the polymer solution at the tip of the capillary tube. Therefore, it has a decisive effect on the performance of electrospinning. Up to a critical limit, it is anticipated that the diameter of the nanofibers would undermine as the applied voltage increases. This is because the polymer solution is subjected to greater stretching when the voltage increases. Nevertheless, the diameter of the nanofibers may overturn, and beads or beaded nanofibers may develop if the applied voltage is increased further beyond the critical limit. Ahmed et al. [18] revealed the influence of voltage on the diameter of electrospun polyacrylonitrile nanofibers. The SEM images of the electrospun nanofibers and their respective diameter distribution at varying applied voltages are shown in Figure 5.3. The SEM images and the diameter distribution of nanofibers clearly attested to the drop in the fiber diameter by raising the applied voltage up to 40 kV. In addition to the lowest fiber diameter, it can be seen that the electrospun nanofibers had the most uniform fiber diameter when the voltage was at 40 kV. However, the findings of fiber diameter were reversed when the voltage was further raised to 45 kV.

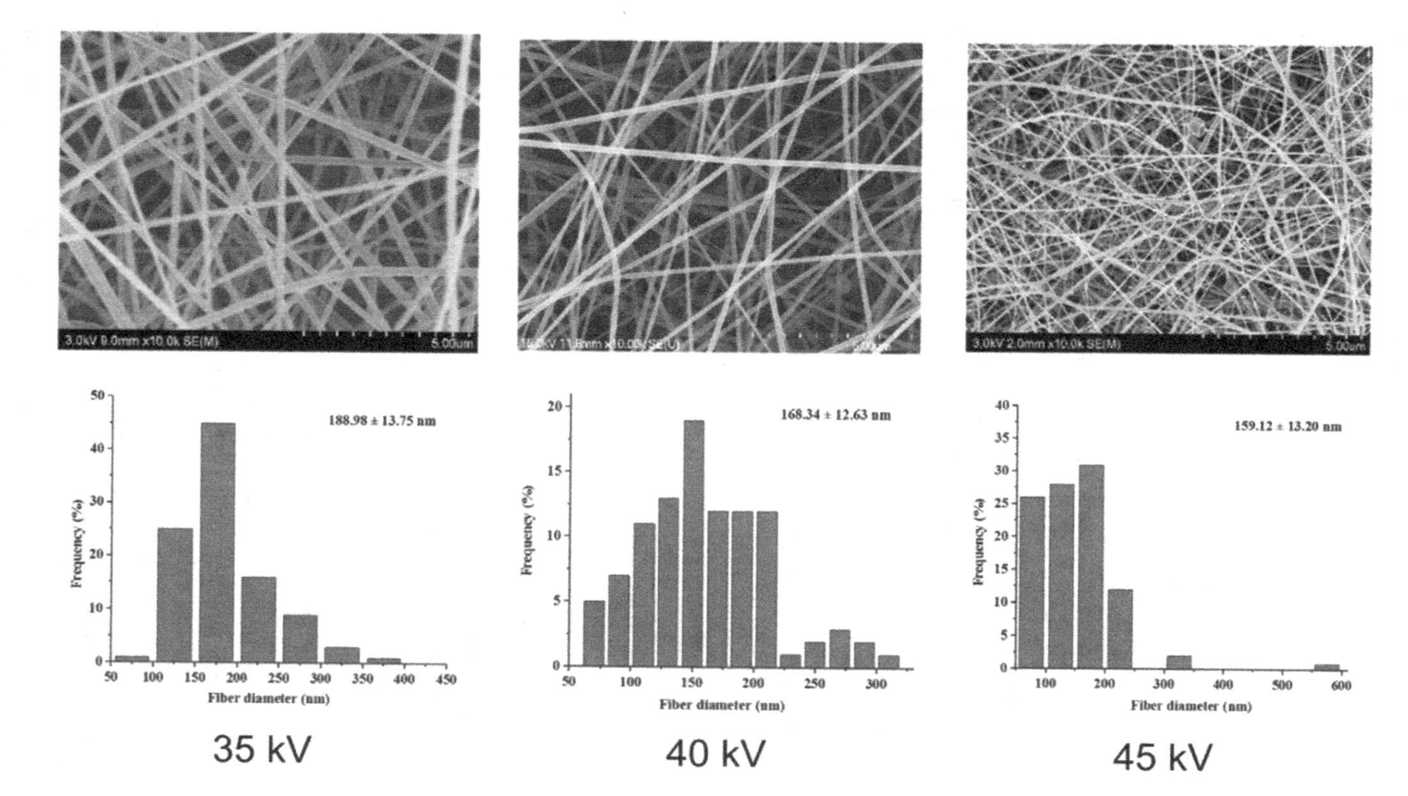

FIGURE 5.3 SEM images associated with the fiber diameter distribution at varying applied voltages.

Source: (Reproduced with permission obtained from [18]).

Although several research studies have revealed that an increase in voltage reduces the fiber diameter, some research groups also declare that the increase in voltage could increase the fiber diameter. The increase in the fiber diameter can be explained by the rise in the jet length with respect to the voltage. Therefore, there is an ongoing vigorous debate among academics worldwide regarding the impact of voltage on fiber diameter. The possible reasons for obtaining these contradicting results are the different types of polymer solutions or voltage levels applied in the respective research studies. Based on the literature studies, voltage obviously affects fiber diameter, but the significance level depends on the polymer used and the distance between the syringe tip and the metal collector. When the distance between the nozzle tip and the metal collecting plate is large, the fiber diameter is closely linked to the polymer density, solution flow and the intensity of the applied electrostatic field. Contrarily, when the distance between the nozzle tip and the metal collecting plate is minimal, the fiber diameter is represented as a function of surface tension, solution flow, dielectric permittivity and electric current intensity [19].

Apart from the fiber diameter, applied voltage significantly influences the bead formation. It has been reported that increasing voltage can suppress bead formation as higher voltage leads to a higher level of stretching. Zuo et al. [20] identified the influence of applied voltage on the bead formation of electrospun fibers. They revealed that increasing the voltage suppressed the bead formation due to the higher electrostatic repulsion force that provides higher stretching, making the fiber diameter smaller. Although increasing the voltage is favorable for producing smooth electrospun fibers without beads, increasing the voltage above the critical limit may result in increased fiber diameter and bead formation. This can be explained by the increase in jet velocity and decrease in the Taylor cone size for the same flow rate. Additionally, an increase in applied voltage raises the resistivity of the polymer solution, which may cause the diameter of the fiber to increase.

5.5.2 Solution Flow Rate

Solution flow rate plays a role in determining electrospun fiber morphology. In fact, the process and solution parameters are interdependent. As discussed in the previous sub-section, the intensity of electric charge has a significant influence on the fiber morphology, but it is closely related to the applied voltage and the conductivity of the polymer solution. Besides the intensity of the electric charge, the solution flow rate is greatly influenced by other parameters, including the solution viscosity and applied electric field. Undoubtedly, the flow rate of the polymeric solution has a significant impact on fiber morphology and bead formation. Generally, there is a critical flow rate of the polymer solution to produce smooth and uniform electrospun fibers, and this critical value is highly dependent on the types of the polymer solution. In particular, increasing the flow rate beyond the critical limit might trigger bead formation. This can be ascribed to low stretching and insufficient time to allow the solvent to evaporate before reaching the collector. In this case, ribbon-like electrospun nanofibers are produced rather than fibers with circular cross-sections [21]. Moreover, an increase in the fiber diameter can be noticed with an increase in the flow rate of polymeric solution, mainly due to insufficient stretching. Figure 5.4

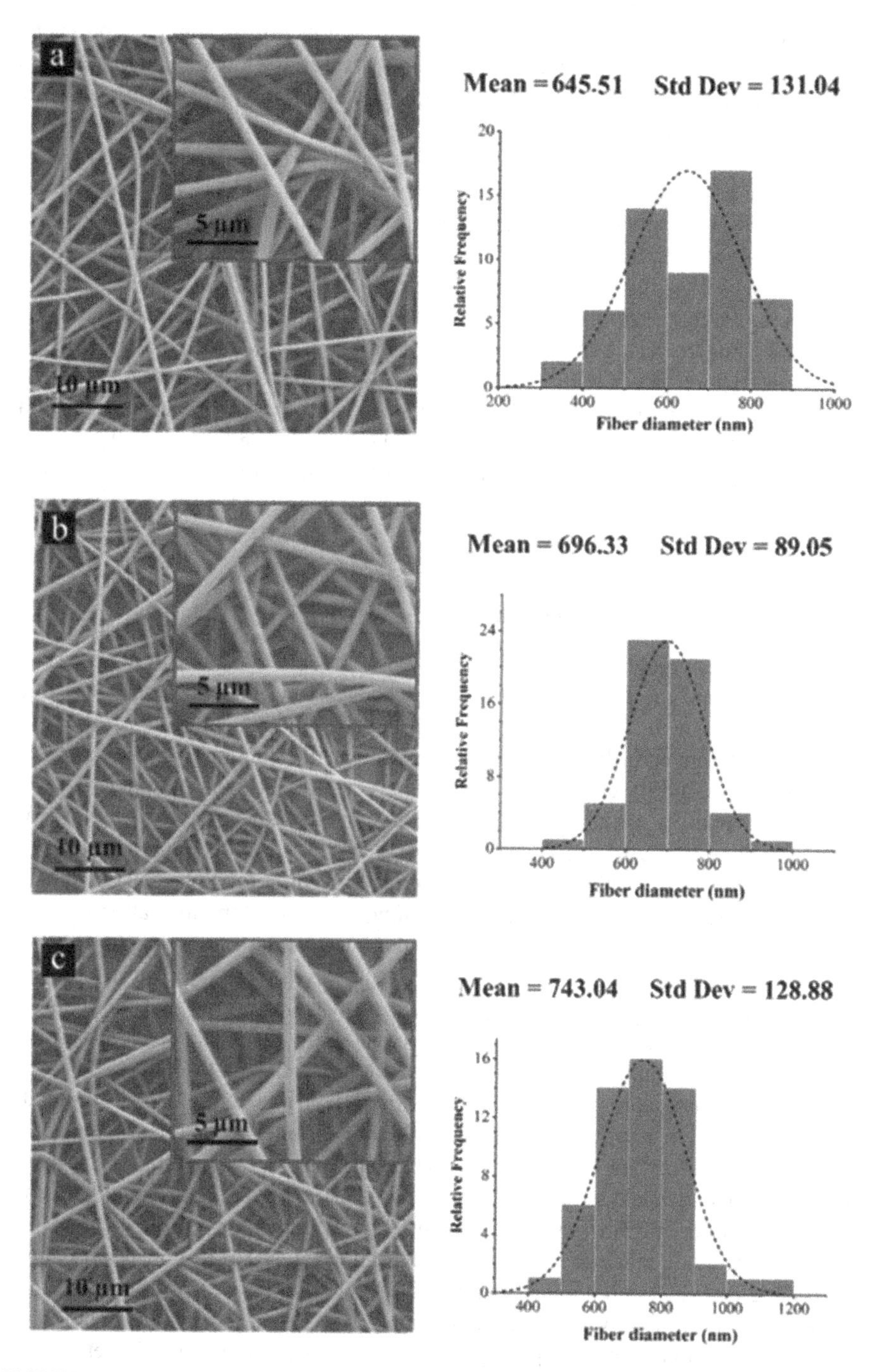

FIGURE 5.4 SEM images and graphs showing fiber diameters with respect to flow rates of poly(methyl methacrylate) solution (a) 0.3 mL h^{-1}, (b) 0.6 mL h^{-1} and (c) 0.9 mL h^{-1} [24].

shows the fiber diameters with respect to the flow rates of poly(methyl methacrylate) solution. It is evident that the increase in the solution flow rate augmented the fiber diameter. With an increase in the flow rate, more polymeric solution will be ejected from the capillary tube without being stretched sufficiently. More specifically, the amount of charged ions is insufficient to stretch the ejected polymeric solution due to the low intensity of the electric field. Any change in the number of charged ions at the polymer surface may affect the fiber morphology. According to the findings reported by Theron et al. [22], an increase in flow rate reduced the surface charge density of different polymers, which led to the fusion of electrospun nanofibers. All these factors contribute to the higher fiber diameter with increasing flow rate. Thus, lowering the solution flow rate has been identified as one way to obtain electrospun fibers with lower diameters. Nonetheless, the solution flow rate should be high enough to replace the ejected polymer to maintain the shape and stability of the Taylor cone. In certain cases, the jet may recede to the needle, leading to jet instability. Consequently, this process leads to inconsistency in fiber diameter [23]. However, this receded jet will be gradually superseded by a cone jet after a certain period of time in electrospinning.

5.5.3 Tip-to-Collector Distance

Apart from the applied voltage and solution flow rate, the distance between the tip of the nozzle and the collector also has a decisive effect on the electrospun fiber morphology. Since the fiber morphology is highly dependent on the deposition time, evaporation extent and whipping or instability interval, changing the tip-to-collector distance can easily affect the fiber morphology [25]. When the polymeric droplet travels from the tip of the nozzle to the collector, the solvent tends to evaporate before reaching the collector. Therefore, a sufficient tip-to-collector distance is necessary to allow the complete evaporation of the solvent to take place. The primary aim of ensuring a proper tip-to-collector distance is to avoid the formation of beads during electrospinning. Like other process parameters, the critical tip-to-collector distance varies with polymer types. When examining the effect of tip-to-collector distance on the diameter of electrospun fibers, it was discovered that the diameter of the electrospun fibers decreased with increase in the distance between the tip of the nozzle and the collector. This is because extending the tip-to-collector distance allows the polymer jet to stretch for a longer period of time before depositing on the collector. However, increasing the tip-to-collector distance might also increase the fiber diameter in certain circumstances. This is due to the weakening of the intensity of electrostatic field strength, causing insufficient stretching of the electrospun fibers. In addition, no fibers will be deposited on the collector if the distance is too large. Hence, it is vital to identify and maintain the critical tip-to-collector distance to produce fine and smooth electrospun fibers. The impact of tip-to-collector distance on the morphology of electrospun fibers was examined by Mazoochi et al. [26]. The SEM images and graphs showing the distribution of fiber diameters with respect to tip-to-collector distances are shown in Figure 5.5. On average, increasing the distance decreased the fiber diameter. This is due to the extended flight time required for the fiber to be stretched before reaching the collector. Moreover, the

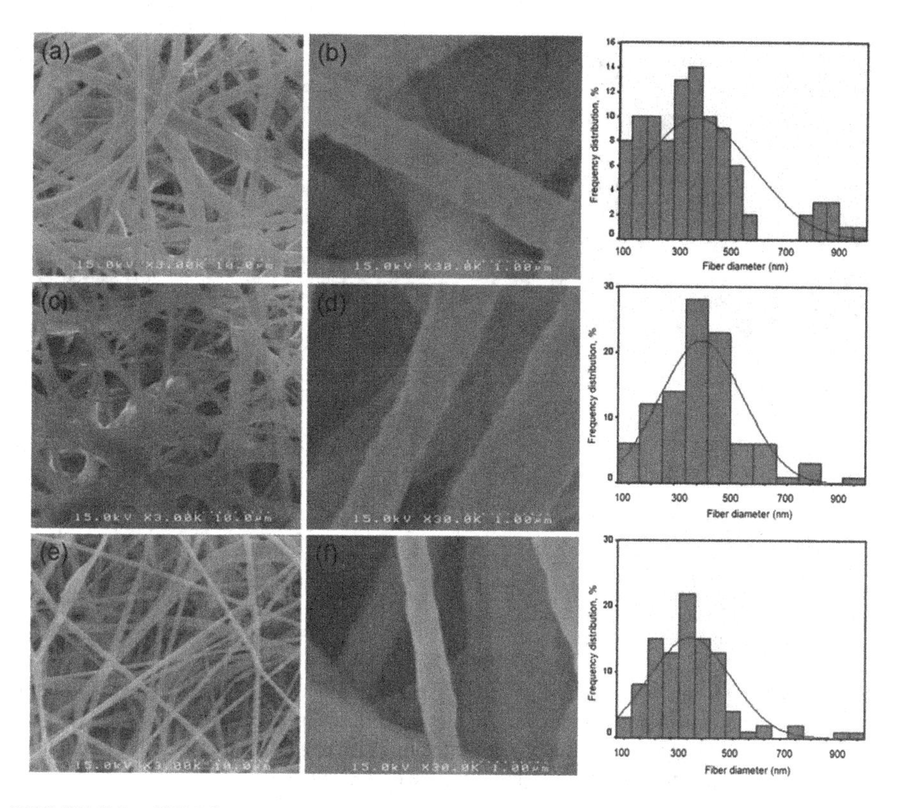

FIGURE 5.5 SEM images and graphs showing the distribution of fiber diameters with respect to tip-to-collector distances (a) 5 cm, (b) 10 cm and (c) 15 cm [26].

results also revealed that decreasing the distance promoted the formation of beads on the fibers. This phenomenon could be attributed to the increase in the electrostatic field strength which results in jet instability. It is worth mentioning that the formation of beads will be restricted at an optimal tip-to-collector distance, considering that the electrostatic field strength offers sufficient stretching force to generate smooth and fine electrospun fibers.

5.5.4 Needle Diameter

The fiber diameter is undoubtedly affected by the needle diameter. As expected, it was found that the decrease in the needle diameter drives a reduction in the diameter of electrospun nanofibers. The needle diameter controls the amount of polymer solution that may be expelled from the capillary tube and the force necessary to start the electrospinning. Essentially, a smaller needle diameter is desirable as it can produce electrospun nanofibers with a smaller diameter and hence a high aspect ratio. Nevertheless, the polymer viscosity should be considered when reducing the needle diameter, as the highly viscous polymer solution cannot be ejected if the needle diameter is too small. Therefore, an appropriate needle diameter which is

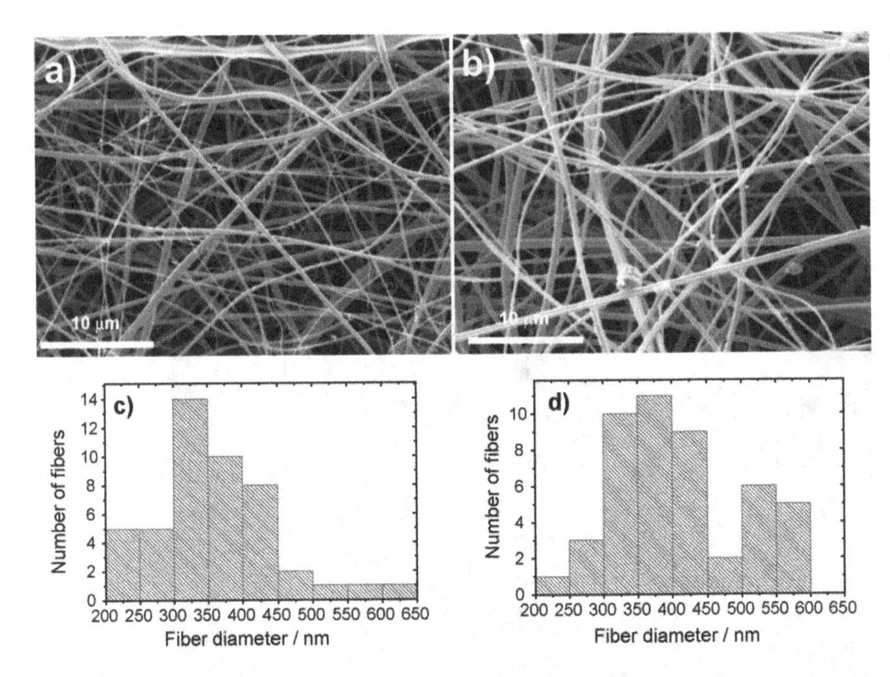

FIGURE 5.6 Morphology and diameter of electrospun nanofiber obtained from the electrospinning with varying inner needle diameters (a) 0.5 mm, (b) 1.7 mm, (c) nanofiber diameter with 0.5 mm inner needle diameter and (d) nanofiber diameter with 1.7 mm inner needle diameters.

Source: (Reproduced with the permission obtained from [27]).

compatible with the polymer viscosity should be used in the electrospinning process. Sencadas et al. [27] evaluated the influence of needle diameter on the diameter of electrospun chitosan nanofibers. Figure 5.6 displays the morphology and diameter of electrospun nanofiber obtained from the electrospinning with different inner needle diameters. They reported that reducing the inner needle diameter decreased the average electrospun nanofiber diameter. These results are in agreement with the findings revealed by Kuchi et al.[28] in which reducing the inner needle diameter (330 µm, 500 µm, 570 µm and 720 µm) decreased the diameter of electrospun titanium dioxide/poly-vinyl pyrrolidone nanofibers. They concluded that the inner needle diameter significantly influenced the diameter of electrospun nanofibers.

5.6 SOLUTION PARAMETERS

The viscosity of the polymer solution might change with the solution concentration. Viscosity is one of the physicochemical properties of the polymer solution, which can influence the morphological and physical properties of the resulting electrospun nanofibers. Other physicochemical properties of polymer solution that might impact the behaviors of electrospun fibers include surface tension and conductivity.

However, the surface tension and conductivity highly depend on the solution composition, such as concentration and additives. Among these three physicochemical properties, viscosity was found to have the most significant impact on the behavior of electrospun fibers. The viscosity of the polymer solution is primarily governed by the molecular weight, processing temperature, polymer structure and concentration. Reducing the molecular weight of the polymers may drop the polymer viscosity, hence promoting the formation of beads on the fibers. In contrast, smooth fibers can be obtained through the increase in the molecular weight of the polymers. In addition, changing the processing temperature may also alter the viscosity of the polymer solutions. It has been revealed that the processing temperature is inversely proportional to the viscosity, implying that the viscosity tends to be lower with high processing temperatures, leading to thinner electrospun fibers [29]. Figure 5.7 shows the morphology of bead-on-string nanofibers with varying solution viscosity. It is obvious that the beads were suppressed by increasing the solution viscosity, indicating that high viscosity favors the formation of smooth fibers without beads. The solution viscosity can be altered by changing the polymer concentration to obstruct the bead formation. Unlike viscosity, high surface tension promotes the formation of beads during electrospinning. However, the surface tension of the polymer solution can be changed by controlling the ratio of polymers and solvents.

5.7 CONCLUSIONS

Nanofibers have found their widespread applications, particularly in the biomedical sectors owing to their large specific surface area, good biocompatibility and high porosity. Hybrid nanofibers with the addition of inorganic nanoparticles were found to have improved mechanical properties and biological activity compared to non-hybrid nanofibers. The main concern in developing hybrid nanofibers is related to the dispersion of nanoparticles in the polymer solution prior to electrospinning. However, a dedicated technique such as *in situ* reaction has been proven to enhance the dispersion of nanoparticles. Uniform dispersion of nanoparticles in the polymer solution is essential to guarantee the optimum performance of hybrid nanofibers.

Electrospinning is the current emerging technology to produce smooth and uniform nanofibers. However, fiber morphology and density are greatly influenced by the processing parameters of electrospinning. Any changes in the applied voltage, solution flow rate and tip-to-collector significantly impact the fiber morphology and density. It should be noted that the diameter of nanofibers is closely linked to their performance. The diameter of electrospun nanofibers must be reduced in order to obtain electrospun nanofibers at their optimum performance. Overall, increasing the voltage and tip-to-collector distance tends to undermine the nanofiber diameter and suppress the formation of beads as the nanofiber is well-stretched before reaching the collector. Nonetheless, it has been reported in several literature studies that the increase in the voltage and tip-to-collector distance over a critical level might overturn the nanofiber diameter and promote bead formation. Unlike the applied voltage and tip-to-collector distance, augmenting the solution flow rate increases the fiber diameter and bead formation. This is mainly due to the low stretching

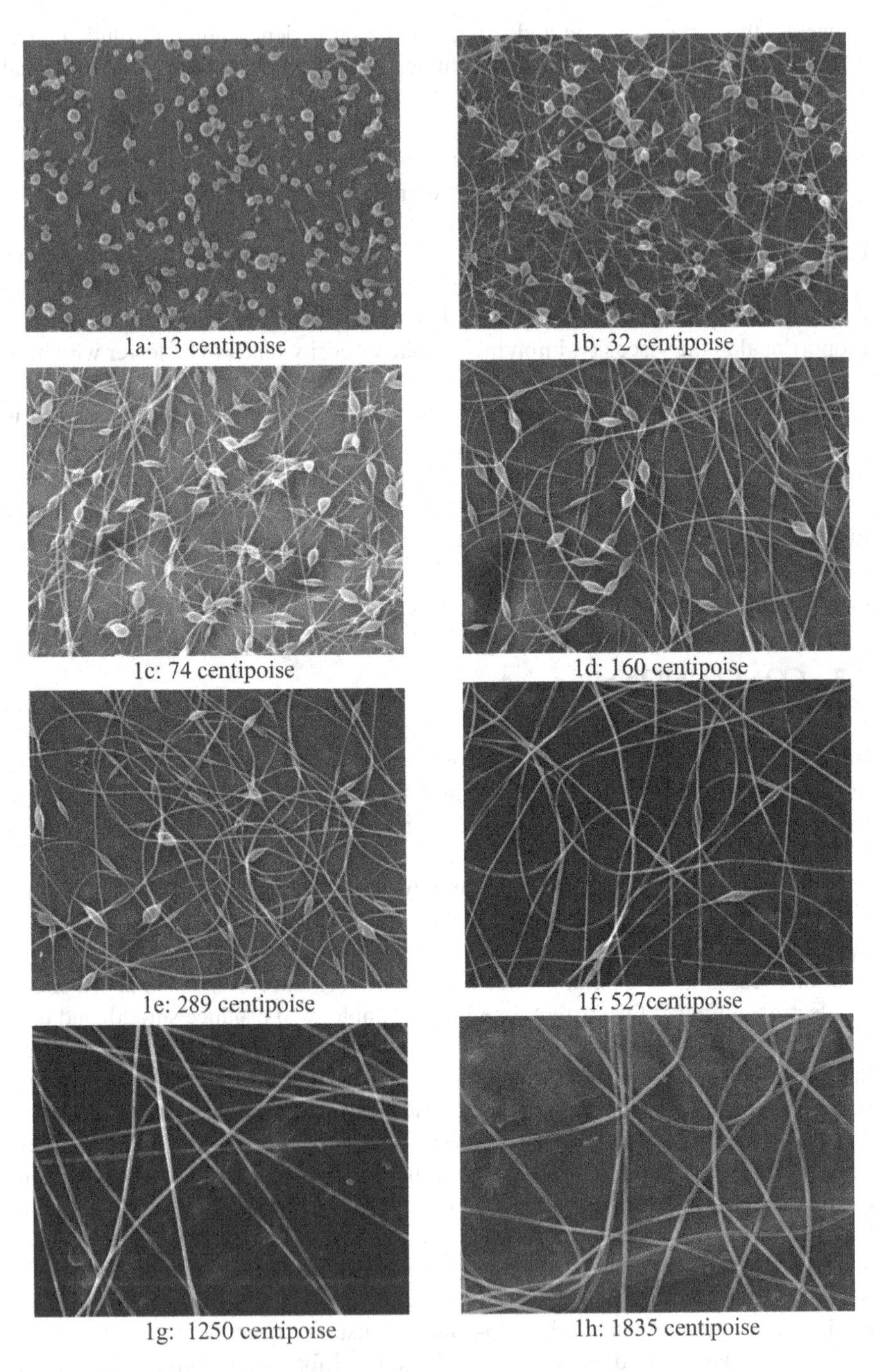

FIGURE 5.7 The morphology of bead-on-string nanofibers with varying solution viscosity (a) 13 cP, (b) 32 cP, (c) 74 cP, (d) 160 cP, (e) 289 cP, (f) 527 cP, (g) 1250 cP and (h) 1835 cP.

Source: (Reproduced with the permission obtained from [30]).

and insufficient time to allow the solvent to evaporate before reaching the collector. Furthermore, the inner needle diameter was also found to be correlated to the diameter of electrospun nanofibers. In particular, decreasing the inner needle diameter reduces the average diameter of electrospun nanofibers. Apart from the process parameters, solution parameters such as the polymer viscosity, conductivity and surface tension play a significant role as well in determining the nanofiber morphology and properties. Continuous exploration of the process parameters is necessary to further optimize the performance of electrospun nanofibers for future applications.

REFERENCES

[1] Ng, L. F.; Yahya, M. Y.; Muthukumar, C. Mechanical Characterization and Water Absorption Behaviors of Pineapple Leaf/Glass Fiber-Reinforced Polypropylene Hybrid Composites. *Polym. Compos.* 2022, *43* (1), 203–214. https://doi.org/10.1002/pc.26367.

[2] Feng, N. L.; Malingam, S. D.; Ping, C. W.; Razali, N. Mechanical Properties and Water Absorption of Kenaf/Pineapple Leaf Fiber-Reinforced Polypropylene Hybrid Composites. *Polym. Compos.* 2020, *41* (4), 1255–1264. https://doi.org/10.1002/pc.25451.

[3] Senthilkumar, K.; Saba, N.; Chandrasekar, M.; Jawaid, M.; Rajini, N.; Siengchin, S.; Ayrilmis, N.; Mohammad, F.; Al-Lohedan, H. A. Compressive, Dynamic and Thermo-Mechanical Properties of Cellulosic Pineapple Leaf Fibre/Polyester Composites: Influence of Alkali Treatment on Adhesion. *Int. J. Adhes. Adhes.* 2021, *106*, 102823. https://doi.org/10.1016/j.ijadhadh.2021.102823.

[4] Feng, N. L.; Malingam, S. D.; Subramaniam, K.; Selamat, M. Z.; Ali, M. B.; Bapokutty, O. The Influence of Fibre Stacking Configurations on the Indentation Behaviour of Pineapple Leaf/Glass Fibre Reinforced Hybrid Composites. *Def. S T Tech. Bull.* 2019, *12* (1), 113–123.

[5] Haider, A.; Haider, S.; Kang, I. K. A Comprehensive Review Summarizing the Effect of Electrospinning Parameters and Potential Applications of Nanofibers in Biomedical and Biotechnology. *Arab. J. Chem.* 2018, *11* (8), 1165–1188. https://doi.org/10.1016/j.arabjc.2015.11.015.

[6] Langer, R. Biomaterials in Drug Delivery and Tissue Engineering: One Laboratory's Experience. *Acc. Chem. Res.* 2000, *33* (2), 94–101. https://doi.org/10.1021/ar9800993.

[7] Sun, Z. B.; Dong, X. Z.; Nakanishi, S.; Chen, W. Q.; Duan, X. M.; Kawata, S. Log-Pile Photonic Crystal of CdS-Polymer Nanocomposites Fabricated by Combination of Two-Photon Polymerization and in Situ Synthesis. *Appl. Phys. A Mater. Sci. Process.* 2007, *86* (4), 427–431. https://doi.org/10.1007/s00339-006-3776-9.

[8] Huang, W.; Xiao, Y.; Shi, X. Construction of Electrospun Organic/Inorganic Hybrid Nanofibers for Drug Delivery and Tissue Engineering Applications. *Adv. Fiber Mater.* 2019, *1* (1), 32–45. https://doi.org/10.1007/s42765-019-00007-w.

[9] Reneker, D. H.; Chun, I. Nanometre Diameter Fibres of Polymer, Produced by Electrospinning. *Nanotechnology* 1996, *7* (3), 216–223. https://doi.org/10.1088/0957-4484/7/3/009.

[10] Theron, S. A.; Yarin, A. L.; Zussman, E.; Kroll, E. Multiple Jets in Electrospinning: Experiment and Modeling. *Polymer (Guildf).* 2005, *46* (9), 2889–2899. https://doi.org/10.1016/j.polymer.2005.01.054.

[11] Ma, Z.; Kotaki, M.; Inai, R.; Ramakrishna, S. Potential of Nanofiber Matrix as Tissue-Engineering Scaffolds. *Tissue Eng.* 2005, *11* (1–2), 101–109. https://doi.org/10.1089/ten.2005.11.101.

[12] Wang, S.; Zhao, Y.; Shen, M.; Shi, X. Electrospun Hybrid Nanofibers Doped with Nanoparticles or Nanotubes for Biomedical Applications. *Ther. Deliv.* 2012, *3* (10), 1155–1169. https://doi.org/10.4155/tde.12.103.

[13] Bhardwaj, N.; Kundu, S. C. Electrospinning: A Fascinating Fiber Fabrication Technique. *Biotechnol. Adv.* 2010, *28* (3), 325–347. https://doi.org/10.1016/j.biotechadv.2010.01.004.
[14] Li, M.; Mondrinos, M. J.; Gandhi, M. R.; Ko, F. K.; Weiss, A. S.; Lelkes, P. I. Electrospun Protein Fibers as Matrices for Tissue Engineering. *Biomaterials* 2005, *26* (30), 5999–6008. https://doi.org/10.1016/j.biomaterials.2005.03.030.
[15] Li, D.; Xia, Y. Fabrication of Titania Nanofibers by Electrospinning. *Nano Lett.* 2003, *3* (4), 555–560. https://doi.org/10.1021/nl034039o.
[16] Luu, Y. K.; Kim, K.; Hsiao, B. S.; Chu, B.; Hadjiargyrou, M. Development of a Nanostructured DNA Delivery Scaffold via Electrospinning of PLGA and PLA-PEG Block Copolymers. *J. Control. Release* 2003, *89* (2), 341–353. https://doi.org/10.1016/S0168-3659(03)00097-X.
[17] Wu, H.; Zhang, R.; Liu, X.; Lin, D.; Pan, W. Electrospinning of Fe, Co, and Ni Nanofibers: Synthesis, Assembly, and Magnetic Properties. *Chem. Mater.* 2007, *19* (14), 3506–3511. https://doi.org/10.1021/cm070280i.
[18] Ahmed, A.; Yin, J.; Xu, L.; Khan, F. High-Throughput Free Surface Electrospinning Using Solution Reservoirs with Different Radii and Its Preparation Mechanism Study. *J. Mater. Res. Technol.* 2020, *9* (4), 9059–9072. https://doi.org/10.1016/j.jmrt.2020.06.025.
[19] Cramariuc, B.; Cramariuc, R.; Scarlet, R.; Manea, L. R.; Lupu, I. G.; Cramariuc, O. Fiber Diameter in Electrospinning Process. *J. Electrostat.* 2013, *71* (3), 189–198. https://doi.org/10.1016/j.elstat.2012.12.018.
[20] Zuo, W.; Zhu, M.; Yang, W.; Yu, H.; Chen, Y.; Zhang, Y. Experimental Study on Relationship between Jet Instability and Formation of Beaded Fibers during Electrospinning. *Polym. Eng. Sci.* 2005, *45* (5), 704–709. https://doi.org/10.1002/pen.20304.
[21] Sill, T. J.; von Recum, H. A. Electrospinning: Applications in Drug Delivery and Tissue Engineering. *Biomaterials* 2008, *29* (13), 1989–2006. https://doi.org/10.1016/j.biomaterials.2008.01.011.
[22] Theron, S. A.; Zussman, E.; Yarin, A. L. Experimental Investigation of the Governing Parameters in the Electrospinning of Polymer Solutions. *Polymer (Guildf).* 2004, *45* (6), 2017–2030. https://doi.org/10.1016/j.polymer.2004.01.024.
[23] Zargham, S.; Bazgir, S.; Tavakoli, A.; Rashidi, A. S.; Damerchely, R. The Effect of Flow Rate on Morphology and Deposition Area of Electrospun Nylon 6 Nanofiber. *J. Eng. Fiber. Fabr.* 2012, *7* (4), 42–49. https://doi.org/10.1177/155892501200700414.
[24] Jafarpour, M.; Aghdam, A. S.; Koşar, A.; Cebeci, F. Ç.; Ghorbani, M. Electrospinning of Ternary Composite of PMMA-PEG-SiO2 Nanoparticles: Comprehensive Process Optimization and Electrospun Properties. *Mater. Today Commun.* 2021, *29*, 102865. https://doi.org/10.1016/j.mtcomm.2021.102865.
[25] Matabola, K. P.; Moutloali, R. M. The Influence of Electrospinning Parameters on the Morphology and Diameter of Poly(Vinyledene Fluoride) Nanofibers- Effect of Sodium Chloride. *J. Mater. Sci.* 2013, *48*, 5475–5482. https://doi.org/10.1007/s10853-013-7341-6.
[26] Mazoochi, T.; Hamadanian, M.; Ahmadi, M.; Jabbari, V. Investigation on the Morphological Characteristics of Nanofiberous Membrane as Electrospun in the Different Processing Parameters. *Int. J. Ind. Chem.* 2012, *3*, 2. https://doi.org/10.1186/2228-5547-3-2.
[27] Sencadas, V.; Correia, D. M.; Areias, A.; Botelho, G.; Fonseca, A. M.; Neves, I. C.; Gomez Ribelles, J. L.; Lanceros Mendez, S. Determination of the Parameters Affecting Electrospun Chitosan Fiber Size Distribution and Morphology. *Carbohydr. Polym.* 2012, *87* (2), 1295–1301. https://doi.org/10.1016/j.carbpol.2011.09.017.
[28] Kuchi, C.; Harish, G. S.; Reddy, P. S. Effect of Polymer Concentration, Needle Diameter and Annealing Temperature on TiO2-PVP Composite Nanofibers Synthesized by Electrospinning Technique. *Ceram. Int.* 2018, *44* (5), 5266–5272. https://doi.org/10.1016/j.ceramint.2017.12.138.

[29] Mit-Uppatham, C.; Nithitanakul, M.; Supaphol, P. Ultrafine Electrospun Polyamide-6 Fibers: Effect of Solution Conditions on Morphology and Average Fiber Diameter. *Macromol. Chem. Phys*. 2004, *205* (17), 2327–2338. https://doi.org/10.1002/macp.200400225.
[30] Fong, H.; Chun, I.; Reneker, D. H. Beaded Nanofibers Formed during Electrospinning. *Polymer (Guildf)*. 1999, *40*, 4585–4592.

6 Development and Characterization of the Cyclodextrin Nanofibers

M. Ramesh and N. Vigneshwari

6.1 INTRODUCTION

Ibuprofen is a poorly water-soluble non-steroidal anti-inflammatory drug, but, the water solubility of ibuprofen can be significantly enhanced by inclusion complexation with cyclodextrins. This new generation of highly porous membranes exhibits great prospect to be used in various separation applications due to their distinguished features such as remarkably high porosity (90%) and interconnected 3D pore structure. As compared with the conventional techniques, electrospinning has been highlighted for developing unique porous membranes. Electrospun nanofibrous membranes have been more and more investigated to a lot of advanced water treatment purposes [1]. The quick dissolving oral films should have bound mechanical integrity not broken throughout handling and transportation, yet, they ought to properly disintegrate within the mouth [2]. Therefore, the fast dissolving oral films should be automatically sturdy and yet, they should be soft, elastic and flexible. Recently, the employment of electrospinning technique is additionally shown to be a really promising approach for developing controlled drug delivery systems and fast dissolving nanofibrous mats in pharmaceutics. Fast dissolving oral drug delivery systems have gotten a lot of attention in pharmaceutics [3, 4]. The fast dissolving oral drug delivery systems are ready as films or strips of edible and water soluble hydrophilic biopolymeric materials which will apace dissolve in mouth and thus can deliver drugs, vitamins and refreshing flavor compounds [5]. The quick dissolving oral films offer the benefit of delivering active compounds in the oral cavity while not requiring water to swallow, and therefore fast dissolving oral films may be advantageous to tablets and pills [6]. The simplicity of the electrospinning setup and the comparatively high production rate of nanofibers make this method extremely engaging for each domain and industry. In the electrospinning technique, endless filament is electrospun from chemical compound solutions (most common) or polymer melts (very limited) below a really high electrical field (Figure 6.1) leading to ultrafine fibers (1,000 times smaller than one human hair) starting from tens of nanometers to a few microns in diameter [7].

This area of research has attracted the interest of a great number of academicians and industrials as such materials turn out to be an alternative solution to the ever-depleting non-renewable sources, environmental pollution, global warming,

 DOI: 10.1201/9781003333814-6

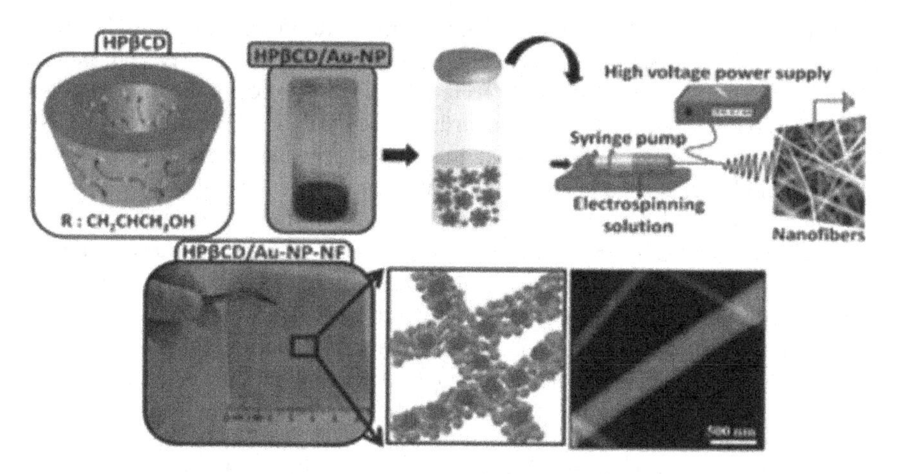

FIGURE 6.1 Nanoparticles incorporated into electrospun cyclodextrin nanofibers.

and energy crisis [8]. In this context, cellulose, starch, alginate, chitin, chitosan, and gelatin have been revealed to be promising candidates with regards to their abundant availability from various resources. Among them, cellulose is by far the most abundant renewable compound obtained from the biosphere and it can be found in plants, algae, tunicates, and some bacteria. This fascinating polymer, seen as an inexhaustive source of raw materials, has potential to be modified and functionalized with several available industrial uses and there is still plenty to discover and celebrate in cellulose. The benefit of cellulose can be further extended when cellulose chains are bundled together, generating highly ordered regions that can be subsequently isolated as nano-particles, known as cellulose nanomaterials or nanocelluloses, considered as a useful class of futuristic materials owing to their physicochemical features. In addition to being renewable and abundant, they combine chemical inertness, excellent stiffness, high strength, low coefficient of thermal expansion, low density, dimensional stability, and ability to modify its surface chemistry [9].

Typically, nano-cellulose can be categorized into two major classes: (i) nanostructured materials (cellulose microcrystals and cellulose microfibrils) and (ii) nanofibers (cellulose nanofibrils, cellulose nanocrystals, and bacterial cellulose). A number of nanocellulose forms can be produced using different methods and from various cellulosic sources. The morphology, size, and other characteristics of each nanocellulose class depend on the cellulose origin, the isolation, and processing conditions as well as the possible pre- or post-treatments. The opportunity of producing nanocellulose with various features is considered a fairly exciting topic, which can promote the exploration of unexplored biomass. The benefits of the 3D hierarchical nanostructure of nanocellulose and its physicochemical characteristics at nano-scale open new prospects in several applications. According to Markets and Markets, the nanocellulose market is forecasted to achieve US $783 million by 2025. The rising demand and the employment of new applications have driven the researchers and the industry to exploit even more the employment of nanocellulose. In addition, the number of papers published is increasing year after year, reflecting the high concern in this type of nanomaterial. This attention expresses itself by the new International Organization for Standardization (ISO), Technical Association of the Pulp and Paper Industry (TAPPI), and Canadian Standards Association (CSA) Standards on CNCs that are being developed and published, highlighting the market interest. This nanofibrous cyclodextrins used in emergency ambulance patient bed [10].

Nanocellulose, which can currently be produced in industrial scale at tons per day, can be employed in several fields in our life, such as nanocomposite materials, biomedical products, wood adhesives, supercapacitors, template for electronic components, batteries, catalytic supports, electroactive polymers, continuous fibers and textiles, food coatings, barrier/separation membranes, antimicrobial films, paper products, cosmetic, cements, and many more emerging uses. The search for novel applications and improving the properties of the current nanocellulose-based materials are crucial driving forces for research and development in various research groups and increasingly in companies [11]. It can be seen that several literature review articles have been published during the last few years and most of them focused on the production of nanocelluloses, their modification and applications. Certain recent findings and advances have not been addressed enough in previous publications, while here, we concisely provide some of the most recent applications of optimized nanocellulose (ONC), cellulose nanocrystals (CNCs). The aim of this chapter is to make a brief summary on the study of nanocelluloses, with a special focus on nanocrystals, as well as their recent applications.

At first, a brief introduction on cellulose, nanocellulose nomenclature, its isolation from several feedstocks, properties, and functionalization are presented. Important challenges related to their production and new directions are addressed. In the subsequent sections, we shed light on current trends and recent research on the use of nanocellulose with special emphasis on nanocomposites, medical, pickering emulsifiers, wood adhesives, adsorption, separation, decontamination, and filtration applications, to provide readers with a comprehensive overview of the advanced science and engineering of nanocellulose-based emerging materials and uses. Other emerging applications of nanocellulose such as paper making, oil and gas drilling and cementing, energy storage systems, sensors and biosensors, which have been extensively reviewed [12], are excluded and they are beyond the scope of the present review. It is expected that this chapter will forge new directions for the preparation of nanocellulose as well as the design and production of nanocellulose-based materials for widespread applications.

6.2 METHODS FOR FABRICATION OF NANOFIBERS

There are many ways to fabricate nanofibers, such as template synthesis, self-assembly, electrospinning, and phase separation. Since the method of template synthesis is not able to produce continuous fibers and in drawing process only visco-elastic materials can be used which tolerate applied tensions, the other alternate methods are used to produce nanofibers [13]. They are self-assembly, electrospinning, and phase separation.

6.2.1 Phase Separation

The phase separation is considered to be one of the simplest methods for production of nano-porous foams. Since this method requires a long time to complete the entire process, it is not the best method for fabrication of nanofibers. The polymer solution quenched below the freezing point of solvent is freeze-dried to produce a porous structure. Various nanoporous foams can be produced by following this process by modifying thermodynamic and kinetic factors. The fabrication of foam scaffolds can be done under five steps: suspension of polymer, phase separation and gelation, extraction of solvent from the gel by means of water, freezing, and then freeze-drying under vacuum [14].

6.2.2 Self-Assembly

In this fabrication method, the molecules and atoms sort out and assemble themselves in the course of fragile and non-covalent forces, for example hydrophobic forces, electrostatic interactions, and hydrogen bonding, and create a stable construction. This fabrication method can be used to make different structures, for example unilamellar and multilamellar vesicles, bilayer, nanoparticles, membranes, fibers, films, micelles, tubes, and capsules [15]. The nanofibers obtained by self-assembly method can be much thinner than those produced by electrospinning, but complication of procedure with low productivity is the major problem associated with self-assembly method [16].

6.2.3 ElectroSpinning

Electrospinning possesses some of the unique properties such as simplicity, affordability, high porosity (good pore size distribution), and yielding continuous fibers. The variety of biomaterials can be used to produce nanofibers, and very low amounts of initial solutions are needed using this method. The fibers which are produced by using the electrospinning method have the diameter in range of 3 nm to several micrometers, whereas fibers obtained using other procedures have the diameter in range of 500 nm up to a few microns. Because of these extremely appreciable properties, electrospinning is a most popular technique for the production of nanofibers. Nanofibers that fell on the stationary collector harvests randomly arranged nanofiber (125–600 nm) matrices, although aligned nanofiber (750–850 nm) mats are synthesized by means of rotatory or disk collector with high-pitched edge [17]. By using different types of electrospinning, hollow fibers, core-shell fibers, nanoparticles, or drug-incorporated fibers, etc. can be produced.

6.2.3.1 Traditional Electrospinning

In traditional electrospinning, three main components are needed: (i) a high voltage source, (ii) syringe pump (nozzle), and (iii) a grounded collector. The nozzle is preferably a metallic needle with a blunt tip for proper observation of the Taylor cone [18]. In this process, first a required amount of polymeric solution (preferably dissolved in a volatile solvent with a specified w:v ratio) is placed into a proper syringe and then to the syringe pump. Then high voltage is applied to the tip of the nozzle, and the elongating conical shape of the droplet is observed. The electrostatic force has to overcome the surface tension of the droplet to form the nanofibers [18, 19]. The structure of the formed fibers can be controlled by various factors such as flow rate of the syringe pump, concentration of the polymer solution, collector type, solution viscosity, applied voltage, distance between the collector and the nozzle, diameter of the nozzle, etc. [19]. Each of these factors affects the fiber morphology significantly. For example, by increasing the voltage, fiber diameter can be decreased, low polymer concentrations can cause electrospraying rather than electrospinning, or increasing the flow rate can reduce the fiber diameter [19].

6.2.3.2 Multi-Jet Electrospinning

The multi-jet electrospinning is developed to improve the productivity and produce composite fibers that cannot be dissolved in regular solvents. Needle diameter, needle number, and configuration play an important role in this approach compared to other electrospinning methods [19]. This method is otherwise known as multi-needle electrospinning. Unfortunately, this method holds one major drawback, which is a strong repulsion among the jets because of the multi needle system [19]. This repulsion, which is generated by the Coulomb force, may cause reduced fiber deposition and poor fiber quality. In order to overcome this problem, needles must be oriented at an appropriate distance.

6.2.3.3 Coaxial Electrospinning

Coaxial method is used to form core-shell nanofibers by using multiple syringe pumps or one syringe pump with various feeding systems. Mostly, a polymer and a composite solution-one is to form shell and the other is to form core parts-it can be used individually, or two different polymer solutions can be employed as forerunner solutions. Directed by the electrostatic repulsions between the surface charges, the polymer solution, which will form the shell part of the composite nanofibers, will be lengthened and will create viscous stress. After that, this stress will be delivered to the core layer, and the polymer solution, which will form the core part, will be promptly stretched [20]. As a result, composite jets will be formed, which will have coaxial structures.

6.2.3.4 Melt Electrospinning

The melt electrospinning technique requires a heating device such as heat guns, lasers, or electrical heating devices to produce nanofibers. A constant heat source is used to keep the polymer solution in its molten state [20]. The main difference between melt electrospinning and conventional electrospinning method is the process of fiber formation. In melt electrospinning, instead of a solution, a molten polymer is used, and the desired product is obtained on cooling; however, in conventional electrospinning, fibers are formed with the help of solvent evaporation [20]. This method can be used with the polymers that do not have a suitable solvent at room temperature.

6.2.3.5 Needless Electrospinning

The needless electrospinning technique has been introduced to avoid the limitations caused by capillaries and needles [21]. The main principle behind this technique is the waves of an electrically conductive liquid self-organize on a mesoscopic scale and form jets when the intensity of the applied electrical field rises above a critical value.

6.3 METHODOLOGY

6.3.1 Nanofibers for Detection of CB Agents

Nanomaterials have high values of specific surface areas, approximately one to two orders of magnitude larger than flat films. This makes them excellent candidates for potential applications in detection of agents. The direct application of electrospun multi-layered mats to garments is one of the main attractive specifications in the area of pH sensitive materials. Up to now, different types of pH detectors have been introduced for various applications; however, direct application of pH sensitive electrospun nanomaterials to garment systems would eliminate costly manufacturing steps and solve seam-sealing problems that have been limiting factors in protective garments. Recently a multi-layer elecrospun nanomaterial with a chemical agent sensor function was developed [21]. The produced multi-layered structure equipped with pH sensitive dyes shows good performance in pH detecting and opens new possibilities for use in protective garments as the chemical and welfare agent detector.

6.3.2 Electrospun Nanofibrous Membranes

A membrane that is prepared by electrospinning is called an electrospun nanofibrous membrane (ENM) and has attractive features. For this reason, it has attracted a lot of attention recently. These features include: high porosity, pore size ranging from tens of nanometers to a few micrometers, high permeability for gases, interconnected

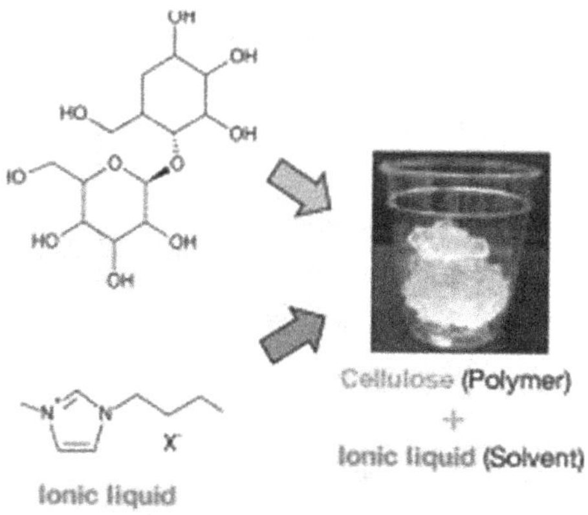

FIGURE 6.2 Multi-layer electrospun nanomaterials.

open pore structures, and a large surface area per unit volume [22]. The electrospinning technique was used to increase the hydrophobicity of the MD membrane [22]. To this end, TiO_2 functionalized with 1H, 2H-perfluorooctyltriethoxysilane was added to the dope solution. Varying amounts of TiO_2 (1%, 5%, and 10%) were applied for fiber production and different concentrations (10%, 15%, and 20%) from PH (PVDF-HFP) were used in order for electrospinning to take place. The presence of TiO_2 not only increases the hydrophobicity of the membrane, but also the membrane pore size was reduced because of the reduction of the fiber's size and it affects the membrane performance. The highest hydrophobicity was in ENM that was made from 20% PH with 10% TiO_2 because it had a good scattering of TiO_2. Its contact angle was reported at about 149 degrees. However, ENM which was made from 10% PH with 10% TiO_2 had the highest LEP because its pore size was decreased. On the basis of the results, they found that ENMs with 10% TiO_2 had better flux and salt rejection stability compared to the commercial membrane and the membranes that do not have TiO_2. When TiO_2 concentration is the same in the two membranes, then the morphology of TiO_2 on fiber surfaces influences the concentration of the polymer, the mass of particles, and volatility of solvents. It was observed that there was no wetting in EMN constructed from 20% PH with 10% TiO_2 after one week of operation, though the feedwater contained high concentrations of salt (7 wt. % NaCl). The flux of this membrane was reported at 40 L m^{-2} h^{-1}.

A nanofiber membrane was constructed and characterized for use in the modified process via electrospinning method [22]. They were able to optimize the structure and properties of the membrane made with the control of polymeric dope composition and parameters related to spinning. Therefore, the performance of the membrane was improved. In addition, they evaluated the effect of the hot-press post treatment on the performance of the MD membrane. Based on the laboratory results, they reported that the membrane structure is highly influenced by factors such as electrospinning process parameters and properties of the dope solution. They

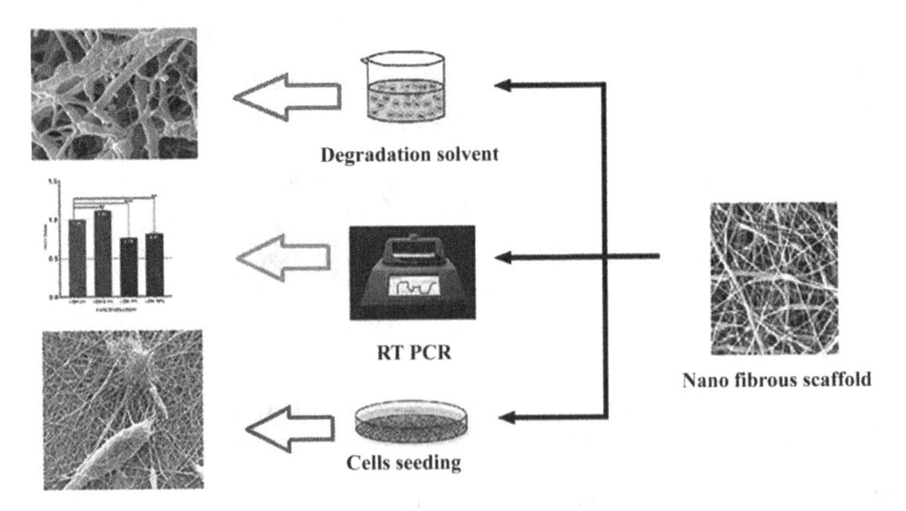

FIGURE 6.3 Morphology of nano-fibrous scaffolds prepared by the electrospinning method.

were controlled polymer concentration and adding appropriate additives in the dope solution to manufacture the nanofiber with small diameter and formation of a membrane with small pore sizes by nanofibers. In addition, the sprayer moving speed can be slowed for membranes with a small pore size and reduced humidity in the spinning chamber. Rough surface with a high hydrophobicity was verified for electrospun membranes by measuring the contact angle of the surface. The hot-press post treatment is essential to be able to increase the permeate flux of water, avoid wetting in membrane pores, and improve fresh nanofiber membrane integrity. The permeate flux of the post-treated nanofiber membrane was reported at 21 kg m^{-2} h^{-1} after a 15 hour test that was higher than the flux of the untreated fresh membrane. Ultimately, electrospun membrane nanofiber (EMN) was introduced as a suitable membrane with a high potential for the modified process. The spectroscopy analysis of the nanofiber-based membrane are shown in Figure 6.4.

The figure shows that the prominent absorption bands (C-H and C-O stretching vibration) of CD are also present for poly-CD nanofibers. The broad absorption band at 2600–2700 cm^{-1} corresponds to the vibration of H-bonded carboxylic OH groups of BTCA, and disappears after TT along with the formation of ester linkage. The strong peak of C = O stretching at 1703 cm^{-1} shifts to higher wave numbers for poly-CD nanofibers suggesting the esterification reaction between CD and BTCA molecules. (A) The other characteristic peaks of BTCA also disappear after TT due to crosslinking. (B) The high resolution XPS spectra of pure CD nanofiber and poly-CD nanofiber after TT. The spectra of the pure CD nanofiber is deconvoluted into three peaks (assigned to C–(C, H, C-O and O-C-O). A new peak is observed in the case of poly-CD nanofiber/after TT which is belong to O = C-O of carboxyl/ester groups. (C) TGA thermograms of BTCA, pure CD nanofiber, poly-CD nanofiber before and after TT. There is an additional degradation step of BTCA at 325 °C before TT, and it becomes indistinct. The degradation step of CD moiety (365 °C) appears after TT as the crosslinker units form the finalstructure of poly-CD network.

Electrospun nanofibrous membranes have attracted much attention from researchers because of its high versatility. Electrospun nanofibrous membranes have become the next-generation filtration media that have promising features and offer good opportunities for advanced filtration techniques in the near future. This versatile technique can be used to fabricate high performance electrospun nanofibrous membranes with a high surface area, a high surface area-to-pore volume ratio, high pore interconnectivity, and uniform pore distribution. Recently, many researchers have focused on the functionalities of electrospun nanofibers to improve their applicability on an industrial scale. For this purpose, nanoparticles have been incorporated into the electrospun nanofibrous membrane to improve its performance. Furthermore, properties such as high porosity, uniform pore size with a narrow pore-size distribution, and a large surface area-to-pore volume ratio make electrospun nanofibrous membranes most desirable as MD membranes to generate a high water vapor flux [23]. Non-woven nanofibrous membranes are widely used in the removal of small particles by the MF process and in desalination by the MD process. The performance of electrospun nanofibrous membranes can be enhanced by considering features such as pore-size distribution, hydrophilicity or hydrophobicity, mechanical strength, and stability. To increase the permeability and permeate

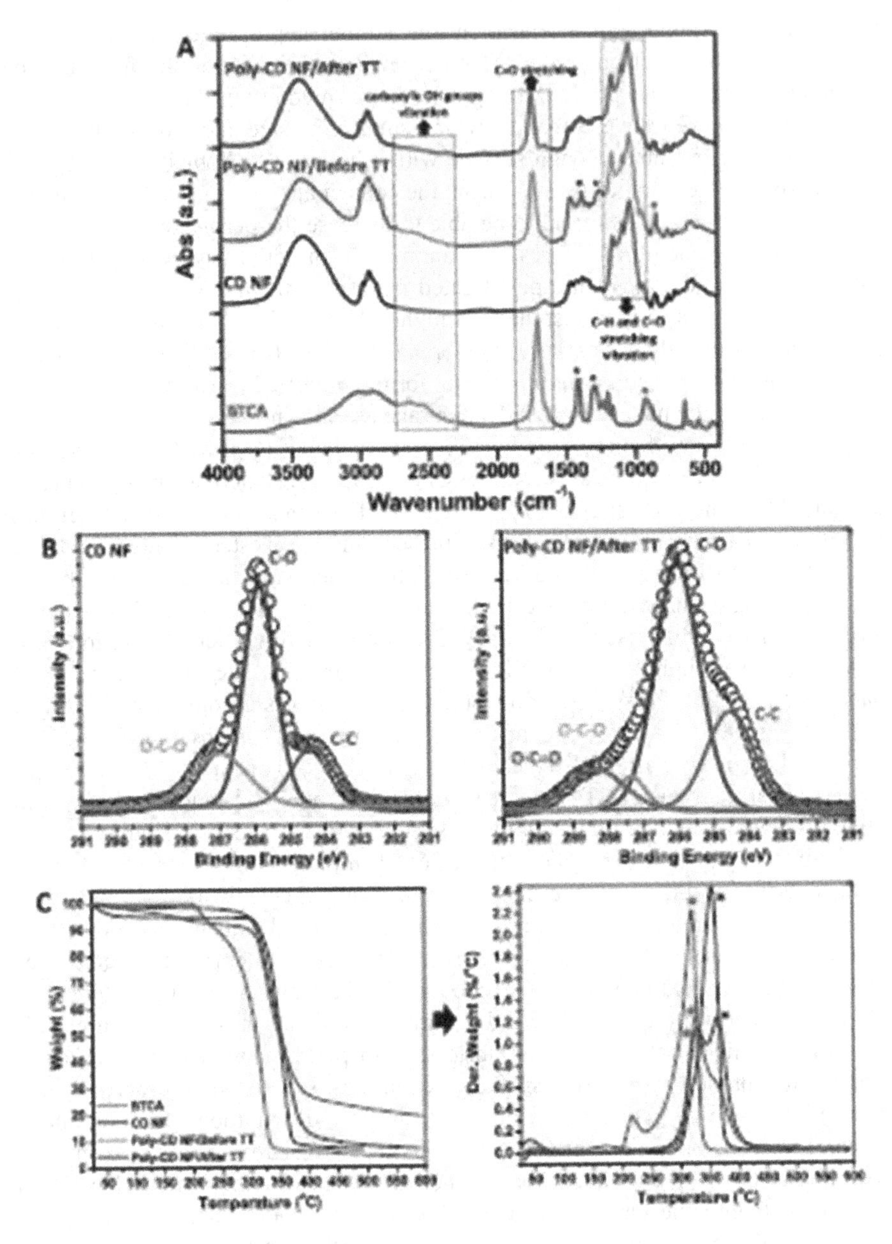

FIGURE 6.4 Spectroscopy analysis of the nanofiber-based membrane.

flux, properties such as membrane thickness and pore size should be optimized. The performance of electrospun nanofibrous membranes can further be optimized with a deeper understanding of how operating parameters and solution parameters can control membrane characteristics in different polymeric solution. Unfortunately, upgrading the electrospinning technique to an industrial scale for commercialization remains a challenge. Therefore, more attention has been given to the high durability and stability of the electrospun nanofibrous membranes to eradicate this problem. Currently, electrospinning is one of the crucial, versatile processes that have influenced the research on water treatment applications. To improve the morphological and topographical features of electrospun nanofibers, various methodologies, such as molecular bonding, insitu polymerization, and addition of molecular dopants, are used in conjunction with electrospinning. Strategies for surface modification, such as nanoparticle coating, treatment with chemicals or heat, grafting, and interfacial polymerization, have been found to be highly effective in enhancing the filtration performance of electrospun nanofibrous membranes [23]. In addition, electrospun nanofibrous membranes are effective in oily wastewater treatment. Thus, considering features such as tunable selectivity, extraordinary permeability, and energy/cost efficiency, it can be concluded that the new-generation membranes used for environmental applications will be based on cost-effective nanofibrous materials.

6.4 METHODS OF NANOFIBERS CHARACTERIZATION

6.4.1 Imaging Methods

The most widely used methods for the evaluation of the structure and characterization of nanofibers are various types of imaging methods. The group of imaging methods involves particularly optical microscopy in the visible range, scanning electron microscopy (SEM), transmission electron microscopy (TEM), and atomic force microscopy (AFM). A prodigious benefit of imaging methods is that the structure can be directly visualized at various places of the nanofibrous sample [24]. So, the attained images provide the beneficial data to compare the local structures within the whole sample. Imaging methods also play an important role in the estimation of in vitro biomedical experiments, depicting the cell cultivation process on various synthetic substrates. The characterization techniques are required to examine the quality, composition, morphology, and structure of the nanofibers.

6.4.2 Optical Microscopy

Optical (light) microscopy has a greater number of advantages such as that preparation of sample is simple and the instrumentation is relatively cheap. The imaging takes place under the atmospheric pressure, and the samples do not need to be dried. Therefore, the polymer samples can be monitored even in the swollen state, the same as they appear in vitro and in vivo experiments [24]. Together with digitization of the signal, optical microscopy permits the monitoring of the changes of polymer sample structures during swelling or drying. Optical microscope is used for introductory examinations of nanofibrous materials during the manufacturing process.

6.4.3 Scanning Electron Microscopy

In general, microscopic imaging techniques are commonly used to observe the parameters such as fiber diameters, alignment, porous structure, fiber morphology, and orientation. With SEM imaging, high-resolution images of a scaffold surface can be obtained and surface properties such as roughness, porosity, smoothness, etc. can be determined [25]. In order to obtain a high-resolution image from scaffolds, samples have to be conductive, so sputtering with a thin layer of a conductive metal such as gold or titanium is a common modification for non-conductive samples. Later, an electron gun is used to produce beams as a cathode source and focuses by electro-magnetic lenses to an exact spot on the sample. The deflection coils are used to shape the selected spot by sampling the whole surface of the sample. This procedure depends on the interaction between the beam and the secondary electrons, which are produced from the sample. Interaction between the secondary electrons from the surface of the sample and the electron beam is monitored and amplified to form an image on the surface.

6.4.4 Transmission Electron Microscopy

TEM technique is considered to be one of the most significant characterization techniques because of its ability to evaluate the interior structure of the samples. With TEM, the pore structure of the scaffolds can be clearly seen. This technique also yields two-dimensional (2D) images of nanofibers and pores similar to that of SEM [26]. When the incoming electron beam fall away from the microscope column, it intermingles with the sample fluorescent screen. A high amount of radiation is emitted from the sample when the electron beam strikes the sample. This interaction causes the elastic and inelastic scattering of the emitted electrons.

6.4.5 Atomic Force Microscopy

AFM technique is frequently used for the evaluation of surface topography. The analytical capabilities of AFM are limited to the uppermost atomic layer of a sample because its operation is based on the interactions with the electron clouds of atoms at the surface [25]. This technique also gives information about morphology, surface roughness, fiber orientation, and particle/grain distribution from the surface of the samples.

6.5 IMPLEMENTATION

The electrospinning process was successfully performed to produce bead-free and uniform cyclodextrin/ibuprofen-IC nanofibers having ~200 nm fiber diameter. The percent loading of drug could be adjusted since cyclodextrin/ibuprofen-IC solutions having different molar ratios (e.g.; 1:1 and 2:1, cyclodextrin:ibuprofen) can be electrospun into nanofibers in the form of self-standing and flexible nanofibrous webs [28]. The cyclodextrin/ibuprofen-IC nanofibrous webs have shown very fast dissolving character when contacted with water or when wetted with artificial saliva suggesting that such electrospun cyclodextrin/ibuprofen-IC nanofibrous webs have shown potential as a fast dissolving oral drug delivery system. It is also noteworthy to mention that the electrospinning of cyclodextrin/ibuprofen-IC nanofibers was performed in water since ibuprofen become water soluble by cyclodextrin [29]. The use of only water provides a great advantage in terms of the industrial processing aspect for the development of such fast dissolving oral drug delivery systems based on cyclodextrin/drug inclusion complex nanofibers. In brief, cyclodextrins can form inclusion complexation with a variety of drug molecules; so, the study with ibuprofen can be extended with other drug molecules in order to develop fast dissolving oral drug delivery systems based on electrospun nanofibrous webs of cyclodextrin/drug inclusion complex nanofibers [30]. The estimation is above 90% removal efficiency for highly concentrated solutions of MB pollutant (40 mg/L) under extremely high flux (3840 $Lm^{-2}h^{-1}$).

6.6 CONCLUSION

Cyclodextrins are very effective for water solubility enhancement for poorly water-soluble drugs by forming inclusion complexation. The electrospinning of nanofibers from cyclodextrin/drug inclusion complexes is a very promising approach to produce fast dissolving nanofibrous webs for oral drug delivery systems. From the literature it is found that the hydroxypropyl-betacyclodextrin is a highly water soluble cyclodextrin derivative which is being used for drug formulations, in order to function both as a nanofiber matrix and complexation agent in order to enhance water solubility and fast dissolution of poorly water-soluble ibuprofen. Essentially, these poly-cyclodextrin nanofibrous webs demonstrate quite rapid uptake of MB from a liquid environment. In general, a bio-based flexible electrospun poly-cyclodextrin nanofiber membrane is a highly efficient molecular filter for wastewater treatment.

REFERENCES

[1] Abouzeid, R. E., Khiari, R., El-Wakil, N., and Dufresne, A. (2018). Current state and new trends in the use of cellulose nanomaterials for wastewater treatment. *Biomacromolecules* 20, 573–597.

[2] Abdul Khalil, H., Chong, E., Owolabi, F., Asniza, M., Tye, Y., Rizal, S., et al. (2019). Enhancement of basic properties of polysaccharide-based composites with organic and inorganic fillers: a review. *J. Appl. Polym. Sci.* 136, 47251.

[3] Abitbol, T., Kam, D., Levi-Kalisman, Y., Gray, D. G., and Shoseyov, O. (2018). Surface charge influence on the phase separation and viscosity of cellulose nanocrystals. *Langmuir* 34, 3925–3933.

[4] Abitbol, T., Rivkin, A., Cao, Y., Nevo, Y., Abraham, E., Ben-Shalom, T., et al. (2016). Nanocellulose, a tiny fiber with huge applications. *Curr. Opin. Biotechnol.* 39, 76–88.

[5] Afrin, S., and Karim, Z. (2017). Isolation and surface modification of nanocellulose: necessity of enzymes over chemicals. *Chem. Biol. Eng. Rev.* 4, 289–303.

[6] Agate, S., Joyce, M., Lucia, L., and Pal, L. (2018). Cellulose and nanocellulose-based flexible-hybrid printed electronics and conductive composites–a review. *Carbohyd. Polym.* 198, 249–260.

[7] Agbor, V. B., Cicek, N., Sparling, R., Berlin, A., and Levin, D. B. (2011). Biomass pretreatment: fundamentals toward application. *Biotechnol. Adv.* 29, 675–685.

[8] Akhlaghi, M. A., Bagherpour, R., and Kalhori, H. (2020). Application of bacterial nanocellulose fibers as reinforcement in cement composites. *Constr. Build. Mater.* 241, 118061.

[9] Alavi, M. (2019). Modifications of microcrystalline cellulose (MCC), nanofibrillated cellulose (NFC), and nanocrystalline cellulose (NCC) for antimicrobial and wound healing applications. *e-Polymers* 19, 103–119.

[10] Almeida, A. P., Canejo, J. P., Fernandes, S. N., Echeverria, C., Almeida, P. L., and Godinho, M. H. (2018). Cellulose-based biomimetics and their applications. *Adv. Mater.* 30, 1703655.

[11] Anderson, S. R., Esposito, D., Gillette, W., Zhu, J., Baxa, U., and Mcneil, S. E. (2014). Enzymatic preparation of nanocrystalline and microcrystalline cellulose. *TAPPI J.* 13, 35–41.

[12] Arof, A., Nor, N. M., Aziz, N., Kufian, M., Abdulaziz, A., and Mamatkarimov, O. (2019). Investigation on morphology of composite poly (ethylene oxide)-cellulose nanofibers. *Mater. Today Proc.* 17, 388–393.

[13] Bala, R., Khanna, S., Pawar, P., and Arora, S. (2013). Orally dissolving strips: a new approach to oral drug delivery system. *Int. J. Pharma. Investig* 3, 67.

[14] Bacakova, L., Pajorova, J., Bacakova, M., Skogberg, A., Kallio, P., Kolarova, K., et al. (2019). Versatile application of nanocellulose: from industry to skin tissue engineering and wound healing. *Nanomaterials* 9, 64.

[15] Bacakova, L., Pajorova, J., Tomkova, M., Matejka, R., Broz, A., Stepanovska, J., et al. (2020). Applications of nanocellulose/nanocarbon composites: focus on biotechnology and medicine. *Nanomaterials* 10, 196.

[16] Bai, L., Xiang, W., Huan, S., and Rojas, O. J. (2018). Formulation and stabilization of concentrated edible oil-in-water emulsions based on electrostatic complexes of a food-grade cationic surfactant (ethyl lauroyl arginate) and cellulose nanocrystals. *Biomacromolecules* 19, 1674–1685.

[17] Bertsch, P., and Fischer, P. (2019). Adsorption and interfacial structure of nanocelluloses at fluid interfaces. *Adv. Colloid Interface Sci.* 35, 571–588.

[18] Cheng, H., Yang, X., Che, X., Yang, M., and Zhai, G. (2018). Biomedical application and controlled drug release of electrospun fibrous materials. *Mater. Sci. Eng. C*, 90, 750–763.

[19] Cheng, H., Kilgore, K., Ford, C., Fortier, C., Dowd, M. K., and He, Z. (2019). Cottonseed protein-based wood adhesive reinforced with nanocellulose. *J. Adhes. Sci. Technol.* 33, 1357–1368.

[20] Gowtham, P., and Arunachalam, V. P. (2020) An efficient monitoring of real time traffic clearance for an emergency service vehicle using IOT. *Int. J. Parallel Prog*. 48,786–812.

[21] Uyar, T., and Kny, E. (2017) *Electrospun materials for tissue engineering and biomedical applications: research, design and commercialization.* Woodhead Publishing.

[22] Tipduangta, P., Belton, P., Fabian, L., Wang, L. Y., Tang, H., Eddleston, M., and Qi, S. (2015) Electrospun polymer blend nanofibers for tunable drug delivery: the role oftransformative phase separation on controlling the release rate. *Mol. Pharm.* 13, 25–39.

[23] Taylor, G. (1969) Electrically driven jets. *Proc. Roy. Soc. A. Math. Phys. Eng. Sci.* 313(1515), 453–475.

[24] Yu, D. G., Li, J. J., Williams, G. R., and Zhao, M. (2018). Electrospun amorphous solid dispersions of poorly water-soluble drugs: a review. *J. Control. Release*. 292, 91–110.

[25] Zhang, Y., Zhou, Y., Cao, S., Li, S., Jin, S, and Zhang, S. J. (2015). Preparation, release and physicochemical characterization of ethyl butyrate and hexanal inclusion complexes with β-and γ-cyclodextrin. *J. Microencap*. 32(7), 711–718.

[26] Bayrak, E. (2022) Nanofibers: Production, characterization, and tissue engineering applications. DOI: 10.5772/Intechopen.102787.

[27] Celebioglu, A., and Uyar, T. (2019) Fast dissolving drug delivery system based on electrospun nanofibrous webs of cyclodextrin/ibuprofen inclusion complex nanofibers. *Mol. Pharm.* DOI: 10.1021/acs.molpharmaceut.9b00798.

[28] Damodaran, P., Subramanian, K. S., and Malaichamy, K. (2019) Development and characterization of cellulosic nanofibre matrix loaded with hexanal. *Int. J. Chem. Stud.* 7(4), 71–74.

[29] Uyar, T. (2017) Electrospinning of functional nanofibers with cyclodextrins. *Cyclodextrin News* 31(2).

[30] Chabalala, M. B., Seshabela, B. C., Van Hulle, S. W. H., Mamba, B. B., Mhlanga, S. D., and Nxumalo, E. N. (2018). Cyclodextrin based nanofibers and membranes: Fabrication, properties and applications. *Intechopen*. DOI: 10.5772/intechopen.74737.

7 Preparation and Characterization of the Electrospun Nanofiber Meshes

Lin Feng Ng, Mohd Yazid Yahya, Syed Mohd Saiful Azwan Syed Hamzah, and Pui San Khoo

7.1 INTRODUCTION

Since a few decades ago, cutting-edge nanotechnology has been continuously growing to develop nanosized structures with a drastic improvement in the material properties. Nanotechnology may yield nanoscale materials with enhanced physicochemical properties by introducing various kinds of nanoscale components, such as nanofibers, nanotubes, nanosheets and nanoparticles, with different dimensions or morphologies into the polymer matrix. The development of nanofiber meshes is considered a remarkable and historic milestone reached by nanotechnology. Nanofibers are primarily formed through the electrospinning process, and the nanofibers are able to form highly porous and thin meshes, which have been employed for a wide range of applications where high porosity is required. Over the years, nanofibers have found various applications in many sectors [1]–[3]. Figure 7.1 summarizes the potential applications of electrospun nanofiber meshes produced from the electrospinning process. Unlike the typical rigid porous structures, electrospun nanofiber meshes are actually dynamic systems where the pore size and shape can be altered. However, a rigid porous structure can also be formed by linking the nanofibers if this is required for certain applications.

The advancement in nanotechnology has initiated the development of several methods to yield nanofibers from various kinds of synthetic polymers. In general, the methods for producing nanofibers can be grouped into electrospinning and non-electrospinning. Even though there are plenty of methods for producing nanofibers, electrospinning has been considered the most promising and popular technique to yield large-scale electrospun nanofibers. This is because some of these non-electrospinning methods have some demerits, such as high processing costs, a limited range of polymer selections and restricted fiber assembly. Electrospinning is considered a simple yet effective method to generate nanofibers with diameters ranging from several micrometers to tens of nanometers [4],[5]. The rationales behind the

DOI: 10.1201978100333814-7

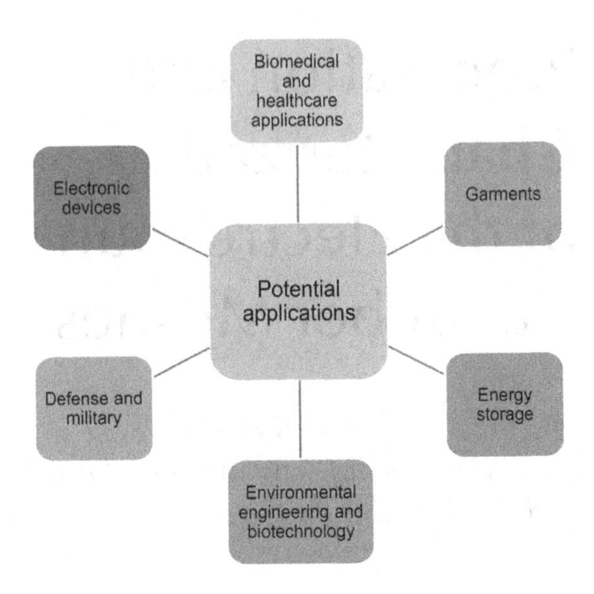

FIGURE 7.1 Potential applications of electrospun nanofiber meshes.

popularity of electrospinning are high production rate, simplicity, reproducibility, less consumption of polymer solution, freedom to control the fiber diameter, convenience to the process, excellent cost-performance ratio and technical advantage in the scale-up process. The development of this technique was initiated in the 1900s, and continuous improvement and growth in this technique have made the production of high-quality nanofibers possible. The electrospinning method is realized by stretching the nanofibers from either synthetic or natural polymer solutions through the electrostatic force. However, processing and solution parameters could affect the morphology and performance of electrospun nanofibers. In addition, the surrounding conditions, such as temperature and humidity, may also affect the electrospinning process, which could have a profound impact on the nanofiber properties. Therefore, the processing and solution parameters should be optimized in order to produce electrospun nanofibers at their exemplary performance.

To date, innumerable research studies have been performed to improve the overall performance of the electrospinning method and the miscellaneous properties of electrospun nanofiber meshes. Given the advancement in electrospinning technology, almost all the polymer solutions with ample molecular weight can be electrospun to form nanofiber meshes. It has been demonstrated that electrospun nanofiber meshes made of natural and synthetic polymers, hybrid polymers and nanoparticle-incorporated polymers have been successfully developed. Different fiber morphologies such as beaded, ribbon, porous and core-shell fibers can be obtained by changing the processing and solution parameters. Since nanofiber meshes have found multiple applications and are closely linked to our daily life, improving the electrospinning technique and miscellaneous properties of nanofiber meshes is essential. This chapter intends to discuss various preparation methods for nanofiber meshes, which can be grouped into electrospinning and non-electrospinning techniques. The working principle of each technique is clearly explained. Apart from that, characterizations of nanofiber meshes to explore their morphological and mechanical properties are uncovered in this chapter.

7.2 PREPARATION METHODS OF NANOFIBER MESHES

Over the years, nanofibers have become an essential part of our daily lives since such materials have found a broad range of applications. The nanofibers derived from natural resources are particularly fascinating for biomedical and agricultural sectors [6]. Due to the increasing demand for green materials, there is a growing interest in biocomposite materials. On this note, it can be seen that numerous research studies are focusing on composites based on eco-friendly fibers [7]–[13]. The large-scale applications of nanofiber meshes have resulted in both environmental and economic benefits. Due to the high demand for nanofiber meshes, it is pivotal to enhance their production rate and functional properties. There are several methods to yield nanofiber meshes, including electrospinning and non-electrospinning techniques. Examples of non-electrospinning techniques are self-assembly, phase separation, drawing, template synthesis, melt-blown technology, melt spinning, etc. The aim of developing non-electrospinning techniques is mainly to improve the production yield of nanofibers. However, electrospinning is still the leading technology among all the preparation methods for nanofiber meshes.

7.2.1 Electrospinning

Although there are innumerable methods to yield nanofiber meshes, electrospinning is regarded as the most versatile and feasible technique to produce ultrathin nanofibers. Remarkable progress has been witnessed in the electrospinning technology to produce high-performance nanofibers with very small diameters. By tuning the processing parameters, it is possible to tailor the shape and morphological properties of the electrospun nanofibers. The fiber diameter, roughness, porosity, fiber alignment and pore interconnectivity are manipulated properties that determine the overall performance of the nanofiber meshes. Apart from the controllable fiber morphology, electrospinning is able to produce continuous nanofibers and form various fiber assemblies such as nonwoven fiber mesh, aligned fiber mesh, patterned fiber mesh, etc. [14]. Lately, the electrospinning method has been utilized to develop three-dimensional (3D) structures as these structures have been shown to offer some benefits over conventional two-dimensional (2D) electrospun nanofibers [15].

Electrospinning is actually an old technology which can be traced way back to the late 1890s. In fact, electrospinning was transformed from electrospraying in 1900, and it was investigated in detail by John Zeleny in 1914. Then, the invention of the electrospinning instrument was patented by Formhals in 1934. Over the years, electrospinning has been regarded as a mature and cutting-edge technology that has gained wide acceptance among researchers and engineers worldwide. The popularity of this technique can be manifested by the fact that more than 200 universities and research institutes around the globe are exploring numerous aspects of the electrospinning process and the nanofibers yielded from this technique [16]. In addition, the number of patents and applications of electrospun nanofiber meshes has been continuously growing in recent years. The electrospinning apparatus is sophisticated, yet the working principle to produce ultrathin nanofibers is simple. With the expansion of this technology, this technique can be further classified into different categories after years of effort to improve the efficiency of electrospinning. Vibration electrospinning, magneto-electrospinning, bubble electrospinning and siro-electrospinning are some examples of electrospinning techniques. Basically, an electrospinning setup is formed by three major components, including a high-voltage power supply, a syringe with a metal needle and a collector in either a horizontal or vertical position. Figure 7.2 illustrates the electrospinning setup and its process for producing electrospun nanofiber meshes. Originally, the electrospinning instrument was based on a single nozzle, but this instrument has the disadvantage of a low production rate. In order to resolve this problem, electrospinning based on multiple nozzles has been introduced with a significant improvement in the production rate. Apart from increasing the number of nozzles, another alternative way to increase the production rate of electrospinning is to increase the voltage level. However, it should be noted that altering the voltage level might affect the morphology of nanofibers.

The working principle of electrospinning revolves around the stretching of the nanofibers from the polymer solution through the electrostatic force. During electrospinning, a high DC voltage is applied to overcome the surface tension of the

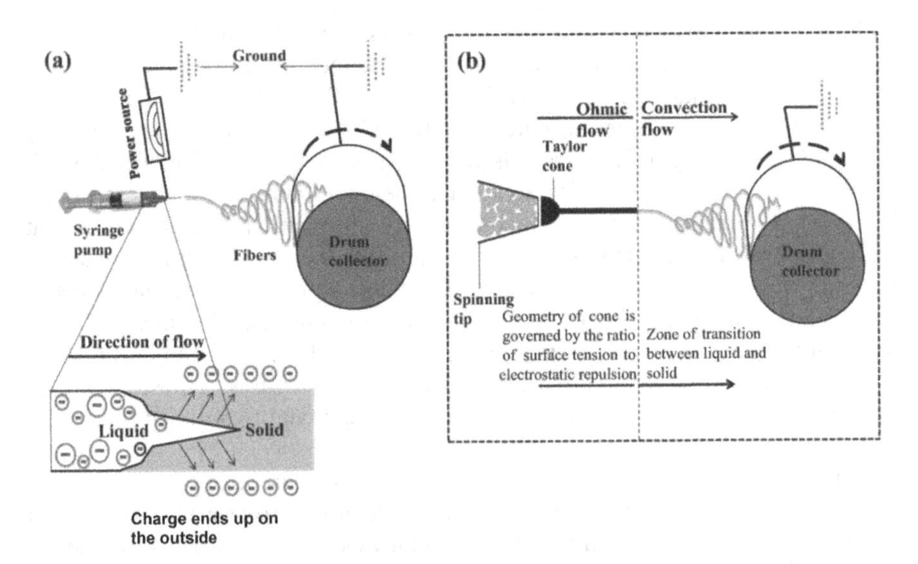

FIGURE 7.2 Electrospinning setup and its process for producing electrospun nanofiber meshes.

Source: (Reproduced with permission obtained from [17]).

polymer droplet at the tip of the metal needle. After applying the electrical charge, the polymer droplet emerges out and elongates, eventually forming a cone-like structure commonly known as a Taylor cone. It is worth mentioning that the DC voltage should be sufficiently higher than the threshold voltage to allow the repulsive electrical force to stretch out the polymer droplet. However, the voltage level required highly depends on the types of polymer solutions. On this note, the critical voltage required to initiate the electrospinning can be determined via Equation (7.1) [18].

$$V_c^2 = 4\frac{H^2}{L^2}\left(\ln\frac{2L}{R} - \frac{3}{2}\right)(0.117\pi\gamma R) \tag{7.1}$$

Where V_c is the critical voltage, H is the distance between the capillary and the collector, L is the length of the capillary, R is the radius of the capillary and γ is the surface tension of the polymer solution.

Subsequently, the jet of the polymer solutions continuously flows in the direction of the electric field and finally reaches the metal collector. Meanwhile, the solvent of the polymer will evaporate very quickly in the pathway, leaving only the nanofiber meshes lying on the metal collector. When the Taylor cone travels from the spinneret to the metal collector, the jet is stretched on the way toward the metal collector, resulting in the production of ultrathin nanofiber meshes. Generally, the electrospinning setup can be in a horizontal or vertical arrangement. When comparing the working principle of these two arrangements, the only difference between these two electrospinning arrangements is the types of the driving force that pull the fibers out of the syringe. Specifically, the effective forces involved in the horizontal electrospinning setup are the electrostatic force from the voltage supply and an opposite attractive force of the collector, which pull the nanofibers. In contrast, the opposite attractive force of the collector and the gravitational force are the two pulling forces that draw the nanofibers from the tip of the metal needle and form the Taylor cone in the vertical electrospinning setup [19].

Several types of solvents are commonly applied in the electrospinning process, such as *N,N*-dimethylformamide (DMF), methanol, acetic acid, isopropanol, 2-methoxyethanol, etc. However, DMF is among the most widely used solvent, as seen in several research studies [20]–[23]. When discussing the electrospinning technique, the solvent used in this technique is one of the essential parts that cannot be neglected. This is because the initial step in the electrospinning process involves the dissolution of the polymer in a certain solvent. In other words, the solvent plays an important role in preparing the polymer solution, and it eventually impacts its spinnability as the solvent determines the viscosity of the polymer solution. Aside from the viscosity, the solvent might also influence the surface tension of the polymer solution to a certain extent. Yang et al. [24] reported that the surface tension of the polymer solutions varied with different types of solvents. Moreover, they revealed that it is possible to convert beaded to smooth electrospun nanofibers by dropping the surface tension of the polymer solution without altering its concentration. Towards an optimum spinnability of the polymer solution, the solvent should have extraordinary volatility, boiling point, vapor pressure and ability to maintain the integrity of the polymer solution [16]. On this matter, the volatility of the solvent should be considered the top priority as it can significantly affect the formation and morphology of the electrospun nanofiber meshes [25]. Therefore, selecting an appropriate solvent for successful electrospinning is one of the critical steps. Overall, electrospinning is considered the most widely employed technique to produce nanofiber meshes. However, there are still some challenges associated with the electrospinning process, which needs more effort to overcome its shortcomings, such as the high cost of technology and low production yield. Furthermore, vapors from certain solvents emitted into the air might raise occupational health issues.

7.3 NON-ELECTROSPINNING

As of today, over 100 kinds of polymers have been used to produce nanofibers, including natural and synthetic polymers. The high demand for nanofibers has triggered the necessity to develop more nanofiber fabrication methods. Even though producing sufficient nanofiber meshes for research purposes is rather simple, there is difficulty in scaling up the production yield for commercial use. Over the past few decades, various fabrication methods have been developed to elevate the productivity of nanofiber meshes. Aside from electrospinning, non-electrospinning methods are other viable options for obtaining nanofiber meshes to boost the production yield. As mentioned in the previous section, non-electrospinning methods include self-assembly, phase separation, drawing, template synthesis, melt-blown technology, melt spinning, etc. The majority of these techniques generate nanofibers in the form of randomly oriented fiber meshes [26]. Analogous to electrospinning, the processing parameters of non-electrospinning fabrication methods can also be optimized to enhance the functional properties of nanofiber meshes. For example, the morphology, dimension and porosity can be altered by changing the processing parameters. The following sub-sections will discuss different types of non-electrospinning methods and their respective working principles.

7.3.1 Self-Assembly

Self-assembly is one of the non-electrospinning techniques which yields nanofibers by using small molecules as building blocks to form nanofibers [27]. In the biomedical sector, self-assembly is often used to prepare 3D scaffolds for tissue regeneration, engineering nerve tissue and cartilage tissue [28],[29]. In this technique, convergence synthesis is often applied in synthesizing complex molecules. However, convergence synthesis requires dedicated laboratory instruments, and this process is only limited to certain polymers [26]. In fact, self-assembly is a bottom-up nanofiber fabrication technique in which the small molecules organize and arrange themselves concentrically through intermolecular forces such as hydrophobic forces, hydrogen bonding and electrostatic reactions. In self-assembly, the morphology and functional properties of nanofibers are governed by the non-covalent forces that hold the molecules together. It is actually a very efficient technique to yield nanofibers with diameters lower than a few nanometers. Compared to electrospun nanofibers, those nanofibers yielded from self-assembly possess even smaller diameters [30]. Although it is efficient in producing nanofibers, it is undeniable that self-assembly encompasses several shortcomings, such as complicated process, high cost, time-consuming, low production yield, difficulty in controlling the fiber dimensions and a very limited range of polymer selections. These shortcomings make self-assembly less attractive to researchers and engineers.

7.3.2 Phase Separation

Phase separation is another quite simple technique applied to produce various types of nanofibers from either natural or synthetic polymers. It can be further divided into several categories, including emulsion freeze-drying, hydrogel freeze-drying and solution freeze-drying, based on the types of solvent used in the phase separation procedure. This technique involves several steps, including dissolution, gelation, extraction and freezing. These steps eventually lead to the formation of porous nanofiber meshes. The working principle of phase separation is the separation of polymer- and solvent-rich domains to obtain the nanoscale fiber network. In particular, the polymer at the required concentration is firstly mixed with the solvent either at room temperature or elevated temperatures to obtain a gel of a polymer. Thereafter, the gelation temperature is reached and maintained to trigger the phase separation to produce nanofibrous matrices. It should be noted that the gelation temperature varies according to the concentration of the polymer solution. Furthermore, the gelation duration also highly depends on the polymer concentration and gelation temperature [31]. From here, it can be seen that gelation is the most challenging and critical step to control, as it could affect the morphology of nanofibers. The platelet-like structures and nanofiber network with a fiber diameter ranging from 50–500 nm are formed during the gelation. Finally, the matrix is dried, and the solvent is extracted from the mixture through the applied vacuum using a freeze dryer, leading to the formation of nanofibers. Figure 7.3 illustrates the phase separation process to yield nanofibers.

Nevertheless, the morphology of the nanofibers can be affected by both the processing and solution parameters. By changing the types of solvent, types of polymer, gelation temperature and gelation duration, the fiber morphology and the

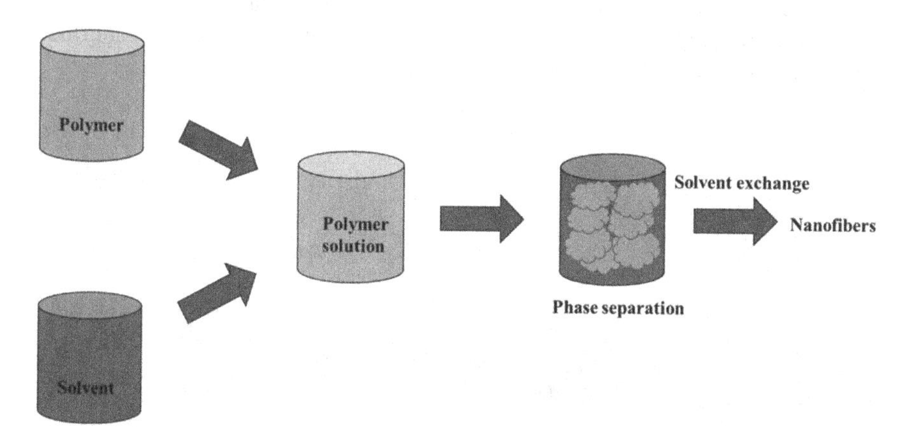

FIGURE 7.3 Phase separation process.

Source: (Reproduced with permission obtained from [32]).

performance of the nanofiber meshes will be altered. Overall, the phase separation technique offers several fascinating advantages in producing nanofibers. Simplicity, low cost and the freedom to design the fiber morphology are those advantages provided by the phase separation technique. The freedom to control the fiber morphology is especially attractive as it allows the researchers and engineers to tailor the mechanical properties of the nanofiber meshes by modifying the processing and solution parameters. However, despite being an affordable and simple technique, it still encompasses some limitations which retard the expansion of this technique to a wide range of applications. Currently, phase separation is only applicable to a very limited number of polymers since this technique requires gelation capability. Polylactide acid (PLA), polyglycolide, poly(methyl methacrylate), polystyrene and epoxy are examples of polymers that can yield nanofibers using the phase separation technique. Other drawbacks of phase separation include time consumption, low structural stability and difficulty in maintaining porosity [33].

7.3.3 Drawing

Drawing is another non-electrospinning technique which is similar to the dry spinning process. It is commonly used in the fiber-forming industry to produce long and smooth nanofibers. During the drawing process, a device known as micromanipulator is used to yield the nanofibers. This method is fascinating as it only requires the sharp tip of the micropipette to produce nanofibers. Alternatively, a hollow glass micropipette can be used to ensure a continuous polymer extraction to avert the volume shrinking problem, which could influence the diameter of nanofibers [34]. The drawing process consists of mainly three steps, including (1) the deposition of a drop of the polymer solution on the substrate material, (2) moving the micropipette towards the edge of the polymer drop on the substrate surface and (3) dipping and pulling back the micropipette from the polymer droplet. At the end of the third step, the nanofiber is extracted from the polymer droplet at a very slow rate. The liquid fibers solidify after the solvent evaporates during the drawing process. It should be noted that the micropipette must be pulled gently and slowly at a rate of around 10^{-4} m/s. The three steps in the drawing are repeated multiple times to produce nanofibers. Figure 7.4 illustrates the drawing process to yield nanofibers.

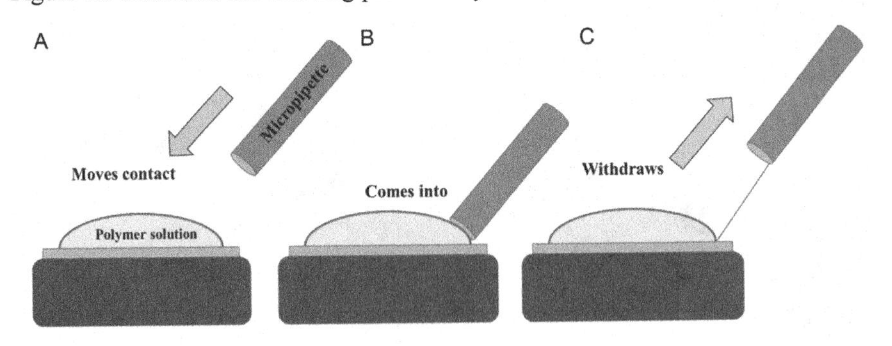

FIGURE 7.4 Drawing process.

Source: (Reproduced with permission obtained from [32]).

For this technique, the morphology of the nanofibers primarily depends on the extraction rate, polymer concentration and solvent evaporation rate. The main advantages of this technique are simplicity, low cost and freedom to modify the parameters. The freedom to modify the key parameters enables better fiber dimension and morphology control. Moreover, due to its simplicity, this technique does not require highly trained technicians to handle the entire process. However, it is not a demerit-free technique to yield nanofibers. The nanofibers produced from the drawing have a diameter higher than 100 nm, which is larger than the nanofibers produced from other viable techniques. To date, the drawing method is only limited to laboratory scale as it is a discontinuous process with a low production yield. Moreover, drawing can be time-consuming and has a very limited number of polymers that can produce nanofibers. It only applies to viscoelastic polymers that can sustain the pulling force during drawing. Since the solvent evaporates during drawing, the viscosity of the droplet may augment. The increase in the viscosity may result in the volume shrinkage of the droplet, which in turn, affects the fiber diameter and the flexibility in drawing continuous fibers.

7.3.4 Template Synthesis

Template synthesis is another non-electrospinning approach that utilizes nano-porous membranes consisting of several nanosize pores (5–50 mm thickness) to produce aligned nanofibers, nanowires and nanotubes with controllable dimensions. Figure 7.5 depicts the template synthesis method to yield nanofibers. As the name suggests, this method requires a template or mold to synthesize the desired nanofiber. As can be seen

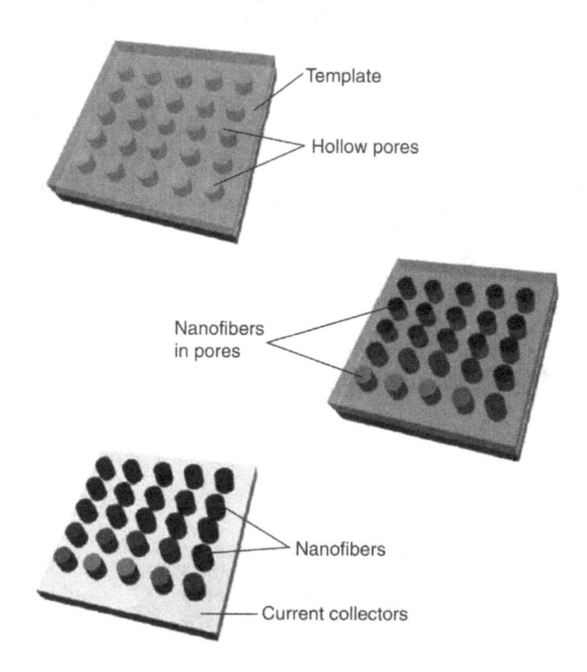

FIGURE 7.5 Template synthesis method.

Source: (Reproduced with permission obtained from [35]).

in Figure 7.5, the template is placed on the collector. Once the nanofibers have been formed, the template can be removed from the collector, leaving the nanofibers lying on the collector. The working principle of this technique involves the pressurized water exerting a certain level of pressure on the polymer solution, which allows the polymer solution to pass through the membrane with numerous nanosized pores. Nanofibers are formed and solidified once the polymer extrusion is in contact with the solidifying solution. The main virtue of this technique is that the diameter of the nanofibers can be easily controlled by changing the templates, as the pore size of the template determines the fiber diameter. However, it should be noted that this method is incapable of producing long nanofibers. The maximum fiber length can only be up to a few micrometers.

7.3.5 Melt-Blown Technology

In comparison with other non-electrospinning methods, melt-blowing is a relatively simple yet versatile and cost-effective technique to generate continuous nanofiber, forming a web structure of nanofiber with random orientation. Additives and binders are not required in melt-blowing, making this technique attractive to researchers and scientists. The nanofiber meshes produced from melt blowing exhibit high specific surface area, decent stiffness and tunable permeability [36]. In melt-blown technology, the nonwoven nanofiber meshes are yielded in a single step where the molten polymers are extruded through an orifice of the die with the aid of a high-velocity hot air stream [37]. The melt-blowing process to generate a nanofiber web structure is shown in Figure 7.6. The nonwoven nanofiber meshes possess very low diameters as the hot air stream reduces the fiber diameter drastically. Subsequently, the nanofiber with a low diameter is solidified by the cooling air, and the web structure with

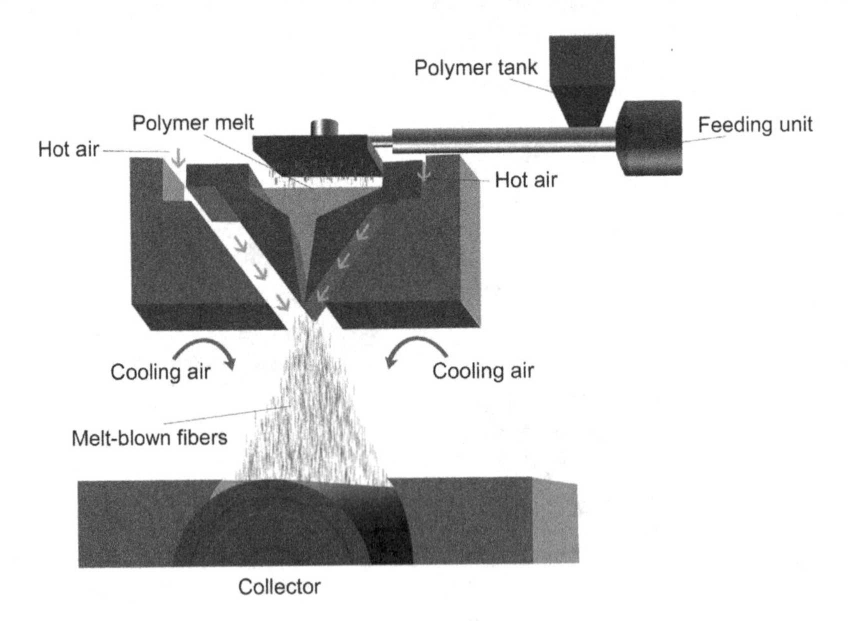

FIGURE 7.6 Melt-blowing process.

Source: (Reproduced with the permission obtained from [38]).

random fiber orientation is deposited on the collector. In general, this technique is able to generate nanofibers with a diameter ranging from 1–10 μm, and all thermoplastic polymers can be processed through melt-blown technology [38]. Since nanofiber meshes are formed without controlled stretching, it offers a lower processing cost and higher production rate than other nanofiber-generating techniques.

The fiber diameter of the meshes could be affected by several factors, including the die-to-collector distance, extrusion rate, polymer viscosity, temperature of the polymer melt and velocity of the hot air stream. Yesil and Bhat [39] revealed that it is possible to reduce the fiber diameter by increasing the die-to-collector distance and air pressure. Furthermore, it was also found that the fiber diameter slightly decreased with an increase in the die temperature. This is expected since elevated temperatures help to reduce the viscosity of the polymer solution, which leads to a drastic drop in the fiber diameter when the fibers exit the orifice. Although melt blowing is a simple technique to produce nanofiber meshes, there are a few hurdles which could limit its use for large-scale applications. On top of that, melt blowing has difficulty in producing very thin nanofibers due to the inability to design and fabricate the sufficiently small orifice of the die. Moreover, the high polymer viscosity is another key factor which retards the use of melt blowing to produce nanofiber meshes. In this regard, several attempts have been made to tackle the hurdles mentioned earlier in the melt-blowing techniques. Modification of the melt-blowing setup, special design of the die and reduction in the polymer viscosity are those possible ways to improve the performance of melt-blowing technique.

7.3.6 Melt Spinning

Electrospinning, melt blowing and spinning are three methods commonly applied to produce nonwoven fiber meshes [37]. Unlike the electrospinning that stretches the fibers through electrostatic force, melt spinning draws the fibers from the polymer solution using a spinneret containing innumerable pores and a take-up wheel. The melt spinning setup is depicted in Figure 7.7. In melt spinning, the

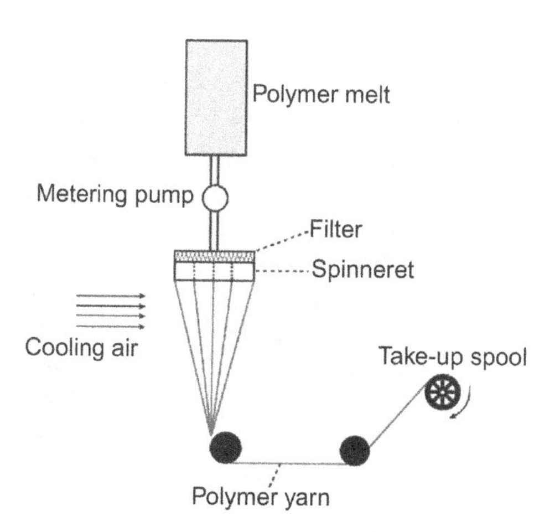

FIGURE 7.7 Melt spinning setup.

Source: (Reproduced with the permission obtained from [40]).

polymers, in their pristine condition, are heated to polymer melts with an appropriate viscosity until they can be extruded to form continuous nanofiber filaments. The continuous nanofibers are then solidified by cooling air during the drawing process, leading to highly oriented strong nanofibers with a high production rate. The average fiber diameter in melt spinning is typically larger than 10 μm [41]. To date, there are around 100 kinds of polymers, including synthetic and natural polymers, which can yield nanofibers through either solvent or melt spinning [17]. Melt spinning is commonly regarded as the most economical spinning method as it does not involve the use of solvent and has a relatively simple process. Table 7.1 summarizes the merits and demerits of various nanofiber-generating methods.

TABLE 7.1
Merits and Demerits of Various Nanofiber-Generating Methods

Nanofiber-generating methods	Flexibility to design fiber diameter	Merits	Demerits
Electrospinning	Yes	• A broad range of polymer selection • Yield long and continuous nanofibers • Form various fiber assemblies • The simplicity of the process	• High cost • Low production yield • Jet instability • Certain kinds of solvents are toxic
Self-assembly	No	• Obtain smaller fiber diameter easily	• Sophisticated process • Complicated process • High cost • Time-consuming • Low production yield • Difficult to control the fiber dimensions • Limited range of polymer selections
Phase separation	No	• The simplicity of the process • Low cost • Freedom to design the fiber morphology	• Limited range of polymer selections • Time-consuming • Low structural stability • Hard to maintain the porosity
Drawing	No	• The simplicity of the process • Low-cost • Freedom to design the fiber morphology • No highly trained technicians are required	• Relatively large fiber diameter • Discontinuous process • Low production yield • Time-consuming • Limited range of polymer selections
Template synthesis	Yes	• Controllable fiber diameter	• Incapable of producing long nanofibers • Sophisticated process

Nanofiber-generating methods	Flexibility to design fiber diameter	Merits	Demerits
Melt-blown technology	Yes	• Yield long and continuous nanofibers • High production rate • No special additives and binders are required • Low processing cost	• Difficulty in producing very thin nanofibers • Limited range of polymer selections
Melt spinning	Yes	• Yield long, continuous and highly oriented nanofibers • High production rate • A broad range of polymer selection • No solvent recovery process is required	• Only applicable for thermoplastic polymers

7.4 CHARACTERIZATION OF NANOFIBER MESHES

Thanks to the comprehensive research studies focusing on the process parameters of various nanofiber-generating methods, it is now proven that the process parameters have a decisive effect on the morphological properties of the nanofibers and the overall performance of nanofiber meshes or scaffolds. From this standpoint, the morphological properties of nanofibers can be tailored by altering the process parameters of certain nanofiber-generating methods. Therefore, it indicates that the deeper the understanding of the process parameters and manufacturing technology, the better the control of the final properties of the nanofiber meshes. It is worth mentioning that other factors, such as fiber alignment and fiber mesh thickness, also have a major impact on the final properties of the nanofiber meshes. Therefore, aside from the processing parameters, solution parameters, fiber alignment and fiber mesh thickness should not be neglected to ensure the development of nanofiber meshes at their optimum performance.

The research studies exploring the miscellaneous properties of nanofiber meshes and scaffolds are on an upward trend. Yesil and Bhat [39] investigated the effect of die temperature, die-to-collector distance (DCD) and air pressure on the tensile and tear strengths of melt-blown nonwovens based on polyethylene nanofibers. They reported that the tensile strength dropped with an increase in DCD due to the loosening of the web structure. However, the tensile strength increased when the air pressure was increased up to 35 kPa. This trend overturned when the air pressure was further increased to 70 kPa. It is worth noting that the die temperature was found not to affect the tensile strength of the nonwovens. On the other hand, the tear strength of nonwovens was found to increase with an increase in DCD up to a critical level. Any further increase in the DCD may deteriorate the tear strength. The same trend was noticed for the effect of air pressure on the tear strength of nonwovens. However, the increase in die temperature was found to increase the tear strength of nonwovens due to enhanced fiber bonding. Neves et al. [42] explored the morphological and tensile properties of electrospun nanofiber meshes prepared using three different patterned collectors. The electrospun nanofiber meshes were based on polyethylene oxide and poly(ε-capro-lactone). They revealed that the electrospun nanofiber meshes produced using different patterned collectors did not show any unique characteristics of a specific material. In terms of mechanical properties, the tensile properties of the nanofiber meshes were highly governed by the fiber alignment. The nanofiber meshes produced on the longitudinal screw collector showed higher tensile properties due to a higher degree of alignment toward the loading direction. Conte et al. [43] identified the effect of fiber density and strain rate on the mechanical properties of electrospun polycaprolactone nanofiber meshes. They reported that increasing the fiber density increased the tensile strength but undermined Young's modulus. The fiber density was observed to have no significant impact on the tensile strain. However, reducing the strain rate to 0.5 mm/min increased the tensile strength, modulus and strain. Butcher et al. [44] conducted a fascinating study where the effect of concentration and bloom strength of gelatin solution on the morphological and tensile properties of electrospun gelatin nanofiber meshes was identified. The findings showed that an increase in the gelatin concentration or bloom strength led to an increase in fiber diameter. The fiber diameters and orientations of the electrospun nanofibers are shown in Figure 7.8. According to Figure 7.8, the diameter variability also increased with an increase in the concentration and bloom strength. The tensile properties of the electrospun nanofiber meshes with respect to bloom

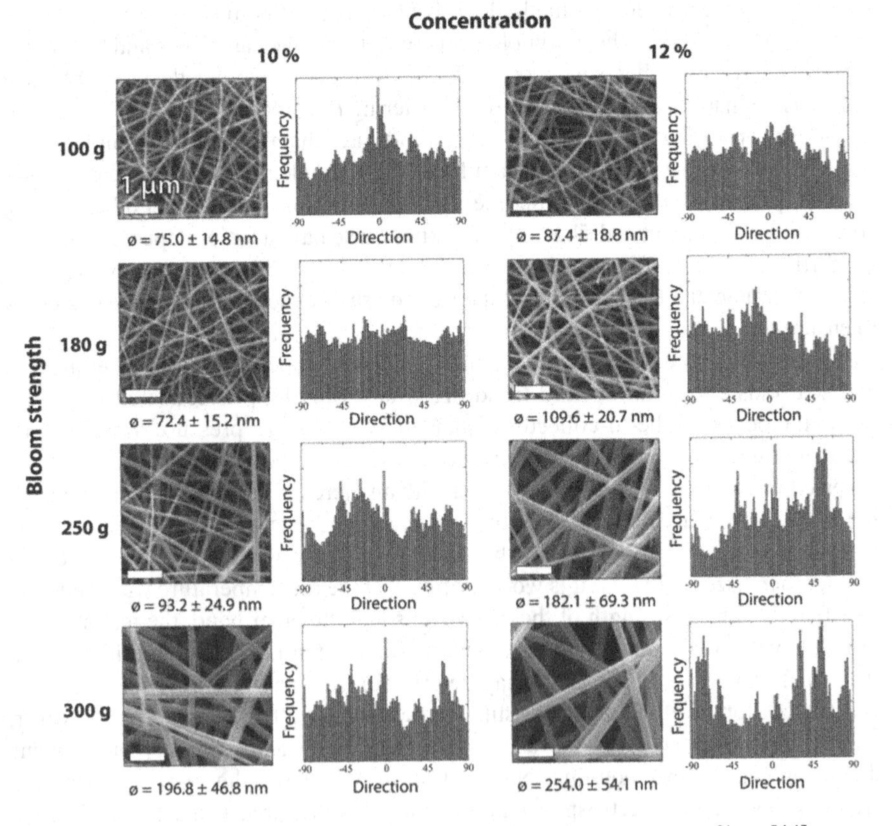

FIGURE 7.8 The fiber diameters and orientations of the electrospun nanofibers [44].

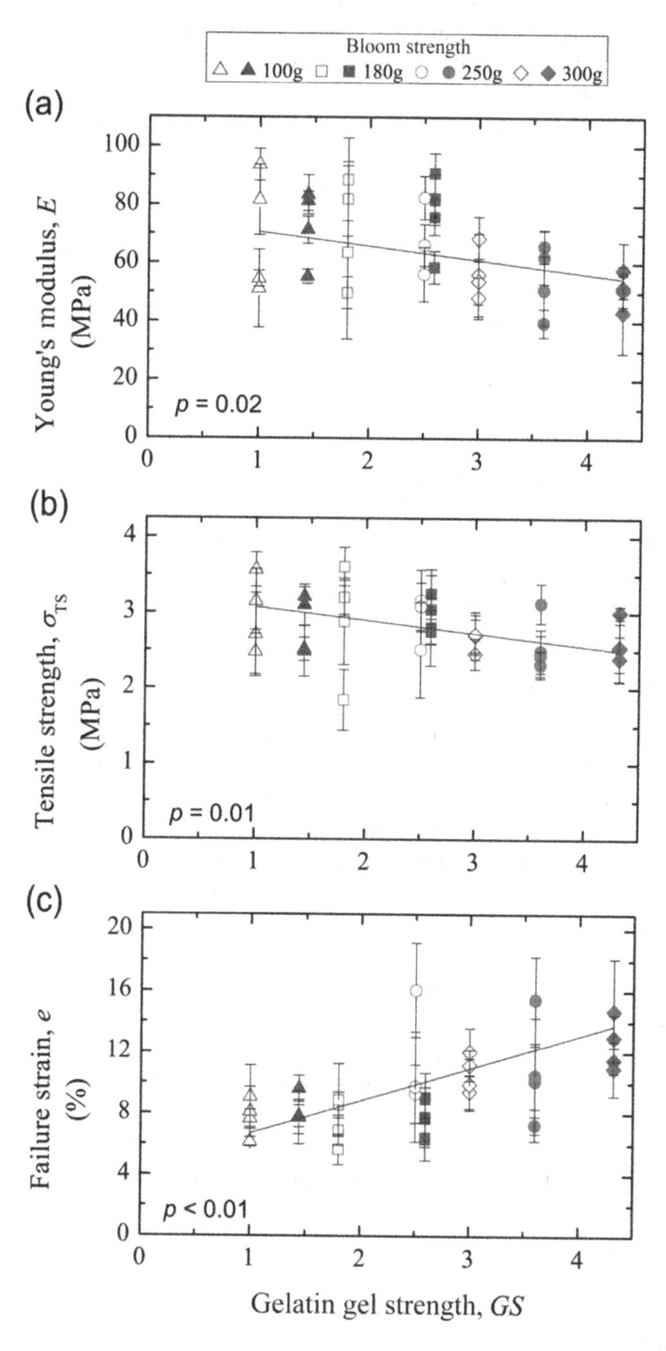

FIGURE 7.9 Tensile properties of the electrospun nanofiber meshes with respect to bloom strength [45].

strength are shown in Figure 7.9. According to the results, it is clear that increasing the bloom strength undermined the tensile strength and modulus but augmented the tensile strain. On the other hand, the viscosity was observed to increase with increase in the gelatin concentration, but it has profound influence only on the tensile strain, while the fiber diameter has a statistical relationship only with tensile strength and strain. Mubyana et al. [45] unveiled the influence of specimen thickness and fiber alignment on the mechanical properties of electrospun polycaprolactone (PCL) nanofiber meshes. The findings evidenced that the failure strain was not correlated to the specimen thickness. Nonetheless, all the tensile strength, modulus and toughness decreased with an increase in the specimen thickness. When scrutinizing the tensile properties of nanofiber meshes with varying fiber alignments, it was shown that the tensile strength, modulus and toughness of aligned fiber meshes were higher than those of random fiber meshes. However, randomly oriented fiber meshes encompassed higher tensile strain than aligned fiber meshes. In accordance with the literature studies, many factors could significantly impact the functionality of nanofiber meshes. The performance of nanofiber meshes is largely dependent on the fiber-generating methods and their processing parameters. Due to the advancement and promising characteristics of electrospinning technique, performing more research studies focusing on the electrospinning is necessary to uplift further the productivity and functional properties of electrospun nanofiber meshes.

7.5 CONCLUSIONS

Nanotechnology is one of the rapidly growing fields nowadays, and it is closely linked to our daily life. The transformation in nanotechnology has led to the development of nanofibers having a diameter in the nanoscale. Since nanofibers have a very small diameter, it implies that nanofibers possess a large surface area-to-volume ratio. This physical property endows nanofiber meshes with superior functional properties. Many nanofiber-generating methods have been developed to yield nanofibers with decent functional properties. However, the versatility of electrospinning has made this method successfully garner significant interest among researchers. Electrospinning is particularly favorable for biomedical applications such as tissue regeneration. Despite the versatility of electrospinning, it is still not completely understood as many processing parameters of this technique can be manipulated, and these parameters may eventually affect the functional properties of the products. In order to uplift the productivity of nanofiber, several non-electrospinning methods have also been utilized to yield high-performance nanofiber meshes. Indeed, each of the nanofiber-generating techniques has its unique features. Meanwhile, these techniques also encompass certain shortcomings that must be carefully addressed. It is, therefore, pivotal to perform more comprehensive research studies to upscale each nanofiber-generating technique to obtain enhanced nanofiber meshes for a broad range of biomedical or industrial applications. When referring to the literature studies, it is obvious that the morphology of nanofibers can be affected by the processing and solution parameters. From this standpoint, it can be considered to have merit as it offers flexibility and freedom to modify and design the fiber morphology so as to fit well with particular applications. By judiciously designing the fiber morphology, the overall functional properties of the nanofiber meshes can be improved and optimized.

REFERENCES

[1] Chang, W. M.; Wang, C. C.; Chen, C. Y. The Combination of Electrospinning and Forcespinning: Effects on a Viscoelastic Jet and a Single Nanofiber. *Chem. Eng. J.* 2014, *244*, 540–551. https://doi.org/10.1016/j.cej.2014.02.001.

[2] Persano, L.; Camposeo, A.; Tekmen, C.; Pisignano, D. Industrial Upscaling of Electrospinning and Applications of Polymer Nanofibers: A Review. *Macromol. Mater. Eng.* 2013, *298*, 504–520. https://doi.org/10.1002/mame.201200290.

[3] Luo, C. J.; Stoyanov, S. D.; Stride, E.; Pelan, E.; Edirisinghe, M. Electrospinning versus Fibre Production Methods: From Specifics to Technological Convergence. *Chem. Soc. Rev.* 2012, *41*, 4708–4735. https://doi.org/10.1039/c2cs35083a.

[4] Reneker, D. H.; Chun, I. Nanometre Diameter Fibres of Polymer, Produced by Electrospinning. *Nanotechnology* 1996, *7* (3), 216–223. https://doi.org/10.1088/0957-4484/7/3/009.

[5] Li, D.; Xia, Y. Electrospinning of Nanofibers: Reinventing the Wheel? *Adv. Mater.* 2004, *16*, 1151–1169. https://doi.org/10.1002/adma.200400719.

[6] Neo, Y. P.; Ray, S.; Easteal, A. J.; Nikolaidis, M. G.; Quek, S. Y. Influence of Solution and Processing Parameters towards the Fabrication of Electrospun Zein Fibers with Sub-Micron Diameter. *J. Food Eng.* 2012, *109*, 645–651. https://doi.org/10.1016/j.jfoodeng.2011.11.032.

[7] Feng, N. L.; Malingam, S. D.; Ping, C. W. Mechanical Characterisation of Kenaf/PALF Reinforced Composite-Metal Laminates: Effects of Hybridisation and Weaving Architectures. *J. Reinf. Plast. Compos.* 2021, *40* (5–6), 193–205. https://doi.org/10.1177/0731684420956719.

[8] Ng, L. F.; Yahya, M. Y.; Muthukumar, C. Mechanical Characterization and Water Absorption Behaviors of Pineapple Leaf/Glass Fiber-Reinforced Polypropylene Hybrid Composites. *Polym. Compos.* 2022, *43* (1), 203–214. https://doi.org/10.1002/pc.26367.

[9] Feng, N. L.; Malingam, S. D.; Subramaniam, K.; Selamat, M. Z.; Juan, W. X. The Investigation of the Tensile and Quasi-Static Indentation Properties of Pineapple Leaf/Kevlar Fibre Reinforced Hybrid Composites. *Def. S T Tech. Bull.* 2020, *13* (1), 117–129.

[10] Feng, N. L.; Malingam, S. D.; Ishak, N. M.; Subramaniam, K. Novel Sandwich Structure of Composite-Metal Laminates Based on Cellulosic Woven Pineapple Leaf Fibre. *J. Sandw. Struct. Mater.* 2021, *23* (7), 3450–3465. https://doi.org/10.1177/1099636220931479.

[11] Sivakumar, D.; Ng, L. F.; Zalani, N. F. M.; Selamat, M. Z.; Ab Ghani, A. F.; Fadzullah, S. H. S. M. Influence of Kenaf Fabric on the Tensile Performance of Environmentally Sustainable Fibre Metal Laminates. *Alexandria Eng. J.* 2018, *57* (4), 4003–4008. https://doi.org/10.1016/j.aej.2018.02.010.

[12] Chandrasekar, M.; Siva, I.; Kumar, T. S. M.; Senthilkumar, K.; Siengchin, S.; Rajini, N. Influence of Fibre Inter-Ply Orientation on the Mechanical and Free Vibration Properties of Banana Fibre Reinforced Polyester Composite Laminates. *J. Polym. Environ.* 2020, *28* (10), 2789–2800. https://doi.org/10.1007/s10924-020-01814-8.

[13] Sanjay, M. R.; Yogesha, B. Studies on Hybridization Effect of Jute/Kenaf/E-Glass Woven Fabric Epoxy Composites for Potential Applications: Effect of Laminate Stacking Sequences. *J. Ind. Text.* 2018, *47* (7), 1830–1848. https://doi.org/10.1177/1528083717710713.

[14] Teo, W. E.; Ramakrishna, S. A Review on Electrospinning Design and Nanofibre Assemblies. *Nanotechnology* 2006, *17* (14), R89. https://doi.org/10.1088/0957-4484/17/14/R01.

[15] Vong, M.; Diaz Sanchez, F. J.; Keirouz, A.; Nuansing, W.; Radacsi, N. Ultrafast Fabrication of Nanofiber-Based 3D Macrostructures by 3D Electrospinning. *Mater. Des.* 2021, *208*, 109916. https://doi.org/10.1016/j.matdes.2021.109916.

[16] Bhardwaj, N.; Kundu, S. C. Electrospinning: A Fascinating Fiber Fabrication Technique. *Biotechnol. Adv.* 2010, *28* (3), 325–347. https://doi.org/10.1016/j.biotechadv.2010.01.004.

[17] Haider, A.; Haider, S.; Kang, I. K. A Comprehensive Review Summarizing the Effect of Electrospinning Parameters and Potential Applications of Nanofibers in Biomedical and Biotechnology. *Arab. J. Chem.* 2018, *11* (8), 1165–1188. https://doi.org/10.1016/j.arabjc.2015.11.015.

[18] Taylor, G. Electrically Driven Jets. *Proc. Roy. Soc. London. A. Math. Phys. Sci.* 1969, *313*, 453–475. https://doi.org/10.1098/rspa.1969.0205.

[19] Thenmozhi, S.; Dharmaraj, N.; Kadirvelu, K.; Kim, H. Y. Electrospun Nanofibers: New Generation Materials for Advanced Applications. *Mater. Sci. Eng. B Solid-State Mater. Adv. Technol.* 2017, *217*, 36–48. https://doi.org/10.1016/j.mseb.2017.01.001.

[20] Shen, W.; Ao, F.; Ge, X.; Ning, Y.; Wang, L.; Ren, H.; Fan, G. Effects of Solvents on Electrospun Fibers and the Biological Application of Different Hydrophilic Electrospun Meshes. *Mater. Today Commun.* 2022, *30*, 103093. https://doi.org/10.1016/j.mtcomm.2021.103093.

[21] Mei, L.; Mao, M.; Chou, S.; Liu, H.; Dou, S.; Ng, D. H. L.; Ma, J. Nitrogen-Doped Carbon Nanofibers with Effectively Encapsulated GeO2 Nanocrystals for Highly Reversible Lithium Storage. *J. Mater. Chem. A* 2015, *3* (43), 21699–21705. https://doi.org/10.1039/c5ta03911h.

[22] Luo, L.; Xu, W.; Xia, Z.; Fei, Y.; Zhu, J.; Chen, C.; Lu, Y.; Wei, Q.; Qiao, H.; Zhang, X. Electrospun ZnO-SnO2 Composite Nanofibers with Enhanced Electrochemical Performance as Lithium-Ion Anodes. *Ceram. Int.* 2016, *42* (9), 10826–10832. https://doi.org/10.1016/j.ceramint.2016.03.211.

[23] Xi, M.; Wang, X.; Zhao, Y.; Zhu, Z.; Fong, H. Electrospun ZnO/SiO2 Hybrid Nanofibrous Mat for Flexible Ultraviolet Sensor. *Appl. Phys. Lett.* 2014, *104* (13), 133102. https://doi.org/10.1063/1.4870296.

[24] Yang, Q.; Zhenyu, L. I.; Hong, Y.; Zhao, Y.; Qiu, S.; Wang, C. E.; Wei, Y. Influence of Solvents on the Formation of Ultrathin Uniform Poly(Vinyl Pyrrolidone) Nanofibers with Electrospinning. *J. Polym. Sci. Part B Polym. Phys.* 2004, *42* (20), 3721–3726. https://doi.org/10.1002/polb.20222.

[25] Juncos Bombin, A. D.; Dunne, N. J.; McCarthy, H. O. Electrospinning of Natural Polymers for the Production of Nanofibres for Wound Healing Applications. *Mater. Sci. Eng. C* 2020, *114*, 110994. https://doi.org/10.1016/j.msec.2020.110994.

[26] Nayak, R.; Padhye, R.; Kyratzis, I. L.; Truong, Y. B.; Arnold, L. Recent Advances in Nanofibre Fabrication Techniques. *Text. Res. J.* 2012, *82* (2), 129–147. https://doi.org/10.1177/0040517511424524.

[27] Hartgerink, J. D.; Beniash, E.; Stupp, S. I. Self-Assembly and Mineralization of Peptide-Amphiphile Nanofibers. *Science (80-.).* 2001, *294* (5547), 1684–1688. https://doi.org/10.1126/science.1063187.

[28] Zhang, S. Fabrication of Novel Biomaterials through Molecular Self-Assembly. *Nat. Biotechnol.* 2003, *21*, 1171–1178. https://doi.org/10.1038/nbt874.

[29] Holmes, T. C.; De Lacalle, S.; Su, X.; Liu, G.; Rich, A.; Zhang, S. Extensive Neurite Outgrowth and Active Synapse Formation on Self-Assembling Peptide Scaffolds. *Proc. Nat. Acad. Sci. Unit. Stat. Am.* 2000, 6728–6733. https://doi.org/10.1073/pnas.97.12.6728.

[30] Kenry; Lim, C. T. Nanofiber Technology: Current Status and Emerging Developments. *Prog. Polym. Sci.* 2017, *70*, 1–17. https://doi.org/10.1016/j.progpolymsci.2017.03.002.

[31] Garg, T.; Rath, G.; Goyal, A. K. Biomaterials-Based Nanofiber Scaffold: Targeted and Controlled Carrier for Cell and Drug Delivery. *J. Drug Target.* 2015, *23* (3), 202–221. https://doi.org/10.3109/1061186X.2014.992899.

[32] Sabzehmeidani, M. M.; Ghaedi, M. Adsorbents Based on Nanofibers. In *Interface Science and Technology*; M. Ghaedi, Ed.; Elsevier: Amsterdam, 2021; pp. 389–443. https://doi.org/10.1016/B978-0-12-818805-7.00005-9.
[33] Tsuboi, Y.; Yoshida, Y.; Okada, K.; Kitamura, N. Phase Separation Dynamics of Aqueous Solutions of Thermoresponsive Polymers Studied by a Laser T-Jump Technique. *J. Phys. Chem. B* 2008, *112* (9), 2562–2565. https://doi.org/10.1021/jp711128s.
[34] Beachley, V.; Wen, X. Polymer Nanofibrous Structures: Fabrication, Biofunctionalization, and Cell Interactions. *Prog. Polym. Sci.* 2010, *35* (7), 868–892. https://doi.org/10.1016/j.progpolymsci.2010.03.003.
[35] Meyer, B.; Croce, F. Materials | Nanofibers. In *Encyclopedia of Electrochemical Power Sources*; J. Garche, Ed.; Elsevier: Amsterdam, 2009; pp. 607–612. https://doi.org/10.1016/B978-044452745-5.00058-7.
[36] Kara, Y.; Molnár, K. A Review of Processing Strategies to Generate Melt-Blown Nano/Microfiber Meshes for High-Efficiency Filtration Applications. *J. Ind. Text.* 2022, *51* (1), 137S-180S. https://doi.org/10.1177/15280837211019488.
[37] Ellison, C. J.; Phatak, A.; Giles, D. W.; Macosko, C. W.; Bates, F. S. Melt Blown Nanofibers: Fiber Diameter Distributions and Onset of Fiber Breakup. *Polymer (Guildf).* 2007, *48* (11), 3306–3316. https://doi.org/10.1016/j.polymer.2007.04.005.
[38] Ramazan, E. Advances in Fabric Structures for Wound Care. In *Advanced Textiles for Wound Care*; S. Rajendran, Ed.; Woodhead Publishing: Duxford, 2019; pp. 509–540. https://doi.org/10.1016/b978-0-08-102192-7.00018-7.
[39] Yesil, Y.; Bhat, G. S. Structure and Mechanical Properties of Polyethylene Melt Blown Nonwovens. *Int. J. Cloth. Sci. Technol.* 2016, *28* (6), 780–793. https://doi.org/10.1108/IJCST-09-2015-0099.
[40] Qu, H.; Skorobogatiy, M. Conductive Polymer Yarns for Electronic Textiles. In *Electronic Textiles: Smart Fabrics and Wearable Technology*; T. Dias, Ed.; Woodhead Publishing: Kidlington, 2015; pp. 21–53. https://doi.org/10.1016/B978-0-08-100201-8.00003-5.
[41] Grafe, T.; Graham, K. Polymeric Nanofibers and Nanofiber Webs: A New Class of Nonwovens. *Int. Nonwovens J.* 2003, *12* (1), 51–55. https://doi.org/10.1177/1558925003os-1200113.
[42] Neves, N. M.; Campos, R.; Pedro, A.; Cunha, J.; Macedo, F.; Reis, R. L. Patterning of Polymer Nanofiber Meshes by Electrospinning for Biomedical Applications. *Int. J. Nanomedicine* 2007, *2* (3), 433–438.
[43] Conte, A. A.; Sun, K.; Hu, X.; Beachley, V. Z. Effects of Fiber Density and Strain Rate on the Mechanical Properties of Electrospun Polycaprolactone Nanofiber Meshes. *Front. Chem.* 2020, *8*, 610. https://doi.org/10.3389/fchem.2020.00610.
[44] Butcher, A. L.; Koh, C. T.; Oyen, M. L. Systematic Mechanical Evaluation of Electrospun Gelatin Meshes. *J. Mech. Behav. Biomed. Mater.* 2017, *69*, 412–419. https://doi.org/10.1016/j.jmbbm.2017.02.007.
[45] Mubyana, K.; Koppes, R. A.; Lee, K. L.; Cooper, J. A.; Corr, D. T. The Influence of Specimen Thickness and Alignment on the Material and Failure Properties of Electrospun Polycaprolactone Nanofiber Meshes. *J. Biomed. Mater. Res.—Part A* 2016, *104* (11), 2794–2800. https://doi.org/10.1002/jbm.a.35821.

8 Design and Characterization of the Electrospun Nanofibers Mats

Paulo A. M. Chagas, Gabriela B. Medeiros, Edilton N. Silva, Sirlene Morais, Gustavo C. Mata, Vádila G. Guerra, Mônica L. Aguiar, and Wanderley P. Oliveira

8.1 HISTORY

In the seventeenth century, the electrostatic attraction of a droplet was first observed by William Gilbert, the Court Physician of Queen Elizabeth I of England and Ireland [19]. In the eighteenth century, higher evaporation rates of charged fluids than uncharged fluids were observed by Giovanni Battista Beccaria [20]. In the nineteenth century Louis Schwabe, the silk manufacturer for Queen Victoria's wedding dress, developed the extrusion spinneret. Other contributions were made in France, Germany, and Switzerland, such as the Nobel prize winning Lord Rayleigh, who studied the theoretical stability of electrically charged water jets [21]. Charles Vernon Boys constructed an electrical apparatus to produce fibers from melts such as shellac, beeswax, and gutta-percha [22]. Professor Charles also quoted "the old but little-known experiment of electrical spinning" [19,23].

Only in the early twentieth century, the technique started to be applied in the industry when John Francis Cooley ordered the first electrospinning patent [24]. The technique was popularized from there, spreading through the world. From 1924 to 1929, Professor Kiyohiko Hagiwara of the Imperial University of Kyoto used electricity to orientate viscose solutions [25]. In 1939 Formhals used alternating currents up to 100 kV to control fiber flights between two parallel electrodes [26]. In 1964 Geoffrey Taylor created a mathematical model of the cone shape formed by a pendant drop under an electrical field, known as Taylor Cone, and used it until the actuality [27]. From there, many contributions were performed worldwide, and the term "electrospinning" was popularized, especially in the '90s [28].

8.1.1 Description of the Electrospinning Process

The electrospinning apparatus is quite simple, resembling a circuit capacitor. Its mechanisms include applying a very high electrical potential difference, usually

 DOI: 10.1201978100333814-8

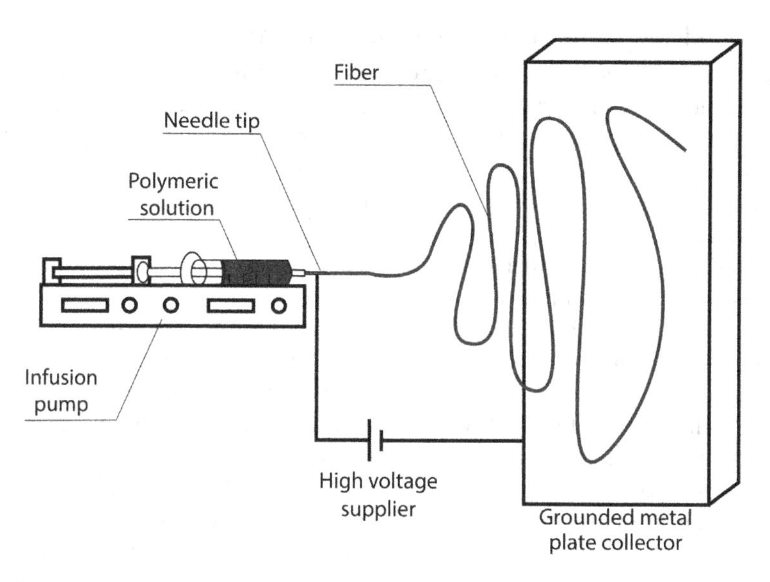

FIGURE 8.1 Traditional electrospinning apparatus using a flat plate as a collector.

ranging from 10 to 40 kV, between a charged syringe needle and a grounded metal plate collector. A polymer is then controllably squeezed through the syringe by an infusion pump, entering the negative pole (needle) of the "capacitor" [29]. Hence, a pendant droplet forms a funnel shape on the needle tip, known as Taylor cone [30,31]. The edge point of the "funnel" elongates from the needle tip towards the grounded metal plate, conducted by the electrical field force. The polymer is then stretched, diminishing its diameter to the nanometer scale. The solvent evaporates during the flight, then the fibers solidify, depositing randomly on the metal plate [31,32]. Figure 8.1 shows a schematic of a typical electrospinning process.

8.2 FUNDAMENTALS OF THE ELECTROSPINNING

Despite the electrospinning set simplicity, the technique has many factors and variables to control and equalize during the production of nanofibers. Additionally, some of those factors are correlated and even intrinsic to one another. In other words, the manipulation of a variable often influences the properties and responses of another one. This section will explore the relationship between those parameters and how to handle them.

8.2.1 Solution Parameters

The solution parameters typically involve the chemical properties of the electrospinning precursor solution. Polymer and solution properties will influence the formation and stability of the jets and, consequently, the resulting electrospun fiber mat. Those properties can be manipulated before the electrospinning process during the polymer solution preparation.

8.2.1.1 Polymer Concentration and Molecular Weight

There are many ways that the polymer concentration can interfere with the fiber formation. When the concentration is low, the fiber can break during the flight or even not form a fiber. Suppose that the polymer chains do not have enough interchain interactions, such as hydrogen bonds or crosslinks. In that case, the solution will not entangle, fragmenting the fiber before it arrives in the collector [6], and causing bead formation [33].

A high concentration can also be a problem. It is impossible to produce chitosan fibers using solutions with less than 2% solid content, while above 2%, the viscosity becomes a hindrance for electrospinning [34]. Above 2%, the viscosity increases due to the hydrogen bonds formed with N and O in the chitosan structure. To solve this issue, some authors make use of adjuvants, such as polyvinyl alcohol (PVA) [35–37] or dimethyl sulfoxide (DMSO) [34]. Solvent's addition diminishes the interaction degree between polymer chains, reaching the entanglement threshold necessary to produce fibers. Some authors have also added several additives, such as active pharmaceuticals to add functional properties to electrospun nanofibers, such as antimicrobials, essential oils, propolis extract, and others. Figure 8.2 shows some fibers of PVA with additives.

The same effect was observed by reducing the molecular weight of CS and, consequently, its entanglement degrees [38]. Koski *et al.* [39] produced PVA electrospun fibers, testing molecular weights varying from 9000 to 186,000 $g.mol^{-1}$. They observed that there are threshold values for the concentration to create fibrous structures and that the fiber size increases with the molecular weight. The fibers tend to be flattened instead of circular at a higher molecular weight. While producing silk fibroin, Park and Um [40] found that the molecular weight also influences the resulting fibers' crystallinity and mechanical properties. Promnil *et al.* [41] tested polylactic acid (PLA) to produce fibers variating the concentration (10, 15, and 20% w/v) and the molecular weight (high and low). They found that the higher concentration with low molecular weight presented higher tensile strength and Young's modulus, while a higher molecular weight presented the highest elongation at break. Ngadiman *et al.* [42] observed a different result when working with PVA. They obtained stronger fibers with higher tensile strength and Young's modulus using higher molecular weight to electrospin them.

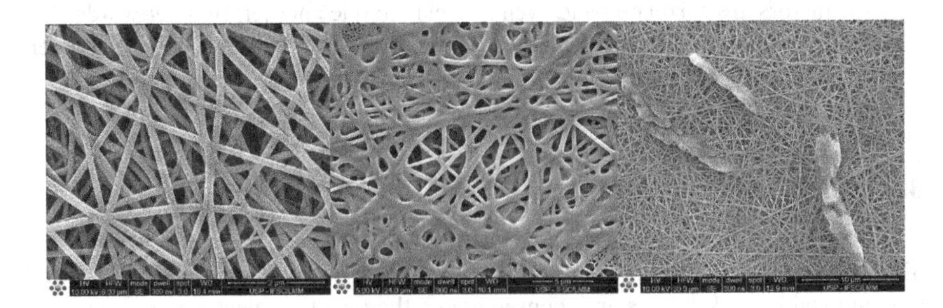

FIGURE 8.2 The three images are PVA fibers loaded with chitosan (left), chitosan and citric acid (middle), and chitosan and CTAB (right).

8.2.1.2 Rheology and Viscosity

The Ostwald de Wale power law is frequently used as a model to describe the rheology of the electrospinning solutions, whose typical behavior ranges from Newtonian to pseudoplastic fluids. Equation 8.1 shows the Ostwald model.

$$\tau = K \cdot \Upsilon^n \tag{8.1}$$

where τ is the shear stress (or applied force/area, Dyn · cm^{-2}), γ is the shear rate (or resulting deformation, 1.s^{-1}),n (flow index, dimensionless), and K (consistency index, Dyn·cm^{-2} · s^n) [31,43]. The parameter K (consistency factor) is the shear strength at a shear rate of 1.0 s^{-1}, and corresponds approximately to the relative viscosity of the composition at γ = 1.0 s^{-1}. If K is high, the fibers also increase in diameter size [44].

It is possible to categorize some pseudoplastic and dilatant fluids based on their time-dependent behavior, in thixotropic or rheopectic, respectively [45]. Thixotropic systems present a high initial resistance to deformation, relaxing with time elapsed. Pseudoplastics with rheopectic behavior show the opposite of the thixotropy, an increased acceptance of initial deformation, creating hindrances along the time. The rheological behavior of the solution can also affect the nanofiber diameter, as can be seen in Figure 8.3.

8.2.1.3 Surface Tension

During the electrospinning process, the pendant droplet turns into a fiber when the attraction force, exerted by the electric field, overcomes the surface tension. The surface tension acts as an initial resistance force that can modulate the Taylor cone (further discussed). Some authors reported that diminishing the surface tension also reduces the formation of beaded fibers [46,47]. When the polymer is in low concentrations in the electrospinning solution, the surface tension minimizes the total surface free energy [48], diminishing the superficial charge and the influence of the electrical field over the pendant droplet. When the surface tension is too low, it favors the

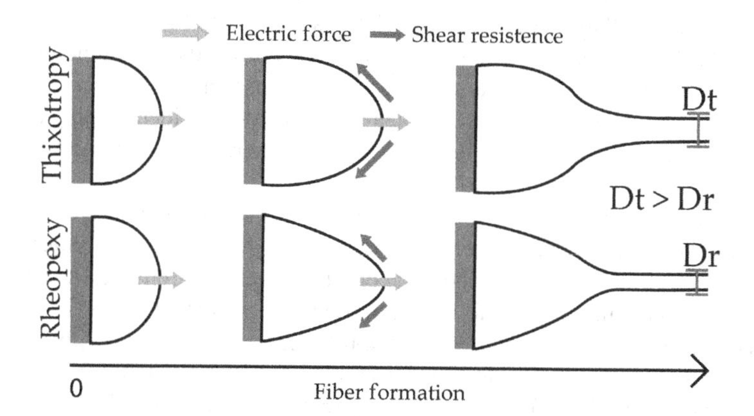

FIGURE 8.3 Demonstration of the Taylor cone, originating by fluids with different rheological properties, and its effect on the fiber size diameter.

process of electrospraying instead of electrospinning [49,50]. Some authors even used those properties to produce electrospraying and electrospinning simultaneously.

8.2.1.4 Solvent System

The solvent used for electrospinning has two significant roles during fiber production. First is the need to completely dissolve the polymer that will be electrospun. The second issue is the pressure vapor of the solvent. If a solvent is too volatile, it will evaporate earlier in the Taylor cone, clogging the needle tip and deforming the Taylor cone. If the solvent has a very low vapor pressure, it will not evaporate during the fiber jet flight [33,51]. If the fiber is not fully dried during the deposition, it can favor fiber fusion and aggregation [52].

Zaarour *et al.* [53] produced polyvinylidene fluoride (PVDF) fibers using the chamber humidity and the volatility of solvents to make fibers with different shapes. Using mixtures in different proportions of acetone and N,N-dimethylformamide (DMF) as the solvent, they could exploit the phase equilibrium between the ambient water, the solvent, and the polymer to induce roughness and grooves over the surfaces of the fibers. Huang *et al.* [54] also used mixtures of solvents to improve their nanofibers. Using formic acid as a solvent, they observed that adding small amounts of pyridine can prevent bead formation when electrospinning fibers of nylon-4,6.

The solvent system is also related to the needle tip-to-collector distance and the humidity in the electrospinning chamber. Depending on the distance between the needle and the collector, the fiber jet will have more or less time to dry, and the humidity because of the excess of water in the air can difficult solvent evaporation, preventing the fiber solidification when deposited over the collector.

8.2.1.5 Electrical Conductivity

Electrical conductivity possesses a significant role during fiber elongation. It is responsible for the fiber attraction towards the collector plate and helps the fiber elongation and diminution of its diameter.

The polymeric solution forms a pendant droplet when extruded inside the needle tip. This droplet is charged by one of the poles of the electrical field generated by the high-voltage supplier. Those charges, applied to the polymeric solution, redistribute themselves over the polymeric surface towards an equipotential distribution alongside the fiber. As polymeric solutions tend to be viscous, charge repulsion among themselves creates a shear force inside the fiber, helping to stretch it and consequently diminishing the fiber size diameter. Figure 8.4 exemplifies the process.

The pendant droplet deforms itself to form a funnel shape, entering the next step of the fiber formation, called the Taylor cone [31,55]. When the polymer surface charges overcome the surface tension, the polymer stretches from the Taylor cone in the direction of the metal plate, following the field lines of the electric field applied. As a viscous material, the polymer solution elongates into fibers with nanoscale dimensions [32]. Figure 8.4, e) exemplifies the Taylor cone formation.

Stranger *et al.* [56] describe a parallel between the solution conductivity and the Taylor cone geometry. Increases in the charge density reduce the curvature and the diameter of the Taylor cone. It is possible to change the conductivity solution by

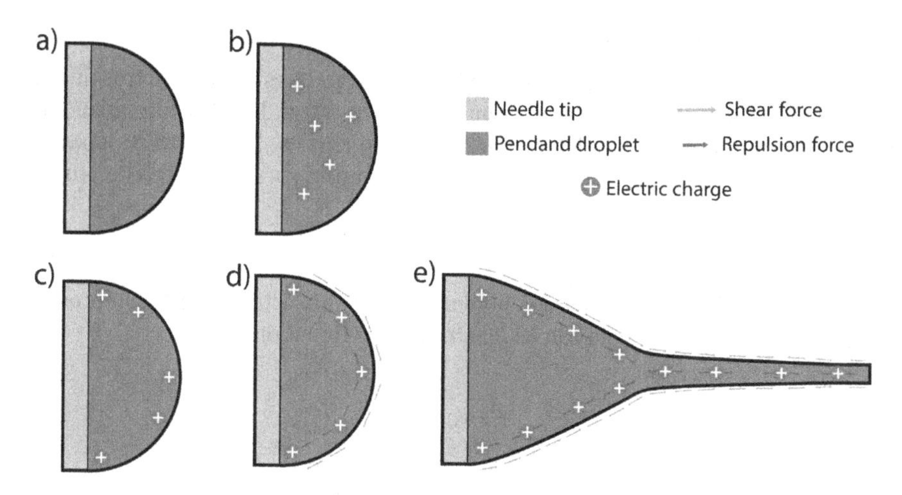

FIGURE 8.4 a) Pendant droplet with no charge; b) initial charges applied to the droplet; c) charges distribute themselves over the surface; d) charges reorganize and shear force is created by the superficial charges' repulsion; and e) Taylor cone formation and elongation.

Source: (Copyright © 2022 Gustavo C. Mata et al.) [37].

adding salts [52,54]; some studies show that trace amounts of salt can affect the polymer solution conductivity [57]. Topuz *et al.* [58] observed that for some samples, the addition of salt (tetraethylammonium bromide) diminished the fiber diameter of polyamide nanofibers and prevented bead formation. Yu *et al.* [59] studied the influence of salt addition in coaxial electrospinning of polyvinylpyrrolidone (PVP), obtaining fibers of 120 ± 40 nm after a slightly increased amount of salt.

The addition of organic acids as polymer solvents can also increase the conductivity of the polymeric solution [37]. Huang *et al.* [54] produced ultrathin electrospun nylons beaded fibers (3 nm) and removed the fiber beads by raising the conductivity of the solution and adding small amounts of pyridine. The influence of salt and acid addition on the conductivity (σ) [60], is described in Equation 8.2.

$$\sigma = \sum n_i q_i \mu_i \tag{8.2}$$

being n_i the number of ions i, q_i the charge of the ions i and μ_i the mobility of the ion i Since the charge of the ions does not change and the ion mobility μ_i is high in acid solutions, the number of ions is the variable able to change the conductivity.

Generally, as conductivity increases, fiber diameter tends to decrease. Angammana and Jayaram [61] studied the effect of solution conductivity by adding NaCl salt to the electrospun polyethylene oxide (PEO)/water system and found that the relationship between conductivity and fiber diameter follows a power law. However, excessively high conductivity can hinder the formation of the Taylor cone, leading to the emergence of multiple fibers jets on the fluid droplet and the induction of protrusions in the resulting nanofiber.

8.2.2 Processing Variables

Processing variables are related to the physical parameters of the electrospinning process. Changing the processing parameters might affect the fiber formation phenomenon and the resulting electrospun fibers' properties. Following is presented an overview of the effects of the main electrospinning parameters on the process performance and fibers properties.

8.2.2.1 Electrical Field

As previously discussed, a charged polymer under an electric field E can produce nanofibers. The electric field can be determined by Equation 8.3.

$$E = k_0 \cdot \frac{Q}{d^2} \tag{8.3}$$

where E is the electric field in N/C; k_0 is the electrostatic constant in the vacuum;Q is the charge that generated the electric field; and d is the distance between the charge of the electric field and the observed point.

High voltages (from 10 to 40 kV) are commonly used during the electrospinning technique as a way to increase the charges (Q) in the poles and, consequently, the electric field (E). The electric field acts as a driven force, pulling and accelerating the polymer across its poles. Those properties make conductive polymers interesting for electrospinning. By retaining more charges, they are subjected to a higher electrical field and an increasing acceleration across the poles, generating fibers with lesser sizes.

Medeiros *et al.* [62] showed that the electrical field was the most influential parameter on zein/polyethylene oxide fibers, in controlling the presence or absence of beads. Dhanalakshmi *et al.* [63] worked on producing fibers of Nylon11 via electrospinning and observed that increasing the voltage potential from 10 to 20kV diminished their average diameter and the bead formation. Other studies show that it is possible to control the fibers' deposition and diameter by variating the electrical field [64–66]. However, a very high voltage can deform the fibers, creating beads and fiber fusion, being that the critical values are polymer dependent [33,51,67].

The direction of the electrical field can influence fiber formation. Typically, the polymer pole carries a positive charge. However, changing the type of charges dispersed on the polymer surface from positive to negative can also influence fiber formation [29]. Moreover, the type of electrical current used – alternate or direct – to create the electrical field [68] or the introduction of a magnetic field over the electric field [69,70] can also affect the morphology and deposition of the fibers.

8.2.2.2 Feed Flow Rate

The feed flow rate of the polymeric composition controls the quantity of material per time extruded through the needle and consequently pulled towards the collector. This parameter is significant because of a flow rate threshold for the formation of the Taylor cone. Lower flow rates imply insufficient material is sent to maintain a stable Taylor cone, resulting in jet dispersion and fiber breakage. The electrical field's force over the fiber overcomes the fiber's resistance to deformation. On the other hand, excess material overcomes the electrical field's capacity to "pull" the fibers, leading to larger fibers, bead formation, and electrospraying [33,71]. Some authors have already reported that the fiber starts forming beads and fiber aggregation above a specific trigger value of flow rate [72,73].

8.2.2.3 Needle Tip and Fibers Collectors

The distance between the needle tip and the collector is an essential parameter in the electrospinning technique. As seen in Equation 8.3, varying the distance (d) will cause a quadratic effect on the electrical field (E). Therefore, the influence of the needle tip distance from the collector is linked to the magnitude of the electrical field created. So, diminishing the distance will increase the electric field and consequently prejudice or enhance the formation of the fibers, as previously discussed. The needle tip-to-collector distance also influenced solvent volatility. Small distances do not give enough time for the fiber to fully dry before the deposition, creating fiber fusion in the fiber mat [30].

Observing the needle as a separate parameter, studies are controversial about the inner diameter of the needle tip and the resulting fiber size diameter. Macossay *et al.* [74] produced poly(methyl methacrylate) electrospun nanofibers and did not observe any correlation between the fiber diameter with the inner needle opening. However, in recent studies, He *et al.* [75] produced polyethylene-oxide nanofibers via electrospinning and found that by diminishing the needle's inner hole, the fiber diameter also decreased. Other authors observed the same trend when using electrospun n-Bi_2O_3/Epoxy-PVA nanofibers [76].

The isolated influence of the collector in the electrospinning process is primarily related to the electric field arrangement. Since the fibers travel through the poles (*i.e.*, the feeding source and the collector), they tend to follow the electric field's density lines [77]. The disposition of the field lines will differ if the collector shape is a flat plate, a drum, a bar, a ring, *etc.* [78–80], as shown in Figure 8.5.

Nguyen *et al.* [81] used dielectric films of polyimide to block part of the electric field between the needle tip and a drum collector. They observed that the rectangular gap

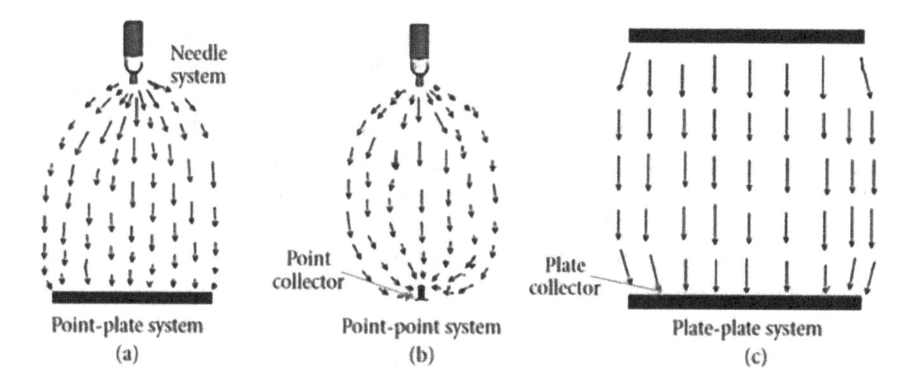

FIGURE 8.5 Distinct electrical field behavior, according to the needle and collector electrospinning apparatus. a) Needle-plate system, b) point-point system, c) plate-plate system.

Source: (Copyright © 2011 Rahul Sahay et al.) [62].

between the films acted as a preferential path to the fibers, aligning the fibers over the drum collector. Lee *et al.* [82] electrospun TiO_2 with polyvinylpyrrolidone (PVP) fibers using grounding collectors of Si electrodes with gaps of Al_2O_3 between the electrodes. The researchers observed that the fibers were stretched on the gaps and more aligned over the electrodes. De Prá *et al.* [83] produced polycaprolactone fibers using a rotating drum, static copper wires, and a rotating mandrel. They compared a rotating drum with a static one, without rotation. They observed that increasing the rotation speed from 0 to 2000 rpm diminished the fiber size from 1142 ± 391 nm to 663 ± 334 nm, improving the organization and the fiber alignment. The collector's material (*e.g.*, aluminum, copper, *etc.*) can also influence the electrospinning process by being more or less conductive.

8.2.3 Environmental Conditions

Returning to the electrical field formula (Equation 8.3), the role of the environmental condition is associated with the constant (K_0). Depending on the electrical permissiveness of the ambient air, the electrical field can change, favoring or hindering the fiber formation. For example, high humidity can disperse the charges on the poles of the electrospinning apparatus or even create a current between the needle tip and the collector.

The humidity also influences the solidification process [33,84]. Içoğlu and Oğulata 85,86 varied the environmental electrospinning conditions when producing nanofibers of polyetherimide (PEI) and poly trimethylene terephthalate (PTT). They observed that increasing the environmental temperature and the relative humidity of the ambient environment resulted in a decrease in the medium-size diameter for both polymers. However, they obtained the inverse response to the previous study when using 1-methyl-2-pyrrolidinone as a solvent to dissolve PEI polymer. As the humidity increased, the fiber diameter also increased [87]. Alswid and Issa [88] obtained the same result using polyvinyl alcohol (PVA) nanofibers. Depending on the solvent's pressure vapor, the excess humidity in the environment hinders solvent evaporation.

The equilibrium phenomena of solvent evaporation were explored by Zaarour *et al.* [53]. They produced polyvinylidene fluoride (PVDF) nanofibers with rough and grooved surfaces by controlling the solvent mixture and humidity in the electrospinning chamber. Figure 8.6 shows the rugged and grooved nanofibers obtained by the author. Barua and Saha [89] produced polyacrylonitrile (PAN) fibers dissolved in DMF to investigate the influence of humidity and temperature. After studies in the ternary diagram H_2O/DMF/PAN, they observed that when the humidity is high, the diffusion of water in the polymer solution turns the system into a thermodynamically unstable phase. The instability leads to a separation in a polymer-rich phase and a polymer-lean phase, controlling the fiber porosity. Depending on the relative humidity, the fibers tend to absorb the water from the air during the elongation step. Water condensation releases latent heat. In response, the fibers absorb the water and expel the solvent earlier, preventing fiber solidification [30].

Another relevant factor is the temperature. Ramazani and Karimi [90] tested the influence of the temperature when producing nanofibers of polycaprolactone (PCL), observing that the fiber diameter decreased with the increase in temperature.

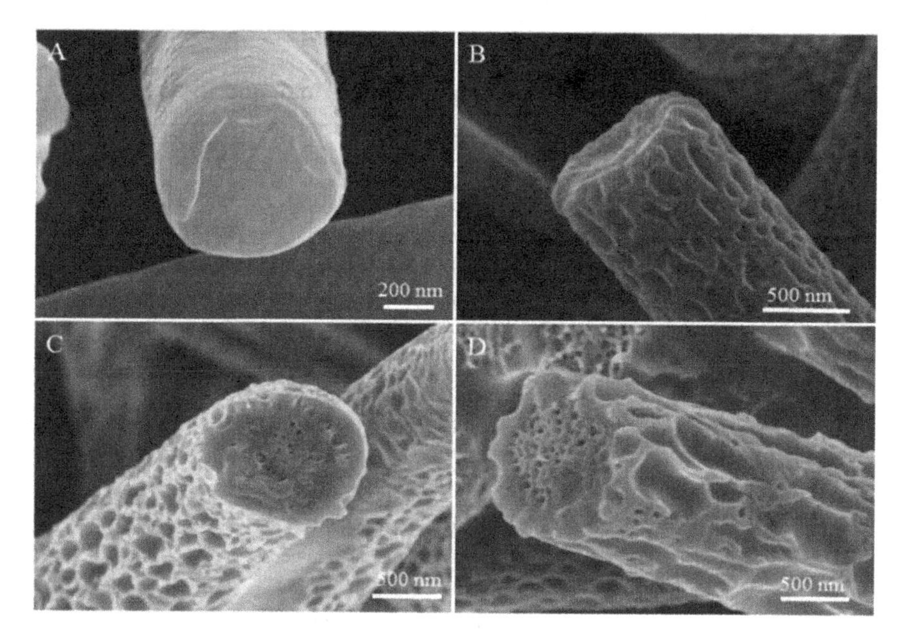

FIGURE 8.6 PVDF nanofibers with rough and grooved surfaces, produced using relative humidity at a) 5%, b) 25%, c) 45%, and d) 65%.

Source: (Copyright © 2018 Bilal Zaarour et al.) [53].

8.3 PROPERTIES OF THE FIBERS MAT AND METHODS OF CHARACTERIZATION

Electrospun nanofibers exhibit properties such as nanoscale size, high surface area for a small volume, and high molecular orientation. The functionalities of nanofibrous materials can be adjusted through the fiber diameter, the surface morphology, and the internal structure of the nanofibers [91,92]. Fiber characterization provides fundamental data to understand the relationship between the structure and properties of nanofibers. One of the limitations in the nanofiber's characterization is the manipulation of the nanofibers since these fibers are small, which makes it difficult to find an appropriate observation mode and difficult to prepare the nanofiber sample. With the evolution of nanotechnology, some of these limitations have already been overcome and new sophisticated techniques have already been developed or are under study [93].

The type of characterization performed to test nanofibers is based on the applicability of the final product. Many common techniques are used to characterize conventional engineering materials, and some not-so-common techniques have been employed in the characterization of nanofibers. Characterization techniques are in constant study and development, seeking more efficient and accurate techniques [93,94]. Therefore, in this section, basic information on the most common types of characterizations applied to analyze the properties of electrospun nanofibers will

TABLE 8.1
Morphological, Physicochemical, Mechanical, Thermal, and Biological Characterization Methods of Electrospun Nanofibers

Characterization type	Methods
Morphological	Optical Microscopy (OM)
	Scanning Electron Microscopy (SEM)
	Field Emission Scanning Electron Microscopy (FESEM)
	Transmission Electron Microscopy (TEM)
	Atomic Force Microscopy (AFM)
	Mercury Intrusion Porosimetry
	Liquid Extrusion Porosimetry
	Capillary Flow Porosimetry
Chemical	Energy-Dispersive X-ray Spectroscopy (EDX)
	X-ray Photoelectron Spectroscopy (XPS)
	X-Ray Diffraction (XRD)
	Contact Angle Measurements
	Fourier Transform Infrared Spectroscopy (FTIR)
	Nuclear Magnetic Resonance (NMR)
Mechanical	Tensile Test
	Dynamic Mechanical Analysis (DMA)
	Atomic Force Microscopy (AFM)
Thermal	Differential Thermal Analysis (DTA)
	Differential Scanning Calorimetry (DSC)
	Thermogravimetric Analysis (TGA)
Biological	*In Vitro* Test
	In Vivo Test

be presented. Morphological, physicochemical, mechanical, thermal, and biological characterizations will be presented in this section. Table 8.1 summarizes the characterization methods that are described in more detail in this section.

8.3.1 Morphology

The morphological properties of nanofibers include fiber diameter, diameter distribution, fiber orientation, pore size, porosity, cross-sectional shape, and surface roughness. Techniques such as Optical Microscopy (OM), Scanning Electron Microscopy (SEM), Field Emission Scanning Electron Microscopy (FESEM), Transmission Electron Microscopy (TEM), and Atomic Force Microscopy (AFM) are used for the characterization of morphological properties. In addition, Mercury Intrusion Porosimetry, Liquid Extrusion Porosimetry, and Capillary Flow Porosimetry, techniques used to measure porosity, are also covered in this section [91,93–95].

Generally, microscopic characterizations are applied as preliminary characterization techniques. The measurement and distribution of the nanofiber's diameter are obtained from OM and SEM images. Using these two techniques for porosity analysis only determines the surface of the voids, as most pores are not cylindrical in morphology [93]. The use of SEM for morphological analysis of the fibers is the most appropriate technique used by those surveyed, as it allows increases greater than that of OM. To perform the SEM, the sample must be electrically conductive; therefore, in the case of nanofibers, the sample must be previously prepared with a conductive coating, usually made with gold or platinum, which can alter the diameter readings at higher magnifications [91,95].

In SEM, the sample surface is irradiated with a high-energy electron beam under a vacuum. The interaction of the incident beam with the sample surface results in the image signal that is collected by the detector modulating the display brightness, generating an image of the surface of the sample. Despite the complexity of the imaging mechanism, it is possible to obtain high-resolution images [93,96,97]. Other techniques used in the qualification and quantification of the morphology of nanofibers are FESEM and TEM.

FESEM is employed to observe the morphological behavior of samples with resolutions from 1 to 20 nm. From FESEM it is possible to obtain topographical information on the surface of the sample and of small whole particles, which is difficult to visualize in SEM images. In general, the functioning mechanism of FESEM is the same as that of SEM. The main difference between these two techniques is the electron generation system. In SEM a thermionic emission is used, while in FESEM a field emission gun (FEG) is used as an electron source, in which a potential gradient is applied to emit the electron beam. Commonly, the FESEM technique is equipped with energy-dispersive X-ray spectroscopy (EDX) used to obtain information about the characteristics or elemental compositions of nanofibers [93,98].

Samples analyzed by SEM must be dry. As an alternative to this limitation, TEM is applied, in which the sample does not necessarily need to be dry for the analysis to be carried out [95]. TEM can provide images with 0.1 nm resolutions [93]. The working principle of TEM is done through the transmission of an electron beam through the ultrathin section of the samples. When the electron beam strikes the sample, a large amount of radiation is transmitted/emitted from the sample. Elastic and inelastic scattering of transmitted electrons occurs as the interaction between the electron beam and the nanofibers occurs. The elastically scattered electrons generate the images, allowing the observation of the structural characteristics and the defects of the sample in high resolution. Despite the high resolution, the electron beam generated by TEM can damage the analyzed sample, especially when they are biological samples [93,97,99].

AFM is a technique used for morphological analysis. A cantilever can tip (usually 15 μm long monolithic silicon) or probe presses and scans the surface of the sample with atomic resolution. The forces of attraction and repulsion resulting from the interaction of the probe/tip with the surface generate signals that represent the topography of the sample and generate the images. Non-conductive samples can be used in this technique. The characterization of nanofibers by AFM is not widely used, as the width of the fibers obtained is not accurate, especially for fibers with smaller dimensions. The displacement of the fibers occurs as the probe moves across the surface of the sample, which impairs the resolution of the images. A solution to this problem is the use of a fixative substrate in the sample [91,93,96]. The images obtained by microscopic techniques represent a small area of the sample. Thus, to represent the sample, several regions of the sample must be selected and, with this information, an analysis of the characteristics of the fibers must be carried out.

Another morphological property of the fiber is porosity, defined as empty spaces in the nanofibers. Characterization of pore size and porosity can be performed by the techniques of mercury intrusion porosimetry, liquid extrusion porosimetry, and capillary flow porosimetry. The mercury intrusion porosimetry is a technique commonly used for assessing materials porosity. This technique consists of the forced passage of goods through the nanofiber sample. The greater the increase in pressure, the smaller voids are reached by the mercury. Very small or determined pores are not measured from this measuring technique. If there are large voids, the technique is unfeasible, as the goods easily flood even before the pressure is applied [91,93].

The pressure applied in the mercury intrusion porosimetry is high, thus it is impracticable for the characterization of nanofibers with a low resistance to pressure. Given this limitation, the liquid extrusion porosimetry technique was developed, in which lower pressures are used. This technique consists of passing a liquid into the voids of the fibers through a pressure generated by the gas over the liquid column. It is also possible to measure the pressure required to introduce the liquid into the pores. The capillary flow porosimetry technique also uses the passage of a liquid under gas pressure through the sample, in which the gas flow is recorded as a function of the differential pressure. The surface tension of the liquid, the surface energy of the fibers, and the diameter of the pores directly influence the required gas flow, which enables the characterization of the fiber [93,97].

8.3.2 Physicochemical Properties

To evaluate the physicochemical properties of the surface of nanofibers, several techniques can be used, for example, energy-dispersive X-ray spectroscopy (EDX), X-ray photoelectron spectroscopy (XPS), X-Ray diffraction (XRD), contact angle measurements, Fourier transform infrared spectroscopy (FTIR), and nuclear magnetic resonance (NMR). As previously mentioned, the EDX technique is used together with microscopy to characterize the chemical composition of nanofibers. The interaction between particles or electromagnetic radiation and the sample emits X-rays that are characteristic of each chemical element, making it possible to identify the chemical composition of the sample [91,93]. XPS is also a technique used along with microscopy to provide chemical information about the surface of materials, such as elemental and molecular composition. The main difference between EDS and XPS is that EDX detects an elemental composition of a specific point on the surface and provides only elemental composition information. XPS, on the other hand, detects existing elements across the surface of the sample. Furthermore, with XPS it is possible to determine the elements using their binding energies [93].

X-Ray Diffraction (XRD) is a technique applied to determine the atomic and molecular structure of a sample, a beam of X-rays incident on crystalline atoms and diffracted in many specific directions. The analyzed sample must be a very fine and homogeneous powder, and its average mass composition must be determined. Thus, one can qualitatively determine the crystalline phases and structure existing in a nanofiber. Phase formation, crystallite size, lattice voltage, lattice parameter, and content spacing of each phase are obtained. The results found in the literature show low levels of crystallinity for most of the studied nanofibers, while the observed molecular orientation is relatively high [93].

The chemical properties of the surface of the nanofibers can be evaluated in terms of their hydrophobicity, which can be measured by analyzing the contact angle between a drop of water and the surface of the nanofibers [91]. This technique evaluates the wettability of a droplet with the surface indicating the quality, roughness, and hydrophobicity/hydrophilicity of the surface. Contact angles smaller than 90° indicate that the material is wetted by the liquid (partially hydrophilic), and angles greater than 90° indicate non-wetting (partially hydrophobic), while 0° and 180° indicate complete wetting (hydrophilic) and complete non-wetting (hydrophobic), respectively [100].

Among the techniques for characterizing the molecular structure of nanofibers are FTIR and NMR. In the mixture of materials in the production of electrospun nanofibers, it is possible with the use of these techniques to determine the intermolecular interaction [91]. The working principle of FTIR is the passage of infrared radiation through a sample. The emitted radiation can be absorbed, passed, and/or transmitted from the sample. This technique provides a spectrum in which information about chemical bonds is obtained and produces the molecular fingerprint of the sample. Thus, FTIR is used to identify unknown materials in the sample, quantify the components, and determine the quality of the analyzed sample. NMR has been used to investigate variations in polymer distribution, in addition to evaluating the presence of carbonaceous materials within the polymeric matrix of nanofibers along with the effects on morphology and physical properties of nanofibers. The interactions of rotating nuclei in a strong magnetic field are analyzed and permit obtaining information of important organic functional groups presented in the electrospun nanofibers since they present similar absorption frequencies, independent of the structure of the molecules. A stationary external magnetic field causes certain nuclei in a molecule to absorb selective radio frequencies. In the NMR spectrum, the absorbed energy that induces a transition in nuclear spins is observed [93,95].

8.3.3 Mechanical Strength

The mechanical properties of the fibers describe their characteristics regarding the application of loads and displacements. The mechanical properties depend on the technique used for measurement, process conditions used, fiber orientation, and bonds, in addition to environmental conditions. Generally, electrospun fibers do not have the necessary mechanical properties for a given application, mainly due to the formation of beads and pores in polymeric nanofibers. Electrospun nanofibers of the same polymer produced from different solvents generally have different mechanical properties. Thus, it is necessary to improve these properties by modifying the solution and process parameters in the production of electrospun fibers [92–94].

The main modes of deformation are tension, compression, bending, shear, and torsion. However, there is a challenge in measuring the mechanical properties of fibers with diameters in the micro to the nano range. Thus, the most suitable form of applied loads for nanofibers is stretching, commonly done by tensile test [92,94]. The tensile test is carried out in testing machines that subject the fiber samples to elongation, measuring the force applied for this purpose, until the sample ruptures. From the data obtained, it is possible to calculate Young's modulus, tensile strength, and stress at break [93,95].

Another technique used is dynamic mechanical analysis (DMA), in which the viscoelastic characteristics of the samples are determined. In DMA tests the specimen is subjected to periodic stress in one of several different modes of deformation (bending, tension, shear, and compression). The modulus, as a function of time or temperature, is measured and provides information about phase transitions. This technique provides information such as applied stress, resultant force or resultant strain, and angle of lag or delays between a maximum stress and maximum strain [93,101]. Elastic properties of electrospun fiber mats can be obtained through AFM. Very light contact between the tip and the sample generates cantilever deflection due to the repulsion of the atomic layers. Thus, it is possible to obtain the flexural modulus and shear modulus of the fibers, and the modulus of elasticity [91,95].

8.3.4 Thermal Analysis

Thermal properties are important for the physicochemical and structural characterization of nanofibers. The most used thermoanalytical methods are differential thermal analysis (DTA), differential scanning calorimetry (DSC), and thermogravimetric analysis (TGA) [94]. The TGA measures the weight loss or gain of the material as a function of the heating temperature, enabling the determining, for example, residual solvent losses, losses of incorporated substances, and material degradation. The DSC comprises testing the cooling or heating of the sample and determining the temperature and heat flux associated with material transitions as a function of time and temperature. In the DTA technique, the temperature difference between a substance and a reference material is measured as a function of temperature while the substance and the reference material are subjected to a controlled temperature program. The basic difference between DSC and DTA is that the former is a calorimetric method in which differences in energy are measured. In DTA, temperature differences are recorded [93,94,102]. DSC is a powerful technique used to determine the characteristic thermal transitions (glass transition temperature, melting temperature, and even characterization of the crystallization structure) of materials. The TGA is used to determine the thermal stability of the analyzed sample.

In carrying out the DSC, a sample is placed in an aluminum or platinum pan, and another aluminum pan is used as a reference (pan without sample), both at the same temperature. As the temperature increases, the energies required for the aluminum reference pan and the sample can be precisely monitored. A thermal imbalance occurs between the pans, which induces a heat flow between the sample and the reference pans. With the temperature and heat flow measurement data, it is possible to build the DSC curve. Through the DSC curve, it is possible to obtain sample transition temperatures such as melting (endothermic peak), crystallization (exothermic peak), and glass transition (baseline change) temperatures, as well as specific heat, enthalpy of fusion, the kinetics of reaction and stability [93,102]. The DSC system should be recalibrated periodically by using pure reference standard substances (e.g., indium, zinc, silver, gold, or salts) to furnish reliable and precise measurements of the heat flow, and consequently the thermodynamic properties of the analyzed material.

In the TGA equipment, the sample is placed on a scale with a platinum crucible that is placed inside an oven. The sample mass variation is recorded as the temperature increases. TGA provides information about the composition of materials through the decomposition of all organic contents from heating the sample to high temperatures. In the TGA analysis, the curve of mass versus temperature or time is generated as a result. From the TG curve, it is possible to analyze the thermal stability data of the sample, polymerization, second-order phase transitions, solid/gas interactions, dehydration, and pyrolysis [93,102]. Usually, the TG and DSC measurements are taken consecutively for a consistent assessment of the material's phase transitions and thermal properties.

8.3.5 Biological Characterization

The field of application of electrospun fibers directly depends on the combination of the physical-chemical, thermal, mechanical, and biological properties of the fibers. Treatments performed during or after electrospinning can improve the functionality of nanofibers [103]. As an example, the addition of components with biological action in the fibers can generate suitable antimicrobial and antiviral properties for their application against contaminating microorganisms (e.g., fungus and bacteria) and viruses (e.g., SARS-CoV-2). However, it is essential to prove the toxicity, biocompatibility, and biological action of these fibers. Some polymers have these characteristics, as is the case with many natural polymers, which are generally biocompatible and non-toxic [104].

In vitro cytotoxicity assays are the first tests to assess the biocompatibility of any material. Once non-toxicity has been proven, the material is evaluated in tests with laboratory animals (*in vivo*). These tests consist of placing the fiber in contact with the cell culture (direct contact, indirect contact for example the Agar overlay technique, and contact through extracts and eluates) and evaluating whether cell colonies are inhibited. Implantation tests, sensitization tests, irritation tests, and toxicity tests can be performed *in vivo* tests [105].

However, the mechanical properties are considered more important for engineering applications [92]. Thus, knowing the basic properties of nanofibers (such as morphology, molecular structure, mechanical properties, and thermal properties) is crucial for the scientific understanding of nanofibers and the effective design and use of nanofibrous materials [94].

8.4 INNOVATIVE ELECTROSPUN FIBERS AND DESIGN

Despite the already-known benefits and characteristics of electrospun nanofibers, technological advances in research have contributed to the development of multifunctional materials with innovative design, further extending the use of electrospun nanofibers and improving mechanical, thermal, and physical-chemical properties by proposing the development of special multi-structures from nanofibers [106]. For example, uniaxial and non-conventional electrospun nanofibers have gained a lot of interest in several applications such as sensors [107,108], human health [109], biosensors [110], air filters [111], artificial muscles [112], wound healing [113], oral medication [114], tissue-regeneration [115], and gas purification [116]. In addition, pre- and post-electrospinning modifications [117] have been employed on the surfaces of nanofibers, contributing to the gain of new properties, such as antiviral and antibacterial properties [118,119]. New designs and structures of nanofibers can be achieved from different approaches to produce non-conventional nanofibers such as Janus [120], multiaxial [121–123], spider-net [124 and structures such as bilayer [125,126], trilayer [127,128], aerogels [129] and to build three-dimensional (3D) structures [130–132]. In an example of morphological engineering, Chagas *et al.* [133] developed a membrane containing two layers aiming at the application as wound dressings. As shown in Figure 8.7 a), the structure has a layer of electrospun polylactic acid (PLA) nanofibers and the other layer from microfibers of the blend of PLA and natural rubber containing curcumin. The multifunctional membrane prevented the proliferation and passage of bacteria from the environment to the wound site. In

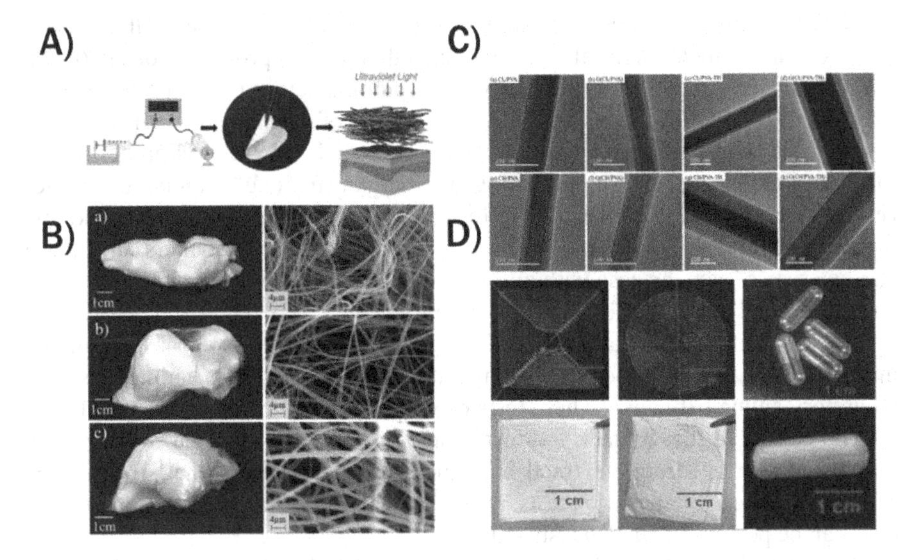

FIGURE 8.7 Different designs and microstructures obtained by the electrospinning technique: a) Bilayer structure being one layer of PLA nanofibers and the other microfibers of the PLA/BN blend containing curcumin. b) Nano and microfibers fluffy 3D structures from polysuccinimide nanofibers. c) Nanofibers obtained by coaxial electrospinning. d) Sandwich structure of nanofibers obtained by electrospinning and electrohydrodynamic printing techniques and subsequently packed in a capsule.

Sources: a) Reprinted with permission from [133]. Copyright 2021, Elsevier B.V. b) Reprinted with permission from [132]. Copyright 2020, Elsevier B.V. c) Reprinted with permission from [122]. Copyright 2020, Elsevier B.V. d) Reprinted with permission from [139]. Copyright 2018, American Chemical Society.

addition, the work presented a strategy to prevent the photodegradation of curcumin under UV light, using the layer of PLA nanofibers, avoiding photodegradation of curcumin for up to 14 days. Electrospinning core-shell nanofibers obtained by coaxial electrospinning are produced with different polymers in the shell and core. It is through the alteration of the polymeric components that it is possible to modulate the release rate of drugs. For example, Dos Santos *et al.* [122] developed a nanofiber core-sheath via coaxial electrospinning using chitosan as the shell layer and PVA with tetracycline hydrochloride as the core layer, as shown in Figure 8.7 b). The effects of the degree of chitosan deacetylation were studied, as well as the physicochemical and biological properties of the post-electrospinning process of crosslinking using genipin. The coaxial nanofibers showed a diameter range of 100–300 nm, sustained release for up to 14 days, improved mechanical properties after crosslinking, and excellent biological histocompatibility. In addition, core-sheath nanofibers showed antibacterial action on bacteria obtained from the human periodontal subgingival pocket of patients with chronic periodontitis.

Zhao *et al.* [134] developed polyurethane (PU) nanofibers via a side-by-side electrospinning method containing two different dyes that were encapsulated separately in Janus nanofibers with potential application in smart fabrics such as face masks, sensors, and packaging. This heterostructure allowed reaching a maximum tensile strength of 6.9 MPa and fracture elongation of 157%, a homogeneous and concentrated distribution of isolated dyes on opposite sides of the nanofiber and thus, without compromising its performance. Taromsari *et al.* [135] developed a biomonitoring sensor with a short response time and high electrical conductivity based on a customizable design of electrospun nanofibers. In this study, MXene, cellulose nanocrystals (CNC), and graphene nanoplatelets (GnP) were employed in styrene–butadiene-styrene (SBS) nanofibers with three new approaches, i) coating of nanofibers using a non-destructive technique that consists of with the simultaneous use of vacuum-assisted filtration and novel ultrasonic atomization; ii) the combination of two nano-additives, GnP and MXene, were used because they presented interesting electrical properties and morphologies; and iii) CNC was used because it is a liquid-phase stabilizer and because it enables the solid-phase compatibilization between GnP/MXene and the nanofiber polymer SBS. As a result, the nanofibers achieved excellent long-term detection stability, high conductivity (54.9 S/m), and response and recovery time of 100 ms, showing their potential use for the next generation of wearable sensors. As illustrated in Figure 8.7 c), Juhasz and co-authors [132] presented a strategy for producing 3D structures from polysuccinimide nano and microfibers. The 3D fluffy structure was obtained from the use of different concentrations of salts ($CaCl_2$, $MgCl_2$, and LiCl) and the presented results show that the structure is achieved from the accumulation of charge in the solution during the electrospinning process and by the ion-solvent interaction.

Liu *et al.* [136] developed a two-dimensional membrane based on an innovative technique to produce electret nanofibers/nets *in situ* using polyvinylidene fluoride (PVDF) polymer at a super high voltage (50 kV). The developed filters showed a high collection efficiency of 99.998% for particles with a size of 300 nm, a low-pressure drop of 93 Pa, high porosity (92.5%), a transmittance of 91%, and a quality factor of 0.085 Pa^{-1}. These results were obtained by combining a chemical modification of the surface of the nanofibers and the formation of nanonets with an average diameter of 21 nm from the adjustments of the properties of the solution.

The versatility of electrospun nanofibers can also be found in the gene therapy area [137], which is considered a revolutionary biotechnology with applications in the biomedical area. For example, Mulholland *et al.* [138] produced PVA nanofibers crosslinked with peptide nanoparticles to release plasmid DNA. The work showed the potential of using nanofibers to release peptides for application in gene therapy, in addition, the results obtained from gene expression were superior to previously published data and did not show cellular toxicity. In this example, Wu *et al.* [139] developed a 3D membrane in a multilayer sandwich structure containing cellulose acetate membranes with ibuprofen and polyvinyl pyrrolidone (PVP) nanofibers with *Ganoderma lucidum* polysaccharide (GLP) using three-dimensional electrohydrodynamic printing and electrospinning technique, as illustrated in Figure 8.7 d). The obtained membrane was packed in a capsule for evaluation of the release profile in a simulation of intestinal (pH 7.4) and gastric (pH 1.7) fluids. The results showed a three-phase release profile, with the immediate dissolution of PVP and the release of GLP, while the greater stability of cellulose acetate nanofibers at different pHs showed a sustained release.

The electrospinning process should probably be the most used to produce nanofibers; despite this, scale-up is still considered a bottleneck concerning this technique [136], and new proposals for improvements have been reached in the already traditional electrospinning technique such as the use of multiple needles [140], needleless systems [141], and multifluidic coaxial electrospinning [142].

8.5 APPLICATIONS

The versatility of nanofibers allows a wide and versatile range of applications for electrospun nanofibers. From medicine to electronics, nanofibers have been helpful in many research and technological areas, with the promise of expansion in their applications.

8.5.1 Air Filtration

Atmospheric particulate matter (PM) pollution is one of the main pollutants harmful to human health, especially smaller PMs, specifically $PM_{2.5}$, which consist of particles smaller than 2.5 μm, due to their ability to penetrate deeply into the lungs and cardiovascular system, causing damage to respiratory cancer and aggravating pre-existing lung and heart diseases [143,144]. In a report made by the World Health Organization (WHO) in 2014, it was estimated that air pollution causes about 7 million deaths per year [145]. In addition, microorganisms such as viruses, bacteria, and fungi are also classified as particulate matter, more specifically, bioaerosols, which can be transported through the air and inhaled by humans, causing various respiratory and infectious diseases such as: Influenza, tuberculosis, and SARS-CoV-2 (COVID-19) [146,147].

To deal with this problem, it is necessary to filter the air to remove these PMs and bioaerosols. Air filters can remove particulate matter suspended in the air. Mainly electrospinning nanofiber filters, due to their large specific surface area, high permeability, good pore interconnectivity, controllable pore size, and low fiber diameter, which make this material highly efficient in air filtration [148]. In addition, they can be manufactured in different configurations and allow the incorporation of functional molecules within the fiber, to better capture particles, viruses, and hazardous gases [149–151]. Some of these incorporations are those made by: Wang *et al.* [152] who manufactured nanofiber filters composed of polyvinyl chloride (PVC) and PU, with robust mechanical properties (12.28 MPa tensile strength), superior air permeability (706.84 mm/s), and efficiency greater than 90%. Song *et al.* [153] reported an electret filtration medium incorporating particles magnetic Fe_3O_4 and Polyhedral oligomeric silsesquioxane (POSS) with PAN nanofibers. Due to the incorporation of magnetic particles, a significant increase in surface charge stability and its greatly improved holding capacity compared to pure PAN nanofiber membranes has been achieved. Nageh *at al.* [154] incorporated ZnO nanoparticles into PVDF nanofibers to achieve inhibition of human adenovirus. Thus, the places where nanofibers can be applied for air filtration are diverse and some are briefly discussed next.

8.5.1.1 Industrial Dust Filtration

Industrial dust commonly arises with industrial manufacturing, processing of cement and ceramic raw materials, kiln operations, and other industrial operations. Industrial dust seriously affects human health and can cause lung diseases such as lung cancer. Therefore, air filters are needed to decrease and control the dust content in the environment.

Electrospun nanofibrous membranes have been successfully applied and commercialized in industrial dust filtration systems now. Zhang *et al.* [155] produced an electrospinning polyamide nanofiber with extremely small pore size, highly open pore structure, and extended front surface. This filter features a filtration performance of 99.996% for ultrafine particles, with a pressure drop of 95 Pa and a quality factor above 0.11 Pa^{-1}. Thus, electrospinning nanoparticle filters can be widely used in industrial applications due to their performance in filtering industrial dust.

8.5.1.2 Locomotive Air Filters

Locomotive filtration is necessary to ensure air quality for the engine and cab. For the engine intake system, dust and other harmful particles must be prevented from being aspired into the engine through air filters. In addition, cabin air filters are designed to improve cabin air quality in vehicles such as cars, trains, and airplanes. It is a matter of great concern that these places are kept free of contamination, especially in recent years with the pandemic caused by the coronavirus, the need to purify the air during travel has become essential.

That said, electrospinning techniques can be very useful in the production of air filters for this purpose, as it is easy to combine with other techniques to prepare multifunctional cabin air filters. For example, antibacterial agents such as Ag nanoparticles and other biocidal agents can be dispersed in the nanofibers, and the fibers not only have good filtration capacity for particles but also can prevent the invasion of bacteria and viruses [154,156]. Therefore, electrospun air filters can efficiently purify the air in the cabin.

8.5.1.3 Indoor Air Filtration

Much of the time in daily life spent in an indoor or enclosed environment makes indoor air filtration a matter of significant concern. According to the Environmental Protection Agency (EPA), most US citizens spend nearly 90% of their time indoors, such as in homes, workplaces, classrooms, and other school environments [157]. Hospitals, being highly sensitive indoor locations, require significant attention when it comes to air filtration. This attention is necessary to prevent the spread of bacteria and viruses through the circulating air. Due to these factors, electrospun nanofibers have gained considerable popularity as a choice for indoor air filter media. This popularity can be attributed to their high filtration efficiency, substantial load capacity, and low-pressure drop. These characteristics stem from their high porosity and small fiber diameter. Moreover, their potential for functionalization with antimicrobial materials further enhances their appeal. This functionalization helps in reducing fungal and bacterial growth on the filter surfaces and curbing their spread into the surrounding environment [158].

8.5.2 Water Treatment

Water quality is a very important issue nowadays, and many of the water problems are caused by urban and industrial pollution, threatening the environment and human life. Contaminants can come from several sources, such as waste treatment plants, pesticides, industries, and stormwater runoff. These pollutants can come in the form of organic, radioactive, heavy metals, and acidic/basic compounds [159]. These pollutants can cause various diseases in humans, such as diarrhea, tuberculosis, dysentery, cholera, etc.

With progressive water scarcity and pollution, there is a demand to develop more effective ways of separating and removing harmful contaminants from water and ensuring safe drinking water. In this context, membrane-based separation technologies have played an important role in water purification due to their effectiveness and energy efficiency [160]. In addition, electrospun nanofiber filters are being increasingly used due to their characteristics, such as a large surface-to-volume ratio and the adjustable functionality of nanofibrous membranes, which make them much more effective for the superficial adsorption of contaminants from polluted water than conventional membranes [161]. The nanofibers have high porosity, and the pores are interconnected, with controllable pore size distribution, which makes this filter suitable for a range of applications [162].

Furthermore, electrospinning nanofibers can be easily functionalized to fulfill desirable functions, such as mixing functional additives in the electrospinning solution [163] or attaching an active species for affinity adsorption [164], surface coating [165], or interfacial polymerization [166]. These advantages give electrospun nanofibers enormous potential to address many emerging health, environmental, energy, and other challenges. Ahead we can see some applications where these innovative filters can be used.

8.5.2.1 Dye Removal

Dyes can end up in freshwater systems due to the residue from the manufacture of many industrial products such as dyes, paper, leather, cosmetics, and textiles. Dyes can be organic or inorganic, may have high turbidity, and contain toxic compounds and oxidants in suspension, causing damage to the environment and human health [167].

Dye removal is usually done using adsorption, coagulation, and biological treatment techniques [168]. Electrospun nanofibers are also gaining space in this area, especially nanofibers produced from biopolymers, due to their environmental and sustainability benefits [169–172]. One example is the method used by Gopakumar *et al.* [173] which uses Meldrum's acid to modify cellulose-based PVDF membranes, which resulted in increased adsorption of crystal violet dye, a cationic dye present in contaminated water.

8.5.2.2 Heavy Metal Removal

Heavy metals are used in various oxidation-reduction reactions. They can be used in the process of various industries such as ceramics, glass, mining, and battery manufacturing; consequently, these metals end up in the wastewater of these industries. Some of the metals that can be found in these waters include arsenic (Ar), lead (Pb), mercury (Hg), cadmium (Cd), chromium, aluminum (Al), and copper (Cu) [174]. When these metals are released into surface waters (rivers, lakes, etc.) or contaminate groundwater, they end up causing serious biological problems in aquatic systems, in addition to severe damage to human health [171]. The most used methods for removing these pollutants are adsorption and filtration, but nanofibers are gaining ground in this application. Vu *et al.* [175] produced mesoporous titanium nanofibers for the adsorption of Cu(II) ions. Already Zhao *et al.* [176] produced PAN nanofibers embedded with branched polyethyleneimine (PEI), used to retain Cr(VI) from aqueous systems.

8.5.3 Catalyst and Biocatalyst

Catalysts are of fundamental importance to our society; it is estimated that 90% of all chemical processes use some type of heterogeneous catalyst [177]. So that the chemical or biological catalyst can maintain high catalytic activity, and stability and simplify the reaction process, it is recommended that it be linked to a substrate. An inert porous material with a large surface area and high permeability to reagents can be an interesting substrate to serve as a base for catalysts. For example, nanoparticles such as platinum (Pt), palladium (Pd), and rhodium (Rh) supported on substrate surfaces (*e.g.*, carbon and oxide) form the basis for widely used industrial catalysts [178]. Thus, for this function, nanofibers are receiving great attention, due to their large surface area that can provide many active sites accessible by reagents, thus increasing the catalytic capacity. The small well-connected pores of the nanofibers ensure effective interaction between the catalyst and the reagent, this interaction is important in continuous flow reactions or biological processes [179]. That said, nanofibers stand out for their versatility during production and can be produced in different configurations; in addition, they can be produced from various organic and inorganic materials, to meet different needs and applications [180]. Next, we will briefly present some applications of catalysts based on nanofibers.

8.5.3.1 Chemical Catalysts

Catalysts produced with nanofibers can be used in chemical reactions such as hydrogenation, dehydrogenation, oxidation, reduction, isomerization, synthesis, aromatization, and other reactions. Due to its good conductivity, chemical stability, high specific surface area, and stable pore structure, it can support a catalyst supporting noble metals such as silver, platinum, palladium, and others [181]. For example, the poly(acrylonitrile-acrylic acid) (PAN-AA) nanofibers, incorporated with palladium (Pd) showed high catalytic activity and a good performance in the hydrogenation of α-olefin at room temperature. Furthermore, the metals attached to the nanofibers have a reduction in the amount of metals that are leached from the catalyst, resulting in less product contamination and environmental pollution. Furthermore, catalysts can be recovered with minimal loss in their performance [182,183].

8.5.3.2 Photocatalysts

Water, air, and soil pollution seriously affect the health and life of people and living beings in general. These pollutants are often treated with filters containing photocatalysts. Photocatalysis harnesses the energy of daylight to initiate a chemical reaction, leading to the complete degradation of organic pollutants carried by water or air into carbon dioxide, water, and inorganic acids—all at room temperature. Most semiconductor photocatalysts are n-type semiconductor materials that have a unique band structure different from that of the parent metal. Electrospun nanofibers, with their extensive surface area, enhance contact with the reactant components. In addition, the photocatalytic material can be supported directly on the nanofibers and used for the treatment of water and waste gases. The photocatalytic materials that can be used in electrospinning are the most diverse, such as ZnO/PAN [184], Cr-$SrTiO_3$ [185], TiO_2-Graphene [186], between others.

8.5.3.3 Biocatalysts

Enzymes are biocatalysts known for their high efficiency in reactions and their selectivity, which are highly appreciated in applications such as pharmaceutical synthesis, food processing, and the production of fine chemicals [187,188]. Furthermore, compared to other catalysts, enzymes are generally non-toxic and easy to handle. To facilitate the enzyme reaction solution, these are usually immobilized on a carrier; moreover, immobilized enzymes have a longer lifetime compared to free enzymes because their chemical structure and conformation are stabilized by the supporting nanofibers. Materials such as nanofibers are used as enzyme carriers due to their large specific surface area and high loading capacity [183].

Enzymes can be immobilized on nanofibers by several mechanisms, such as encapsulation, adsorption, or covalent bonding, depending on the properties of the polymer used 189. Some of the polymers that can be used and that can bring some advantages to the process are described here: cellulose and PVA are hydrophilic and, therefore, are biocompatible with enzymes under physiological conditions [190,191]. PAN and Polyamide, on the other hand, can offer thermal and mechanical resistance [192,193]. In addition, polymers can be combined to give new characteristics to the catalyst, such as the combination of chitosan and poly(lactic acid) that have appropriate solubility and compatibility for electrospinning. By combining polymers with different properties, the obtained nanofibers are biodegradable, mechanically resistant, and thermally stable [194,195]. Examples of enzymes that were successfully immobilized on electrospun nanofibers include the fungal lipases from species such as *Candida rugosa* [191], laccase [194], and *Thermomyces lanuginosus* [193].

8.5.4 Sensors

The role of sensors is to transform physical or chemical responses into a response signal, which in most cases is an electrical signal, which is easy to transfer, convert, process, and display. But this is not the only signal form that can be obtained; there are other forms such as voltage, current, capacitance, resistance, etc. [196]. Sensors can be applied for different purposes such as detecting chemicals for environmental protection, industrial process control, and security and defense applications. For this to be a good sensor it must have small dimensions, low manufacturing cost, selectivity, and reliability. For the sensor to have a high sensitivity and a fast response requires the sensor to have a large specific surface area and a highly porous structure. The characteristics possessed by electrospun nanofibers match these requirements well. Therefore, a nanofibrous structure should be a promising physical structure to form a highly sensitive and fast response sensor [197,198]. In addition, the raw materials used in the production of nanofibers can include a wide range of natural polymers, synthetic and inorganic polymers, and can be combined with different technologies to have different properties [199]. Currently, based on different detection principles, several sensors based on electrospun nanofibers have been developed and are shown following.

8.5.4.1 Electrochemical Sensors

Electrochemical sensors are used as an analytical method to determine the concentration of an analyte based on its electrochemical properties and corresponding changes in the solution. Typically, in an electrochemical sensor, the electrocatalytic reaction produces a quantifiable current (amperometric detection) [200] or a measurable electrical potential or charge buildup (potentiometric detection) [201], or detectably modifies the electrical conductivity of a medium (conductometric detection) between electrodes [202].

Electrospinning can manufacture polymeric nanofibers in a simple process. Being able to produce nanofibers of polymers, metal, metallic oxide, and ceramics facilitates the use of these sensors in different environments. An example of an electrochemical sensor is PVDF/poly (aminophenylboronic acid) (PAPBA) nanofibers which reacted with glucose, and a change in amperometric properties was easily detected [203].

8.5.4.2 Optical Sensors

Optical sensors utilize the fluorescence quenching of the detection material against target chemical molecules. Optical fluorescence sensors were also prepared by a layer-by-layer electrostatic assembly technique to apply a conjugated polymer to the surface of the nanofiber for detection of methyl viologen and cytochrome *c* in aqueous solution [204] and doped silica nanofibers with porphyrin were used to screen TNT vapor [205]. All these nanofiber sensors showed high sensitivity and fast response. In addition to fluorescent properties, electrospun nanofibers embedded in conjugated polymers have also been reported to be able to detect volatile organic compounds (VOCs) based on optical absorption properties [206].

8.5.4.3 Humidity Sensors

Humidity sensors are very important due to their practical applications in monitoring the environment and controlling industrial processes like air conditioning and drying. So far, many moisture sensors based on nanostructured materials have been obtained successfully and with high performance. This is due to the unique structure of electrospun nanofibers: the large specific surface area facilitates the absorption of water molecules; the structure of the nanofibers facilitates the transfer of mass from water molecules to the sensor interaction region. These advantages result in significant gain in the detection signal and good stability [207]. For example, ZnO nanofibers doped with KCl showed a high sensitivity to humidity with high response and response time [208,209].

8.5.4.4 Gas Sensors

The detection of toxic gases such as carbon monoxide (CO), nitrogen dioxide (NO_2), and ammonia (NH_3) is a very serious issue to protect people in homes, offices, and industrial factories. These sensors are responsible for activating alert systems in case of leaks. Thus, it is necessary that gas sensors have a high sensitivity and response time. In this context, detectors produced using nanofibers are particularly notable, particularly those made from semiconductor oxides, and conductive and non-conductive polymers. Pure oxide nanofibers are normally produced by electrospinning a solution containing oxide sol-gel and polymer, followed by calcination treatment to remove the polymer. The detection of gas molecules using oxide nanofibers is based on conductivity changes due to the doping effect of the analyst gases to the oxide. Some oxide nanofibers were evaluated to detect different gases, such as MoO_3 nanofibers for ammonia [210], WO_3 nanofibers for ammonia [165], TiO_2 for NO_2 and H_2 [211]. These sensors exhibited improved sensitivity, faster response, and a lower detection limit than sol-gel based films. Furthermore, the possibility of being able to incorporate metal oxide nanoparticles improves the sensitivity of the sensor [212].

8.5.4.5 Biosensors

Biosensors are devices that can detect or monitor the biological change in the human body. Electrospun nanofibers are suitable materials for this application due to their large surface area, which provides several binding sites available for immobilization of biological recognition elements and various functionalizations. For example, Wang *et al.* [213] used electrosprayed cellulose acetate (CA) nanofibers coated with a fluorescent layer—poly[2-(3-thienyl)ethanol butoxycarbonyl-methyl urethane] (H-PURET)—to detect cytochrome c, a heme-containing respiratory protein associated with myocardial infarction. Du *et al.* [214] have already developed Nylon 6 nanofibers incorporated with gold particles as glucose sensors. Wang *et al.* [215] also reported an easy route to manufacture gold (vinyl alcohol) nanoparticles (Au NPs-PVA) a nanofiber stable in the presence of water and with adjustable densities of Au NPs, and further demonstrate the potential application as efficient biosensors for the detection of H_2O_2.

8.5.5 Medical Applications

Recently, nanofibers have gained attention in the medical and pharmaceutical fields. Their properties can adapt to mimic cellular tissue, and many biocompatible polymers are available. Nanofibers have been used as controlled drug deliverers and scaffolds for tissue regeneration.

8.5.5.1 Electrospun Nanofibers as Drug Delivery Carrier

The use of nanofibers in drug delivery is one application that can improve human health. Due to the high specific surface area and short diffusion passage length, electrospun nanofibers are extensively investigated for the delivery of a number of drugs, including antibiotics [216], anticancer drugs [217], proteins [218], and DNA [219]. By controlling the composition and nanofibers properties, such as mesh size and fiber diameter, it is possible to control the drug release to a specific site in the body, which can improve the treatment of diseases such as AIDS and some forms of cancer [220].

8.5.5.2 Electrospun Nanofibers in Cosmetics

The electrospun nanofibers have several applications in the cosmetic sector, such as face masks, perfumes, deodorants, and antiperspirants. Cosmetic masks can be used for skin care with or without bioactive substances [221]. These skin masks have a high surface area that offers a rapid permeation of the bioactives to the skin [222]. An example of an application is the production of medicinal soaps and anti-aging products containing mangosteen extracts coated with an electrospinning PVA membrane, which maintains the extracts characteristics, and controlling their release from cosmetics [217].

8.5.5.3 Electrospun Nanofibers for Wound Healing

Electrospun nanofibers present excellent properties for wound healing [223], exhibiting three-dimensional properties that mimic the structure of the extracellular matrix of the skin (ECM), and supporting the adhesion and proliferation of new cells. The porous structure is capable of gas and nutrient exchanges, absorbing the excess of exudates and preventing bacterial colonization [223,224].

It is possible to functionalize the electrospun nanofibers with bioactive molecules to improve wound healing. The high ratio of surface per volume and the possibility to use suitable solvents to solubilize the drugs allow high load capacity for the nanofibers. There are many ways to incorporate biomolecules into polymeric nanofibers, modulating the release of the drug during the distinct stages of cicatrization [223–225]. The primary strategy to produce those ECM mimetic materials is combining synthetic polymers with biopolymers, resulting in hybrid materials [226,227].

Incorporating drugs into nanofibers is a viable strategy to accelerate the wound-healing process. A load of natural or synthetic drugs can confer the nanofiber anti-inflammatory, bactericides, fungicides, and analgesic effects. Karuppannan *et al.* [228] incorporated quercetin, a photochemical with high anti-inflammatory and antioxidant activity, on electrospun nanofibers of polycaprolactone (PCL)/gelatin for wound healing. The hydrophobic fibers presented high mechanical resistance, bactericidal activity, and smooth morphology without beads. They also showed biocompatibility with fibroblasts 3T3, stimulating angiogenesis, re-epithelialization, and collagen synthesis. Morais *et al.* [5] developed a hybrid system of polyvinyl alcohol and chitosan (PVA/CS), loaded with green propolis and nystatin as a bactericidal effect, with easy incorporation of the agents. Liu *et al.* [229] prepared nanofiber membranes using cellulose acetate (CA) and polyester urethane (PEU) associated with polyhexamethylene biguanide (PHMB) to test their wound healing process. CA conferred the fiber hydrophobicity and air permeability, favoring the cicatrization, while the PEU preserved the physical and thermal properties of the membranes. PEU also reduced the diffusion of PMHB, increasing its release time. Kenamy *et al.* [230] loaded gelatin and CS nanofibers with cinnamaldehyde, resulting in growth inhibition and antimicrobial activity against gram-positive and gram-negative bacteria. They performed hemocompatibility, biodegradability, and toxicity assays, proving its efficacy as wound-healing electrospun nanofiber.

Nanoparticle incorporation in nanofibers requires specific techniques to be applied for biomedical purposes. The combination of those two systems offers advantages, developing properties that are not present in those systems individually. The main application is to avoid the fast release of the drug, as occurs in the individual systems, to reach the sustained release of hydrophilic drugs [231]. Studying electrospun nanoparticles, Hussein *et al.* [232] synthesized conjugated nanoparticles of Au, CS, and *Punica granatum*. The resulting complex was dissolved in PVA to produce electrospun nanofibers, reticulated later with glutaraldehyde vapor. The hybrid composites provide excellent properties to support fibroblast proliferation. Atashgahi *et al.* [233] produced CS nanoparticles immobilized in epinephrin by electrospraying over gelatin electrospun nanofibers. The results show that the nanofibers are cytocompatible and hemocompatible, providing an interface for red blood cell adhesion, increasing the coagulation rate. This composite is a new approach for safely controlling hemorrhage.

8.6 FINAL REMARKS

This chapter offers a comprehensive overview of various aspects of nanofiber production via electrospinning, covering the history, mechanisms, characterization methods, and innovative applications. The impact of the polymeric solution's chemical properties and electrospinning parameters on the nanofibers' properties is also discussed, along with the techniques for characterizing their morphological, mechanical, chemical, and biochemical properties. Electrospinning has shown its potential applications in diverse areas, ranging from air and water filtration to catalysis, sensors, pollutant removal, and medical sciences.

Since their inception, nanofibers have garnered attention from both academia and industry, demonstrating their efficacy in numerous fields and versatility in diverse applications. However, challenges such as low production rates, the use of harmless solvents, and the need for biodegradable and biocompatible polymers still need to be addressed to expand large-scale nanofiber applications.

REFERENCES

[1] Tomisawa, R. *et al.* Effect of melt spinning conditions on the fiber structure development of polyethylene terephthalate. *Polymer (Guildf)* 116, 367–377 (2017).
[2] Wang, Z. *et al.* Antibacterial and environmentally friendly chitosan/polyvinyl alcohol blend membranes for air filtration. *Carbohydr Polym* 198, 241–248 (2018).
[3] Matulevicius, J. *et al.* Design and characterization of electrospun polyamide nanofiber media for air filtration applications. *J Nanomater* 2014, (2014).
[4] Chagas, P. A. M. *et al.* Bilayered electrospun membranes composed of poly(lactic-acid)/natural rubber: A strategy against curcumin photodegradation for wound dressing application. *React Funct Polym* 163, 104889 (2021).
[5] Morais, M. S., Bonfim, D. P. F., Aguiar, M. L. & Oliveira, W. P. Electrospun Poly (Vinyl Alcohol) nanofibrous mat loaded with green propolis extract, chitosan and nystatin as an innovative wound dressing material. *J Pharm Innov* 1–15 (2022).
[6] Pillay, V. *et al.* A review of the effect of processing variables on the fabrication of electrospun nanofibers for drug delivery applications. *J Nanomater* 2013, (2013).
[7] Augustine, R., Nethi, S. K., Kalarikkal, N., Thomas, S. & Patra, C. R. Electrospun polycaprolactone (PCL) scaffolds embedded with europium hydroxide nanorods (EHNs) with enhanced vascularization and cell proliferation for tissue engineering applications. *J Mater Chem B* 5, 4660–4672 (2017).
[8] Abbas, W. A., Sharafeldin, I. M., Omar, M. M. & Allam, N. K. Novel mineralized electrospun chitosan/PVA/TiO2 nanofibrous composites for potential biomedical applications: computational and experimental insights. *Nanoscale Adv* 2, 1512–1522 (2020).
[9] Linh, N. T. B. & Lee, B. T. Electrospinning of polyvinyl alcohol/gelatin nanofiber composites and cross-linking for bone tissue engineering application. 27, 255–266 (2011).
[10] Teixeira, M. A., Amorim, M. T. P. & Felgueiras, H. P. Poly(Vinyl Alcohol)-based nanofibrous electrospun scaffolds for tissue engineering applications. *Polymers (Basel)* 12, (2020).
[11] Panda, P. K., Sadeghi, K. & Seo, J. Recent advances in poly (vinyl alcohol)/natural polymer based films for food packaging applications: A review. *Food Packag Shelf Life* 33, 100904 (2022).
[12] Fernandez-Saiz, P., Lagaron, J. M. & Ocio, M. J. Optimization of the biocide properties of chitosan for its application in the design of active films of interest in the food area. *Food Hydrocoll* 23, 913–921 (2009).
[13] Aytac, Z., Ipek, S., Durgun, E., Tekinay, T. & Uyar, T. Antibacterial electrospun zein nanofibrous web encapsulating thymol/cyclodextrin-inclusion complex for food packaging. *Food Chem* 233, 117–124 (2017).
[14] Leung, W. W. F. & Sun, Q. Electrostatic charged nanofiber filter for filtering airborne novel coronavirus ({COVID}-19) and nano-aerosols. *Sep Purif Technol* 250, 116886 (2020).
[15] Bian, Y., Zhang, C., Wang, H. & Cao, Q. Degradable nanofiber for eco-friendly air filtration: Progress and perspectives. *Sep Purif Technol* 306, 122642 (2023).
[16] Bonfim, D. P. F., Cruz, F. G. S., Bretas, R. E. S., Guerra, V. G. & Aguiar, M. L. A Sustainable Recycling Alternative: Electrospun PET-Membranes for Air Nanofiltration. *Polymers (Basel)* 13, (2021).
[17] Bortolassi, A. C. C. *et al.* Efficient nanoparticles removal and bactericidal action of electrospun nanofibers membranes for air filtration. *Mat Sci Eng: C* 102, 718–729 (2019).
[18] de Oliveira, A. E., Aguiar, M. L. & Guerra, V. G. Improved filter media with PVA/citric acid/Triton X-100 nanofibers for filtration of nanoparticles from air. *Polym Bull* 78, 6387–6408 (2021).

[19] Tucker, N., Stanger, J. J., Staiger, M. P., Razzaq, H. & Hofman, K. The history of the science and technology of electrospinning from 1600 to 1995. 7, 63–73 (2012).
[20] Morton, W. J. Method of dispersing fluids. US Patent 705691 (1900).
[21] Hurion, A. LORD RAYLEIGH.—The influence of electricity on colliding water drops (Influence de l'électricité sur la rencontre des gouttes d'eau); Proceedings of the Royal Society, t. XXVIII, n° 194, p. 406; mars 1879. *J Phys Théo Appl* 8, 383–384 (1879).
[22] Boys, C. v. On the Production, Properties, and some suggested uses of the finest threads. *Proc Phys Soc Lond* 9, 8 (1887).
[23] Ghosal, K., Agatemor, C., Tucker, N., Kny, E. & Thomas, S. Electrical spinning to electrospinning: a brief history. *RSC Soft Matter* 2018-January, 1–23 (2018).
[24] Cooley, J. F. Apparatus for electrically dispersing fluids. US Patent n. 692,631, (1902).
[25] Kiyohiko, Hagiwara. Process for manufacturing artificial silk and other filaments by applying electric current. *US Patent* 1,699,615. 22 Jan. 1929 (1929).
[26] Attorneys. Method of producing artificial fibers. *A. Form Hals* 2, 415 (1937).
[27] Taylor & Geoffrey. Disintegration of water drops in an electric field. *RSPSA* 280, 383–397 (1964).
[28] Doshi, J. & Reneker, D. H. Electrospinning process and applications of electrospun fibers. *J Electrostat* 35, 151–160 (1995).
[29] Ura, D. P. *et al.* The Role of Electrical polarity in electrospinning and on the mechanical and structural properties of as-spun fibers. *Materials* 13, 4169 (2020).
[30] Reyes, C. G. & Lagerwall, J. P. F. Disruption of Electrospinning due to Water Condensation into the Taylor Cone. *ACS Appl Mater Interf* 12, 26566–26576 (2020).
[31] Yarin, A. L., Koombhongse, S. & Reneker, D. H. Taylor cone and jetting from liquid droplets in electrospinning of nanofibers. *J Appl Phys* 90, 4836 (2001).
[32] Meli, L., Miao, J., Dordick, J. S. & Linhardt, R. J. Electrospinning from room temperature ionic liquids for biopolymer fiber formation. *Green Chem* 12, 1883–1892 (2010).
[33] Haider, A., Haider, S. & Kang, I. K. A comprehensive review summarizing the effect of electrospinning parameters and potential applications of nanofibers in biomedical and biotechnology. *Arab J Chem* 11, 1165–1188 (2018).
[34] Bhattarai, N., Edmondson, D., Veiseh, O., Matsen, F. A. & Zhang, M. Electrospun chitosan-based nanofibers and their cellular compatibility. *Biomaterials* 26, 6176–6184 (2005).
[35] Vu, T. H. N., Morozkina, S. N. & Uspenskaya, M. V. Study of the Nanofibers Fabrication Conditions from the Mixture of Poly(vinyl alcohol) and Chitosan by Electrospinning Method. *Polymers* 14, 811 (2022).
[36] Paipitak, K., Pornpra, T., Mongkontalang, P., Techitdheer, W. & Pecharapa, W. Characterization of PVA-Chitosan Nanofibers Prepared by Electrospinning. *Procedia Eng* 8, 101–105 (2011).
[37] Mata, G. C. da, Morais, M. S., Oliveira, W. P. de & Aguiar, M. L. Composition Effects on the Morphology of PVA/Chitosan Electrospun Nanofibers. *Polymers (Basel)* 14, 4856 (2022).
[38] Homayoni, H., Ravandi, S. A. H. & Valizadeh, M. Electrospinning of chitosan nanofibers: Processing optimization. *Carbohydr Polym* 77, 656–661 (2009).
[39] Koski, A., Yim, K. & Shivkumar, S. Effect of molecular weight on fibrous PVA produced by electrospinning. *Mater Lett* 58, 493–497 (2004).
[40] Park, B. K. & Um, I. C. Effect of molecular weight on electro-spinning performance of regenerated silk. *Int J Biol Macromol* 106, 1166–1172 (2018).
[41] Promnil, S., Numpaisal, P. O. & Ruksakulpiwat, Y. Effect of molecular weight on mechanical properties of electrospun poly (lactic acid) fibers for meniscus tissue engineering scaffold. *Mater Today Proc* 47, 3496–3499 (2021).

[42] Ngadiman, N. H. A., Noordin, M. Y., Kurniawan, D., Idris, A. & Shakir, A. S. A. Influence of Polyvinyl Alcohol Molecular Weight on the Electrospun Nanofiber Mechanical Properties. *Proc Manuf* 2, 568–572 (2015).
[43] Hong, E., Yeneneh, A. M., Sen, T. K., Ang, H. M. & Kayaalp, A. A comprehensive review on rheological studies of sludge from various sections of municipal wastewater treatment plants for enhancement of process performance. *Adv Colloid Interface Sci* 257, 19–30 (2018).
[44] Yördem, O. S., Papila, M. & Menceloğlu, Y. Z. Effects of electrospinning parameters on polyacrylonitrile nanofiber diameter: An investigation by response surface methodology. *Mater Des* 29, 34–44 (2008).
[45] Ionescu, C. M., Birs, I. R., Copot, D., Muresan, C. I. & Caponetto, R. Mathematical modelling with experimental validation of viscoelastic properties in non-Newtonian fluids. *Phil Trans Roy Soc A* 378, (2020).
[46] Higashi, S., Hirai, T., Matsubara, M., Yoshida, H. & Beniya, A. Dynamic viscosity recovery of electrospinning solution for stabilizing elongated ultrafine polymer nanofiber by TEMPO-CNF. *Sci Rep* 10, 1–8 (2020).
[47] Sidaravicius, J. *et al.* Predicting the electrospinnability of polymer solutions with electromechanical simulation. *J Appl Polym Sci* 131, (2014).
[48] Fong, H., Chun, I. & Reneker, D. H. Beaded nanofibers formed during electrospinning. *Polymer (Guildf)* 40, 4585–4592 (1999).
[49] Anu Bhushani, J. & Anandharamakrishnan, C. Electrospinning and electrospraying techniques: Potential food based applications. *Trends Food Sci Technol* 38, 21–33 (2014).
[50] Lim, L. T., Mendes, A. C. & Chronakis, I. S. Electrospinning and electrospraying technologies for food applications. *Adv Food Nutr Res* 88, 167–234 (2019).
[51] Khajavi, R. & Abbasipour, M. Controlling nanofiber morphology by the electrospinning process. *Electros Nanofib*, 109–123 (2017).
[52] Barakat, N. A. M., Kanjwal, M. A., Sheikh, F. A. & Kim, H. Y. Spider-net within the N6, PVA and PU electrospun nanofiber mats using salt addition: Novel strategy in the electrospinning process. *Polymer (Guildf)* 50, 4389–4396 (2009).
[53] Zaarour, B., Zhu, L., Huang, C. & Jin, X. Controlling the secondary surface morphology of electrospun PVDF nanofibers by regulating the solvent and relative humidity. *Nanoscale Res Lett* 13, 1–11 (2018).
[54] Huang, C. *et al.* Electrospun polymer nanofibres with small diameters. *Nanotechnology* 17, 1558 (2006).
[55] Ahn, Y. C. *et al.* Development of high efficiency nanofilters made of nanofibers. *Curr Appl Phys* 6, 1030–1035 (2006).
[56] Stanger, J., Tucker, N., Kirwan, K. & Staiger, M. P. Effect of charge density on the Taylor cone in electrospinning. 23, 1956–1961 (2012).
[57] Zong, X. *et al.* Structure and process relationship of electrospun bioabsorbable nanofiber membranes. *Polymer (Guildf)* 43, 4403–4412 (2002).
[58] Topuz, F., Abdulhamid, M. A., Holtzl, T. & Szekely, G. Nanofiber engineering of microporous polyimides through electrospinning: Influence of electrospinning parameters and salt addition. *Mater Des* 198, 109280 (2021).
[59] Yu, D. G. *et al.* PVP nanofibers prepared using co-axial electrospinning with salt solution as sheath fluid. *Mater Lett* 67, 78–80 (2012).
[60] Zhang, W., Chen, X., Wang, Y., Wu, L. & Hu, Y. Experimental and modeling of conductivity for electrolyte solution systems. *ACS Omega* 5, 22465–22474 (2020).
[61] Angammana, C. J. & Jayaram, S. H. Analysis of the effects of solution conductivity on electrospinning process and fiber morphology. *IEEE Trans Ind Appl* 47, 1109–1117 (2011).

[62] Medeiros, G. B. *et al.* Experimental design to evaluate properties of electrospun fibers of zein/poly (ethylene oxide) for biomaterial applications. *J Appl Polym Sci* 138, 50898 (2021).
[63] Dhanalakshmi, M., Lele, A. K. & Jog, J. P. Electrospinning of Nylon11: Effect of processing parameters on morphology and microstructure. *Mater Today Commun* 3, 141–148 (2015).
[64] Bellan, L. M. & Craighead, H. G. Control of an electrospinning jet using electric focusing and jet-steering fields. *J Vac Sci Technol B: Microelectron Nanomet Struct Process Meas Phenom* 24, 3179 (2006).
[65] Mamtha, V., Narasimha Murthy, H. N., Authade, P. & Sridhar, R. Study of Electrospun fiber diameter using ANSOFT and ANSYS. *Mater Today Proc* 5, 21529–21537 (2018).
[66] Sahay, R., Thavasi, V. & Ramakrishna, S. Design modifications in electrospinning setup for advanced applications. *J Nanomater* 2011, (2011).
[67] Deitzel, J. M., Kleinmeyer, J., Harris, D. & Beck Tan, N. C. The effect of processing variables on the morphology of electrospun nanofibers and textiles. *Polymer (Guildf)* 42, 261–272 (2001).
[68] He, H., Wang, Y., Farkas, B., Nagy, Z. K. & Molnar, K. Analysis and prediction of the diameter and orientation of AC electrospun nanofibers by response surface methodology. *Mater Des* 194, 108902 (2020).
[69] Wu, D. *et al.* Influence of nonionic and ionic surfactants on the antifungal and mycotoxin inhibitory efficacy of cinnamon oil nanoemulsions. *Food Funct* 10, 2817–2827 (2019).
[70] Xu, J. *et al.* Controllable generation of nanofibers through a magnetic-field-assisted electrospinning design. *Mater Lett* 247, 19–24 (2019).
[71] Zargham, S., Bazgir, S., Tavakoli, A., Rashidi, A. S. & Damerchely, R. The effect of flow rate on morphology and deposition area of electrospun nylon 6 nanofiber. 7, 42–49 (2012).
[72] Chowdhury, M. & Stylios, G. K. Analysis of the effect of experimental parameters on the morphology of electrospun polyethylene oxide nanofibres and on their thermal properties. 103, 124–138 (2011).
[73] Yuan, X. Y., Zhang, Y. Y., Dong, C. & Sheng, J. Morphology of ultrafine polysulfone fibers prepared by electrospinning. *Polym Int* 53, 1704–1710 (2004).
[74] Macossay, J., Marruffo, A., Rincon, R., Eubanks, T. & Kuang, A. Effect of needle diameter on nanofiber diameter and thermal properties of electrospun poly(methyl methacrylate). *Polym Adv Technol* 18, 180–183 (2007).
[75] He, H., Kara, Y. & Molnar, K. Effect of needle characteristic on fibrous PEO produced by electrospinning. *Resol Discov* 4, 7–11 (2018).
[76] Abunahel, B. M., Azman, N. Z. N. & Jamil, M. Effect of needle diameter on the morphological structure of electrospun n-Bi2O3/Epoxy-PVA nanofiber mats. *Int J Chem Mater Eng* 12, 296–299 (2018).
[77] Viswanadam, G. & Chase, G. G. Modified electric fields to control the direction of electrospinning jets. *Polymer (Guildf)* 54, 1397–1404 (2013).
[78] Liu, Y. *et al.* Effects of solution properties and electric field on the electrospinning of hyaluronic acid. *Carbohydr Polym* 83, 1011–1015 (2011).
[79] Dalton, P. D., Klee, D. & Möller, M. Electrospinning with dual collection rings. *Polymer (Guildf)* 46, 611–614 (2005).
[80] Kang, D. H. & Kang, H. W. Advanced electrospinning using circle electrodes for freestanding PVDF nanofiber film fabrication. *Appl Surf Sci* 455, 251–257 (2018).
[81] Nguyen, D. N., Hwang, Y. & Moon, W. Electrospinning of well-aligned fiber bundles using an End-point Control Assembly method. *Eur Polym J* 77, 54–64 (2016).
[82] Lee, S. J., Cho, N. I. & Lee, D. Y. Effect of collector grounding on directionality of electrospun titania fibers. *J Eur Ceram Soc* 27, 3651–3654 (2007).

[83] De Prá, M. A. A., Ribeiro-do-Valle, R. M., Maraschin, M. & Veleirinho, B. Effect of collector design on the morphological properties of polycaprolactone electrospun fibers. *Mater Lett* 193, 154–157 (2017).
[84] Mailley, D., Hébraud, A. & Schlatter, G. A Review on the impact of humidity during electrospinning: from the nanofiber structure engineering to the applications. *Macromol Mater Eng* 306, 2100115 (2021).
[85] İçOğLu, H. İ & Oğulata, R. T. Effect of ambient parameters on morphology of electrospun poly (trimethylene terephthalate) (ptt) fibers. *Tekstil Ve Konfeksiyon* 27, 343–354 (2017).
[86] Oğulata, R. T. & İçOğLu, H. İ. Effect of ambient parameters on morphology of electrospun polyetherimide (pei) fibers. *Tekstil Ve Konfeksiyon* 23, 343–354 (2013).
[87] Oğulata, R. T. & İçOğLu, H. İ. Interaction between effects of ambient parameters and those of other important parameters on electrospinning of PEI/NMP solution. 106, 57–66 (2014).
[88] Alswid, S. & Issa, M. Study The effect of conditions of the electro spinning cabin (humidity) on electro-spun polyvinyl alcohol (PVA) nano-fibers. *Al-Nahrain J Eng Sci* 20, 520–525 (2017).
[89] Barua, B. & Saha, M. C. Influence of humidity, temperature, and annealing on microstructure and tensile properties of electrospun polyacrylonitrile nanofibers. *Polym Eng Sci* 58, 998–1009 (2018).
[90] Ramazani, S. & Karimi, M. Investigating the influence of temperature on electrospinning of polycaprolactone solutions. *e-Polym* 14, 323–333 (2014).
[91] Bhardwaj, N. & Kundu, S. C. Electrospinning: A fascinating fiber fabrication technique. *Biotechnol Adv* 28, 325–347 (2010).
[92] Rashid, T. U., Gorga, R. E. & Krause, W. E. Mechanical properties of electrospun fibers—a critical review. *Adv Eng Mater* 23, (2021).
[93] Shojaei, T. R., Hajalilou, A., Tabatabaei, M., Mobli, H. & Aghbashlo, M. Characterization and evaluation of nanofiber materials. in *Handbook of Nanofibers* 1–32 (Springer International Publishing, 2018).
[94] Ko, F. K. & Wan, Y. Characterization of nanofibers. in *Introduction to Nanofiber Materials* 101–145 (Cambridge University Press, 2014).
[95] Islam, M. S., Ang, B. C., Andriyana, A. & Afifi, A. M. A review on fabrication of nanofibers via electrospinning and their applications. *SN Appl Sci* 1, 1–16 (2019).
[96] Bonfim, D. P. F., Medeiros, G. B., De Oliveira, E. A., Guerra, V. G. & Aguiar, M. L. Nanomaterials in the environment: definitions, characterizations, effects, and applications. in *Environmental, Ethical, and Economical Issues of Nanotechnology* 1–30, Jenny Stanford Publishing (2022).
[97] Marquez, A. L., Gareis, I. E., Dias, F. J., Gerhard, C. & Lezcano, M. F. Methods to characterize electrospun scaffold morphology: a critical review. *Polymers (Basel)* 14, 1–22 (2022).
[98] Abd Mutalib, M., Rahman, M. A., Othman, M. H. D., Ismail, A. F. & Jaafar, J. Scanning electron microscopy (SEM) and Energy-dispersive X-ray (EDX) spectroscopy. in *Membrane Characterization*, 161–179, Elsevier (Elsevier, 2017).
[99] Purabgola, A. & Kandasubramanian, B. Physical characterization of electrospun fibers. in *Electrospun Materials and Their Allied Applications*, 71–112, John Wiley (2020).
[100] Kwok, D. Y. & Neumann, A. W. Contact angle measurement and contact angle interpretation. *Adv Coll Interf Sci* 81 (1999).
[101] Bashir, M. A. Use of dynamic mechanical analysis (DMA) for characterizing interfacial interactions in filled polymers. *Solids* 2, 108–120 (2021).

[102] Ozawa, T. Thermal analysis—review and prospect. *Thermochim Acta* 355, 35–42 (2000).
[103] Medeiros, G. B., Lima, F. de A., de Almeida, D. S., Guerra, V. G. & Aguiar, M. L. Modification and functionalization of fibers formed by electrospinning: A Review. *Membranes (Basel)* 12, 861 (2022).
[104] Corradini, E. *et al.* Recent advances in food-packing, pharmaceutical and biomedical applications of zein and zein-based materials. *Int J Mol Sci* 15, 22438–22470 (2014).
[105] Omidi, M. *et al.* Characterization of biomaterials. in *Biomaterials for Oral and Dental Tissue Engineering*, Elsevier (2017).
[106] Ehsani, N. *et al.* Electrospun nanofibers fabricated by natural biopolymers for intelligent food packaging. *Crit Rev Food Sci Nutr.*, (2022).
[107] Mane, P. P., Ambekar, R. S. & Kandasubramanian, B. Electrospun nanofiber-based cancer sensors: A review. *Int J Pharm* 583, (2020).
[108] Ding, B., Wang, M., Yu, J. & Sun, G. Gas Sensors Based on Electrospun Nanofibers. *Sensors* 9, 1609–1624 (2009).
[109] Si, Y., Shi, S. & Hu, J. Applications of electrospinning in human health: From detection, protection, regulation to reconstruction. *Nano Today* 48, 101723 (2023).
[110] Migliorini, F. L. *et al.* Urea impedimetric biosensing using electrospun nanofibers modified with zinc oxide nanoparticles. *Appl Surf Sci* 443, 18–23 (2018).
[111] Bonfim, D. P. F., Cruz, F. G. S., Bretas, R. E. S., Guerra, V. G. & Aguiar, M. L. A sustainable recycling alternative: Electrospun pet-membranes for air nanofiltration. *Polymers (Basel)* 13, (2021).
[112] Ren, M. *et al.* Strong and Robust Electrochemical Artificial Muscles by Ionic-Liquid-in-Nanofiber-Sheathed Carbon Nanotube Yarns. *Small* 17, (2021).
[113] Behere, I. & Ingavle, G. In vitro and in vivo advancement of multifunctional electrospun nanofiber scaffolds in wound healing applications: Innovative nanofiber designs, stem cell approaches, and future perspectives. *Journal of Biomedical Materials Research—Part A* 110, 443–461 (2022).
[114] Santocildes-Romero, M. E. *et al.* Fabrication of electrospun mucoadhesive membranes for therapeutic applications in oral medicine. *ACS Appl Mater Interfaces* 9, 11557–11567 (2017).
[115] Iliou, K., Kikionis, S., Ioannou, E. & Roussis, V. Marine biopolymers as bioactive functional ingredients of electrospun nanofibrous scaffolds for biomedical applications. *Mar Drugs* 20, 314 (2022).
[116] Sun, Y. *et al.* Rational design of electrospun nanofibers for gas purification: Principles, opportunities, and challenges. *Chemical Engineering Journal* 446, 137099 (2022).
[117] Schneider, R. *et al.* Tailoring the surface properties of micro/nanofibers using 0D, 1D, 2D, and 3D nanostructures: a review on post-modification methods. *Adv Mater Interfaces* 8, 2100430 (2021).
[118] Kowalczyk, T. Functional micro- and nanofibers obtained by nonwoven post-modification. *Polymers (Basel)* 12, 1087 (2020).
[119] Sagitha, P., Reshmi, C. R., Sundaran, S. P. & Sujith, A. Recent advances in post-modification strategies of polymeric electrospun membranes. *European Polymer Journal* 105, 227–249 (2018).
[120] Yang, J. *et al.* Electrospun Janus nanofibers loaded with a drug and inorganic nanoparticles as an effective antibacterial wound dressing. *Materials Science and Engineering: C* 111, 110805 (2020).
[121] Yu, D. G., Wang, M. & Ge, R. Strategies for sustained drug release from electrospun multi-layer nanostructures. *Wiley Interdiscip Rev Nanomed Nanobiotechnol* 14, (2022).
[122] dos Santos, D. M. *et al.* Core-sheath nanostructured chitosan-based nonwovens as a potential drug delivery system for periodontitis treatment. *Int J Biol Macromol* 142, 521–534 (2020).

[123] Li, T. *et al.* Multiaxial electrospun generation of hollow graphene aerogel spheres for broadband high-performance microwave absorption. *Nano Res* 13, 477–484 (2020).
[124] Mata, G. C. da, Morais, M. S., Oliveira, W. P. de & Aguiar, M. L. Composition Effects on the Morphology of PVA/Chitosan Electrospun Nanofibers. *Polymers (Basel)* 14, (2022).
[125] Chanda, A. *et al.* Electrospun chitosan/polycaprolactone-hyaluronic acid bilayered scaffold for potential wound healing applications. *Int J Biol Macromol* 116, 774–785 (2018).
[126] Salleh, N. A. B. M. *et al.* Studies on properties and adsorption ability of bilayer chitosan/PVA/PVDF electrospun nanofibrous. *Desalination Water Treat* 206, 177–188 (2020).
[127] Xia, Q. *et al.* A Biodegradable trilayered barrier membrane composed of sponge and electrospun layers: hemostasis and antiadhesion. *Biomacromolecules* 16, 3083–3092 (2015).
[128] Qi, L. *et al.* Unidirectional water-transport antibacterial trilayered nanofiber-based wound dressings induced by hydrophilic-hydrophobic gradient and self-pumping effects. *Mater Des* 201, 109461 (2021).
[129] Awang, N., Nasir, A. M., Yajid, M. A. M. & Jaafar, J. A review on advancement and future perspective of 3D hierarchical porous aerogels based on electrospun polymer nanofibers for electrochemical energy storage application. *J Environ Chem Eng* 9, 105437 (2021).
[130] Jiang, D. H. *et al.* Facile 3D boron nitride integrated electrospun nanofibrous membranes for purging organic pollutants. *Nanomaterials* 9, (2019).
[131] Yoon, Y. *et al.* 3D bioprinted complex constructs reinforced by hybrid multilayers of electrospun nanofiber sheets. *Biofabrication* 11, (2019).
[132] Juhasz, A. G., Molnar, K., Idrissi, A. & Jedlovszky-Hajdu, A. Salt induced fluffy structured electrospun fibrous matrix. *J Mol Liq* 312, 113478 (2020).
[133] Chagas, P. A. M. *et al.* Bilayered electrospun membranes composed of poly(lactic-acid)/natural rubber: a strategy against curcumin photodegradation for wound dressing application. *React Funct Polym* 163, 104889 (2021).
[134] Zhao, S. *et al.* Janus-structural AIE nanofiber with white light emission and stimuli-response. *Small* 18, 2201117 (2022).
[135] Mohseni Taromsari, S., Shi, H. H., Saadatnia, Z., Park, C. B. & Naguib, H. E. Design and development of ultra-sensitive, dynamically stable, multi-modal GnP@MXene nanohybrid electrospun strain sensors. *Chem Eng J* 442, 136138 (2022).
[136] Liu, H., Zhang, S., Liu, L., Yu, J. & Ding, B. High-performance PM0.3 air filters using self-polarized electret nanofiber/nets. *Adv Funct Mater* 30, 1909554 (2020).
[137] Cojocaru, E., Ghitman, J. & Stan, R. Electrospun-fibrous-architecture-mediated non-viral gene therapy drug delivery in regenerative medicine. *Polymers* 14. https://doi.org/10.3390/polym14132647 (2022).
[138] Mulholland, E. J., McErlean, E. M., Dunne, N. & McCarthy, H. O. Design of a novel electrospun PVA platform for gene therapy applications using the CHAT peptide. *Int J Pharm* 598, 120366 (2021).
[139] Wu, S., Li, J. S., Mai, J. & Chang, M. W. Three-Dimensional Electrohydrodynamic Printing and Spinning of Flexible Composite Structures for Oral Multidrug Forms. *ACS Appl Mater Interfaces* 10, 24876–24885 (2018).
[140] Beaudoin, É. J., Kubaski, M. M., Samara, M., Zednik, R. J. & Demarquette, N. R. Scaled-up multi-needle electrospinning process using parallel plate auxiliary electrodes. *Nanomaterials* 12, (2022).
[141] Wei, L. *et al.* Large-scale and rapid preparation of nanofibrous meshes and their application for drug-loaded multilayer mucoadhesive patch fabrication for mouth ulcer treatment. *ACS Appl Mater Interfaces* 11, 28740–28751 (2019).
[142] Chen, H. *et al.* Nanowire-in-microtube structured core/shell fibers via multifluidic coaxial electrospinning. *Langmuir* 26, 11291–11296 (2010).
[143] Su, J. *et al.* Hierarchically structured TiO2/PAN nanofibrous membranes for high-efficiency air filtration and toluene degradation. *J Colloid Interface Sci* 507, 386–396 (2017).

[144] Mannucci, P. M., Harari, S., Martinelli, I. & Franchini, M. Effects on health of air pollution: a narrative review. *Internal and Emergency Medicine* vol. 10, 657–662 (2015).

[145] Schraufnagel, D. E. *et al.* Air pollution and noncommunicable diseases: a review by the forum of international respiratory societies' environmental committee, part 2: air pollution and organ systems. *Chest* 55, 417–426 (2019).

[146] Choi, D. Y. *et al.* Washable antimicrobial polyester/aluminum air filter with a high capture efficiency and low pressure drop. *J Hazard Mater* 351, 29–37 (2018).

[147] Phan, T. L. & Ching, C. T. S. A Reusable mask for coronavirus disease 2019 (COVID-19). *Arch Med Res* 51, 455–457 (2020).

[148] Kadam, V. *et al.* Electrospun bilayer nanomembrane with hierarchical placement of bead-on-string and fibers for low resistance respiratory air filtration. *Sep Purif Technol* 224, 247–254 (2019).

[149] Ma, H. *et al.* High-flux thin-film nanofibrous composite ultrafiltration membranes containing cellulose barrier layer. *J Mater Chem* 20, 4692–4704 (2010).

[150] Ditaranto, N., Basoli, F., Trombetta, M., Cioffi, N. & Rainer, A. Electrospun nanomaterials implementing antibacterial inorganic nanophases. *Appl Sci (Switzerland)* 8 (2018).

[151] Shekh, M. I., Patel, K. P. & Patel, R. M. Electrospun ZnO nanoparticles doped core–sheath nanofibers: characterization and antimicrobial properties. *J Polym Environ* 26, 4376–4387 (2018).

[152] Wang, N. *et al.* Tortuously structured polyvinyl chloride/polyurethane fibrous membranes for high-efficiency fine particulate filtration. *J Colloid Interface Sci* 398, 240–246 (2013).

[153] Song, X., Cheng, G., Cheng, B. & Xing, J. Electrospun polyacrylonitrile/magnetic Fe3O4-polyhedral oligomeric silsesquioxanes nanocomposite fibers with enhanced filter performance for electrets filter media. *J Mater Res* 31, 2662–2671 (2016).

[154] Nageh, H. *et al.* Zinc oxide nanoparticle-loaded electrospun polyvinylidene fluoride nanofibers as a potential face protector against respiratory viral infections. *ACS Omega* 7, 14887–14896 (2022).

[155] Zhang, S. *et al.* A controlled design of ripple-like polyamide-6 nanofiber/nets membrane for high-efficiency air filter. *Small* 13, 1603151 (2017).

[156] Bortolassi, A. C. C. *et al.* Composites based on nanoparticle and pan electrospun nanofiber membranes for air filtration and bacterial removal. *Nanomaterials* 9, 1740 (2019).

[157] LEWIS, Alastair C. et al. *Indoor Air Quality*, Department for the Environment, Food and Rural Affairs, United Kingdom, pp. 142 (2022).

[158] Merenda, A. *et al.* Hybrid polymer/ionic liquid electrospun membranes with tunable surface charge for virus capture in aqueous environments. *J Water Proc Eng* 43, (2021).

[159] Drioli, E. & Giorno, L. *Encyclopedia of Membranes. Encyclopedia of Membranes*, Springer (2016).

[160] Dickhout, J. M. *et al.* Produced water treatment by membranes: A review from a colloidal perspective. *J Coll Interface Sci* 487, 523–534 (2017).

[161] Nasreen, S. A. A. N., Sundarrajan, S., Nizar, S. A. S., Balamurugan, R. & Ramakrishna, S. Advancement in electrospun nanofibrous membranes modification and their application in water treatment. *Membranes* 3, 266–284 (2013).

[162] Wang, X. & Hsiao, B. S. Electrospun nanofiber membranes. *Curr Opin Chem Eng* 12, 62–81 (2016).

[163] Kaur, S. *et al.* Review: The characterization of electrospun nanofibrous liquid filtration membranes. *J Mater Sci* 49, 6143–6159 (2014).

[164] Ma, Z., Kotaki, M. & Ramakrishna, S. Surface modified nonwoven polysulphone (PSU) fiber mesh by electrospinning: a novel affinity membrane. *J Memb Sci* 272, 179–187 (2006).
[165] Wang, G. *et al.* Fabrication and characterization of polycrystalline WO3 nanofibers and their application for ammonia sensing. *J Phys Chem B* 110, 23777–23782 (2006).
[166] Yoon, K., Hsiao, B. S. & Chu, B. High flux nanofiltration membranes based on interfacially polymerized polyamide barrier layer on polyacrylonitrile nanofibrous scaffolds. *J Memb Sci* 326, 484–492 (2009).
[167] Pan, Y., Wang, J., Sun, C., Liu, X. & Zhang, H. Fabrication of highly hydrophobic organic-inorganic hybrid magnetic polysulfone microcapsules: A lab-scale feasibility study for removal of oil and organic dyes from environmental aqueous samples. *J Hazard Mater* 309, 65–76 (2016).
[168] Kandisa, R. V. & Saibaba KV, N. Dye Removal by Adsorption: A Review. *J Bioremediat Biodegrad* 7 (2016).
[169] Hou, D., Hu, X., Ho, W., Hu, P. & Huang, Y. Facile fabrication of porous Cr-doped SrTiO3 nanotubes by electrospinning and their enhanced visible-light-driven photocatalytic properties. *J Mater Chem A Mater* 3, 3935–3943 (2015).
[170] Qureshi, U. A., Khatri, Z., Ahmed, F., Khatri, M. & Kim, I. S. Electrospun Zein Nanofiber as a Green and Recyclable Adsorbent for the Removal of Reactive Black 5 from the Aqueous Phase. *ACS Sustain Chem Eng* 5, 4340–4351 (2017).
[171] Qureshi, U. A. *et al.* Highly efficient and robust electrospun nanofibers for selective removal of acid dye. *J Mol Liq* 244, 478–488 (2017).
[172] Mahmoodi, N. M., Mokhtari-Shourijeh, Z. & Ghane-Karade, A. Synthesis of the modified nanofiber as a nanoadsorbent and its dye removal ability from water: isotherm, kinetic and thermodynamic. *Water Sci Technol* 75, 2475–2487 (2017).
[173] Gopakumar, D. A. *et al.* Meldrum's acid modified cellulose nanofiber-based polyvinylidene fluoride microfiltration membrane for dye water treatment and nanoparticle removal. *ACS Sustain Chem Eng* 5, 2026–2033 (2017).
[174] Jaishankar, M., Tseten, T., Anbalagan, N., Mathew, B. B. & Beeregowda, K. N. Toxicity, mechanism and health effects of some heavy metals. *Interdiscipl Toxicol* 7, 60–72 (2014).
[175] Vu, D. *et al.* Adsorption of Cu(II) from aqueous solution by anatase mesoporous TiO 2 nanofibers prepared via electrospinning. *J Colloid Interface Sci* 367, 429–435 (2012).
[176] Zhao, R. *et al.* Branched polyethylenimine grafted electrospun polyacrylonitrile fiber membrane: a novel and effective adsorbent for Cr(VI) remediation in wastewater. *J Mater Chem A Mater* 5, 1133–1144 (2017).
[177] Fechete, I., Wang, Y. & Védrine, J. C. The past, present and future of heterogeneous catalysis. *Catal Today* 189, 2–27 (2012).
[178] Wegener, S. L., Marks, T. J. & Stair, P. C. Design strategies for the molecular level synthesis of supported catalysts. *Acc Chem Res* 45, 206–214 (2012).
[179] Lu, P. *et al.* Photochemical deposition of highly dispersed pt nanoparticles on porous CeO2 nanofibers for the water-gas shift reaction. *Adv Funct Mater* 25, 4153–4162 (2015).
[180] Rezaee, O., Mahmoudi Chenari, H., Ghodsi, F. E. & Ziyadi, H. Preparation of PVA nanofibers containing tungsten oxide nanoparticle by electrospinning and consideration of their structural properties and photocatalytic activity. *J Alloys Compd* 690, 864–872 (2017).
[181] Zhang, P., Guo, Z. P., Huang, Y., Jia, D. & Liu, H. K. Synthesis of Co3O4/Carbon composite nanowires and their electrochemical properties. *J Power Sources* 196, 6987–6991 (2011).
[182] Ghasemi, E., Ziyadi, H., Afshar, A. M. & Sillanpää, M. Iron oxide nanofibers: A new mag-netic catalyst for azo dyes degradation in aqueous solution. *Chem Eng J* 264, 146–151 (2015).
[183] Shao, L. *et al.* Coupling reactions of aromatic halides with palladium catalyst immobilized on poly(vinyl alcohol) nanofiber mats. *Appl Catal A Gen* 413–414, 267–272 (2012).

[184] Tissera, N. D. *et al.* Photocatalytic activity of ZnO nanoparticle encapsulated poly(acrylonitrile) nanofibers. *Mater Chem Phys* 204, 195–206 (2018).
[185] Hou, D., Hu, X., Ho, W., Hu, P. & Huang, Y. Facile fabrication of porous Cr-doped SrTiO 3 nanotubes by electrospinning and their enhanced visible-light-driven photocatalytic properties. *J Mater Chem A Mater* 3, 3935–3943 (2015).
[186] Qin, N. *et al.* Highly efficient photocatalytic H2 evolution over MoS2/CdS-TiO2 nanofibers prepared by an electrospinning mediated photodeposition method. *Appl Catal B* 202, 374–380 (2017).
[187] Ge, J., Lu, D., Liu, Z. & Liu, Z. Recent advances in nanostructured biocatalysts. *Biochem Eng J* 44, 53–59 (2009).
[188] Li, Y. *et al.* Electrospun polyacrylonitrile-glycopolymer nanofibrous membranes for enzyme immobilization. *J Mol Catal B Enzym* 76, 15–22 (2012).
[189] Wang, G. *et al.* Fabrication and characterization of polycrystalline WO3 nanofibers and their application for ammonia sensing. *J Phys Chem B* 110, 23777–23782 (2006).
[190] Lu, P. & Hsieh, Y. lo. Layer-by-layer self-assembly of Cibacron Blue F3GA and lipase on ultra-fine cellulose fibrous membrane. *J Memb Sci* 348, 21–27 (2010).
[191] Weiser, D. *et al.* Bioimprinted lipases in PVA nanofibers as efficient immobilized biocatalysts. *Tetrahedron* 72, 7335–7342 (2016).
[192] Daneshfar, A., Matsuura, T., Emadzadeh, D., Pahlevani, Z. & Fauzi Ismail, A. Title: Urease-carrying electrospun polyacrylonitrile mat for urea hydrolysis. *Reactive and Functional Polymers*, 87, 37–45 (2015).
[193] Gupta, A. *et al.* Geranyl acetate synthesis catalyzed by Thermomyces lanuginosus lipase immobilized on electrospun polyacrylonitrile nanofiber membrane. *Process Biochemistry* 48, 124–132 (2013).
[194] Maryšková, M. *et al.* Polyamide 6/chitosan nanofibers as support for the immobilization of Trametes versicolor laccase for the elimination of endocrine disrupting chemicals. *Enzyme Microb Technol* 89, 31–38 (2016).
[195] Siqueira, N. M. *et al.* Poly (lactic acid)/chitosan fiber mats: investigation of effects of the support on lipase immobilization. *Int J Biol Macromol* 72, 998–1004 (2015).
[196] Mondal, K. & Sharma, A. Recent advances in electrospun metal-oxide nanofiber based interfaces for electrochemical biosensing. *RSC Adv* 6, 94595–94616.
[197] Hinds, W. C. *Aerosol Technology Properties, Behavior, and Measurement of Airborne Particles Second Edition*, John Wiley (1999).
[198] Su, Z., Ding, J. & Wei, G. Electrospinning: A facile technique for fabricating polymeric nanofibers doped with carbon nanotubes and metallic nanoparticles for sensor applications. *RSC Adv* 4, 52598–52610 (2014).
[199] Ding, R. *et al.* High sensitive sensor fabricated by reduced graphene oxide/polyvinyl butyral nanofibers for detecting Cu (II) in Water. *Int J Anal Chem* 2015, (2015).
[200] Zhao, D., Lu, Y., Ding, Y. & Fu, R. An amperometric L-tryptophan sensor platform based on electrospun tricobalt tetroxide nanoparticles decorated carbon nanofibers. *Sens Actuators B Chem* 241, 601–606 (2017).
[201] Shaibani, P. M. *et al.* The detection of Escherichia coli (E. coli) with the pH sensitive hydrogel nanofiber-light addressable potentiometric sensor (NF-LAPS). *Sens Actuators B Chem* 226, 176–183 (2016).
[202] Li, W. T., Zhang, X. D. & Guo, X. Electrospun Ni-doped SnO2 nanofiber array for selective sensing of NO2. *Sens Actuators B Chem* 244, 509–521 (2017).
[203] Manesh, K. M., Santhosh, P., Gopalan, A. & Lee, K. P. Electrospun poly(vinylidene fluoride)/poly(aminophenylboronic acid) composite nanofibrous membrane as a novel glucose sensor. *Anal Biochem* 360, 189–195 (2007).
[204] Wang, X. *et al.* Electrostatic assembly of conjugated polymer thin layers on electrospun nanofibrous membranes for biosensors. *Nano Lett* 4, 331–334 (2004).

[205] Tao, S., Li, G. & Yin, J. Fluorescent nanofibrous membranes for trace detection of TNT vapor. *J Mater Chem* 17, 2730–2736 (2007).
[206] Yoon, J., Chae, S. K. & Kim, J.-M. Colorimetric sensors for volatile organic compounds (VOCs) based on conjugated polymer-embedded electrospun fibers. *J. AM. CHEM. SOC* 129, 3038–3039 (2007).
[207] Li, Z. *et al.* Highly sensitive and stable humidity nanosensors based on LiCl doped TiO2 electrospun nanofibers. *J Am Chem Soc* 130, 5036–5037 (2008).
[208] Qi, Q., Zhang, T., Wang, S. & Zheng, X. Humidity sensing properties of KCl-doped ZnO nanofibers with super-rapid response and recovery. *Sens Actuators B Chem* 137, 649–655 (2009).
[209] Qi, Q., Feng, Y., Zhang, T., Zheng, X. & Lu, G. Influence of crystallographic structure on the humidity sensing properties of KCl-doped TiO2 nanofibers. *Sens Actuators B Chem* 139, 611–617 (2009).
[210] Gouma, P. I. 47 Nanostructured polymorphic oxides for advanced chemosensors nanostructured polymorphic oxides for advanced chemosensors. *Rev Adv Mater Sci* 5 (2003).
[211] Kim, I. D. *et al.* Ultrasensitive chemiresistors based on electrospun TiO 2 nanofibers. *Nano Lett* 6, 2009–2013 (2006).
[212] Im, J. S., Kang, S. C., Lee, S. H. & Lee, Y. S. Improved gas sensing of electrospun carbon fibers based on pore structure, conductivity and surface modification. *Carbon N Y* 48, 2573–2581 (2010).
[213] Wang, X. *et al.* Electrostatic assembly of conjugated polymer thin layers on electrospun nanofibrous membranes for biosensors. *Nano Lett* 4, 331–334 (2004).
[214] Du, Y., Zhang, X., Liu, P., Yu, D. G. & Ge, R. Electrospun nanofiber-based glucose sensors for glucose detection. *Front Chem* 10, 944428 (2022).
[215] Wang, J., Yao, H. bin, He, D., Zhang, C. L. & Yu, S. H. Facile fabrication of gold nanoparticles-poly(vinyl alcohol) electrospun water-stable nanofibrous mats: Efficient substrate materials for biosensors. *ACS Appl Mater Interfaces* 4, 1963–1971 (2012).
[216] Wang, S., Zhao, X., Yin, X., Yu, J. & Ding, B. Electret polyvinylidene fluoride nanofibers hybridized by polytetrafluoroethylene nanoparticles for high-efficiency air filtration. *ACS Appl Mater Interfaces* 8, 23985–23994 (2016).
[217] Li, Y., Yin, X., Si, Y., Yu, J. & Ding, B. All-polymer hybrid electret fibers for high-efficiency and low-resistance filter media. *Chemical Engineering Journal* 398, (2020).
[218] Castro-Muñoz, R., Ahmad, M. Z. & Fíla, V. Tuning of nano-based materials for embedding into low-permeability polyimides for a featured gas separation. *Front Chem* 7. (2020).
[219] Nanis, L., & Kesselman, W. *Engineering Applications of Current and Potential Distributions in Disk Electrode Systems. Journal of the Electrochemical Society*, 118, 454 (1971).
[220] Zhang, Z. *et al.* Daylight-induced antibacterial and antiviral nanofibrous membranes containing vitamin K derivatives for personal protective equipment. *ACS Appl Mater Interfaces* 12, 49416–49430 (2020).
[221] Wang, B. *et al.* Flexible multifunctional porous nanofibrous membranes for high-efficiency air filtration. *ACS Appl Mater Interfaces* 11, 43409–43415 (2019).
[222] Tiliket, G. *et al.* A new material for airborne virus filtration. *Chem Eng J* 173, 341–351 (2011).
[223] Liu, X., Xu, H., Zhang, M. & Yu, D. G. Electrospun medicated nanofibers for wound healing: review. *Membranes* 11, 770 (2021).
[224] Miguel, S. P. *et al.* An overview of electrospun membranes loaded with bioactive molecules for improving the wound healing process. *Eur J Pharm Biopharm* 139, 1–22 (2019).
[225] Abrigo, M., McArthur, S. L. & Kingshott, P. Electrospun nanofibers as dressings for chronic wound care: advances, challenges, and future prospects. *Macromol Biosci* 14, 772–792 (2014).

[226] Nirwan, V. P. *et al*. Advances in electrospun hybrid nanofibers for biomedical applications. *Nanomaterials* 12, 1829 (2022).

[227] Gul, A., Gallus, I., Tegginamath, A., Maryska, J. & Yalcinkaya, F. Electrospun antibacterial nanomaterials for wound dressings applications. *Membranes (Basel)* 11, 908 (2021).

[228] Karuppannan, S. K. *et al*. Quercetin functionalized hybrid electrospun nanofibers for wound dressing application. *Mat Sci Eng B* 285, 115933 (2022).

[229] Liu, X. *et al*. Antimicrobial electrospun nanofibers of cellulose acetate and polyester urethane composite for wound dressing. *J Biomed Mater Res B Appl Biomater* 100, 1556–1565 (2012).

[230] Kenawy, E., Omer, A. M., Tamer, T. M., Elmeligy, M. A. & Eldin, M. S. M. Fabrication of biodegradable gelatin/chitosan/cinnamaldehyde crosslinked membranes for antibacterial wound dressing applications. *Int J Biol Macromol* 139, 440–448 (2019).

[231] Vargas-Molinero, H. Y. *et al*. Hybrid systems of nanofibers and polymeric nanoparticles for biological application and delivery systems. *Micromachines* 14, 208 (2023).

[232] Hussein, M. A. M. *et al*. Chitosan/gold hybrid nanoparticles enriched electrospun PVA nanofibrous mats for the topical delivery of punica granatum L. Extract: synthesis, characterization, biocompatibility and antibacterial properties. *Int J Nanomedicine* 16, 5133–5151 (2021).

[233] Atashgahi, M. *et al*. Epinephrine-entrapped chitosan nanoparticles covered by gelatin nanofibers: A bi-layer nano-biomaterial for rapid hemostasis. *Int J Pharm* 608, 121074 (2021).

9 Biodegradable Polymeric Nanofibers Prepared via Electrospinning

Anand Gobiraman, N. Santhosh, and S. Vishvanathperumal

9.1 INTRODUCTION

In recent years, research on biodegradable polymeric nanofibers has grown significantly. The surface area, volume, manufacture ability, and uses of these nanofibers all have an impact on how they are electrospun [1–5]. According to some sources, nanofibers are fibers with a length-to-thickness ratio of 1,000 or higher and at least one dimension of 100 nanometers or less. Many different applications, including medicinal applications, filtration, smart textiles, and composite reinforcement, benefit from the enormous surface area of organic fibers. Due to their low cost, simplicity in creation, and ability to be processed in a variety of ways, polymeric materials are frequently used to create nanofibers. As the diameter lowers, the benefits of polymer fibers – such as their high surface area/volume ratio, flexible surface functionality, and outstanding mechanical capabilities – increase. The most crucial technique for processing polymer nanofibers is electrospinning, which uses a top-down strategy to produce materials that are inexpensive and easy to work with. Light weight, controlled pore size, a sizable specific surface area, high porosity, flexibility of surface features, high permeability, amazing mechanical capabilities, and high aspect ratios are the ideal qualities of electrospun nanofibers [6,7].

Since the nanoscale is essentially two-dimensional, the dimensions of the electroless-spun nanofibers depend on the cross-sectional area of each fiber. A 100 nm-diameter nanofiber's specific surface area can reach 1,000 m^2/g [8]. Although large surface areas exceeding 2000 m^2/g may be produced using nano-porous materials, such as adsorbent granules and powders, fibers are preferable for many critical applications because they are simpler to process than powders. The versatility of surface functionality, which can be used in a variety of applications, such as biomedical applications, effective filtration, smart textiles, and enhanced fiber-matrix interaction in composite reinforcement applications, is frequently cited as one of the benefits of large surface areas of organic fibers [9]. Due to their numerous appealing qualities and advantages, such as affordability, simplicity of fabrication, a variety of processing options, adaptability of usage, and recyclability, polymeric materials are frequently employed in the production of nanofibers [10–12]. Electrospun polymer nanofibers are excellent candidates for a wide range of applications because of their

DOI: 10.1201/9781003333814-9

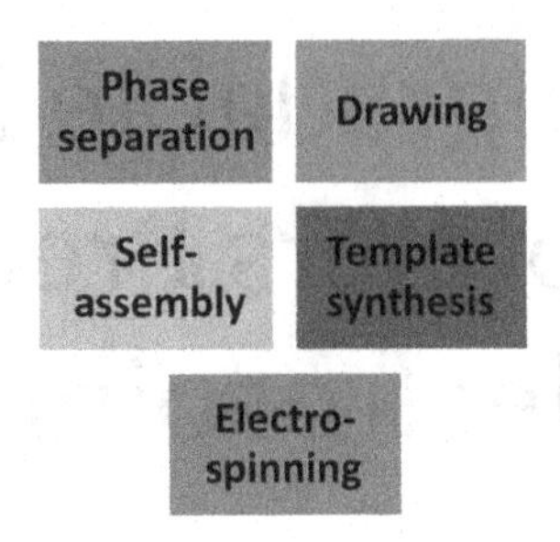

FIGURE 9.1 Schematic of biodegradable polymeric nanofibers.

special qualities [13,14]. Important features of polymer fibers improve as fiber diameter decreases from micrometers to submicrons or nanometers. They have superior mechanical properties compared to other materials, a very high surface area to volume ratio, and flexible surface functionality. Due to their unique qualities, polymeric nanofibers are often used in a variety of applications [15]. Depending on a number of crucial factors that determine whether the criteria for the intended usage are satisfied, nanofibers can be created by selecting the proper blend of polymers and additives and utilizing the appropriate manufacturing techniques [16]. The electrospinning approaches [17–19], based on configurable technical parameters [20], rate of polymerization [21], template synthesis [22], thermal stretching [23], and electrospinning rate [24], are among the different synthesis strategies developed to process nanofibers. Figure 9.1 depicts a schematic of the various production procedures used to create biodegradable polymer nanofibers, including electrospinning methods. The most crucial technique for creating polymer nanofibers is electrospinning [25]. Many nanofibers can be created using electrospinning. Moreover, the electrospinning method can create in-line nanofibers and long continuous nanofibers. The manufacturing capacity for the synthesis of nanofibers is increased by successfully processing a wide variety of polymers to reduce the fiber diameter to the nanoscale range [26–28]. Lightweight design, tunable pore size, extremely high specific surface area, high porosity (typically 90%), flexibility of surface features, high permeability, excellent mechanical properties, high aspect ratios, and lengths up to several centimeters are all desirable characteristics of electrospun nanofibers [29–31]. Using a top-down approach, electrospinning is easier to process and more economical than several bottom-up techniques for creating nanostructures [32]. Because there are fewer material flaws and more molecules are oriented in a certain direction, electrospun nanofibers generally seem to have far higher mechanical qualities than the base material.

The current chapter provides a comprehensive analysis of the electrospinning procedure, the polymers utilized to create biodegradable nanofibers using electrospinning, the specifics of the procedure, and the impact of the factors impacting the procedure. Further, the applications of the electrospun biodegradable polymeric nanofibers are presented with relevant research findings.

9.2 ELECTROSPINNING PROCESS

9.2.1 Overview of Electrospinning Process

The current chapter goes into great detail about the electrospinning process, its operating principle, the variables that affect it, and its scalability. A rapid method for creating extremely tiny fibers from a range of polymer compounds is electrospinning. Over 200 different types of polymers have been subjected to electrospinning by several researchers [33–36]. Several authors have illustrated the experimental setups for producing polymer fibers using electrostatic force [37–39]. A solution of cellulose acetate was placed in an electric field to produce the polymer fibers. One electrode was put on the collector, and the other was submerged in the solution. A metal spinneret with a tiny aperture was used to discharge streams of charged solution. These streams evaporated to create filaments, which were then collected in an electrically grounded collector [40–42]. Figure 9.2 [43] depicts the basic structure of the electrospinning process schematically.

An infusion pump was required for this procedure, which involves suspending a droplet of polymer solution from a stainless-steel capillary tube while controlling the feed rate [44–46]. The polymer solution was suspended from the capillary tube that was wired to a high-voltage DC current, and fibers were collected using a grounded metal screen [47–49]. Electrospinning technique gained greater attention during the 1990s, but notably in recent years because of its ability to create fiber filaments [50–53]. Due to how simple it is to electrospin nanofibers, attention has been paid to many aspects of ultrafine fibers or fibrous structures [54].

Aside from providing an overview of the electrospinning procedure, this chapter also discusses the environmental effects on electrospun polymer nanofibers and the

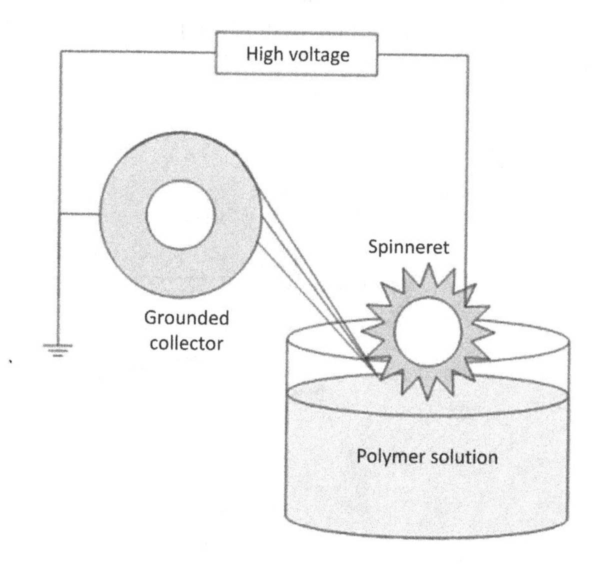

FIGURE 9.2 Basic sketch of the electrospinning process [43].

processing factors linked to electrospinning equipment. The mechanical, electrical, chemical, and optical properties are also examined, along with a number of measurement techniques.

9.2.2 Working Principle of Electrospinning Process

The viscosity of electrospinning polymer solutions must allow them to elongate without separating into droplets and be able to carry charge [55]. The procedure requires a capillary tube with a microscopic needle, a grounded screen, and a high voltage source [56]. The capillary tube is originally filled with a polymer melt or solution during the processing step. However, because solvents are used, the polymer may release offensive odors. Because of this, doing electrospinning in the chamber requires excellent ventilation. When suspended droplets of the polymer solution are present, the liquid surface is charged with a high voltage, generally between 1 and 30 kV. Next, the resulting repercussions of the resulting repercussions are analyzed. A charged jet of solution is also propelled from the apex of the Taylor cone as the electrostatic force surpasses the surface tension of the droplet and the potential nears a critical value. The solvent either hardens or evaporates in less than 10 seconds as the jet moves towards the collector, at which time microscopic fibers are subsequently collected into a web [57–59]. A jet may encounter one of the following three types of instability during this process: bending instability, Rayleigh instability, or whiplash instability [60–62]. Figure 9.3 [43] shows the typical setup for the electrospinning procedure.

The basic block diagram certainly illustrates that when the electrostatic force is larger than the surface tension and the solution stream is evacuated, a Taylor cone appears at the capillary tip. The jet can either form continuous fibers or it can scatter into droplets, depending on the polymer, chain entanglement, and solvent utilized during the process.

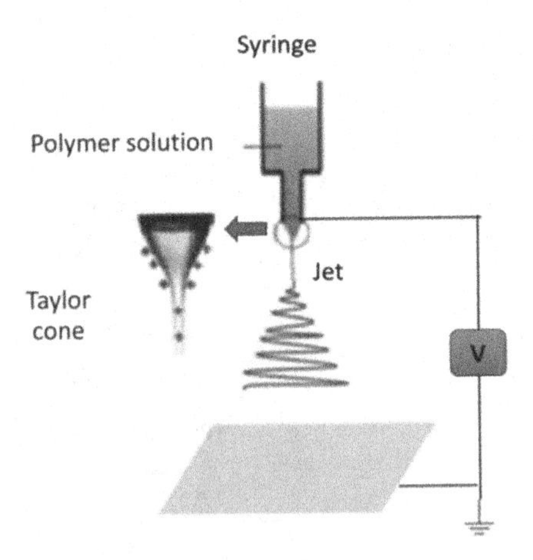

FIGURE 9.3 Block diagram of electrospinning setup [43].

9.3 TYPES OF ELECTROSPINNING TECHNIQUES

Several electrospinning procedures can be categorized according to the equipment's design and the operating principle used. This section goes on to examine a few electrospinning methods.

9.3.1 Melt Electrospinning

Biodegradable polymer nanofibers can be produced using the manufacturing process known as melt electrospinning [63]. The process creates fibrous structures from polymer melts for uses in filtration, textiles, and tissue engineering. The polymer melts or polymer solutions are generally used for electrospinning. When paired with moving collectors, melt electrospinning is a distinct method of 3D printing the polymer nanofibers [64]. Because the fiber collection may be specifically targeted to create the required form and size, melt electrospinning is exceptional. When volatile solvents are not used, there are benefits for some applications where solvent toxicity and buildup during manufacture are a problem [65]. The equipment used in melt electrospinning has an additional provision for melting of polymer. There are numerous advantages for electrospinning and some of them are as follows.

- Melt Electrospinning is an eco-friendly process, without the need for any ventilation system.
- Melt Electrospinning has a high throughput rate and polymeric fiber blends can be easily processed.
- Melt Electrospinning can be effectively used to produce non-soluble polymers as well.

Despite all of these advantages, electrospinning has a substantial disadvantage since it requires a high temperature melting system, has issues with electric discharge, and has poor conductivity with the melt.

9.3.2 Needleless Electrospinning

Needleless electrospinning is a method for creating nanofibers by electrospinning a polymeric solution straight from an uncovered liquid surface (NLES). The NLES process's experimental setup uses a spinneret made of an annular electrode [66]. A magnetic field was added to the experimental setup in order to start an Electroless Spinning (ES) process, which has drawn interest from researchers studying NLES [67]. This modification caused spike formations to appear on the surface of a solution. Furthermore, a spinneret made of a rotating cylinder was added to the apparatus. This spinneret was revolving in a polymeric solution while being partially immersed. Due to this, spikes and a small layer of solution were able to grow on its surface, which caused ES to be activated when an electrical field was applied. Unlike to the spinneret-like spinneret ES, which is started by capillary forces, polymeric jet initiation in NLES is a self-organized process that occurs on a free liquid surface [68]. NLES is therefore a difficult process to manage. Spinnerets have a key impact in the effectiveness and fiber quality of the NLES process, as various researches have shown [69–71]. Both spinning and stationary spinnerets were widely used in NLES. A second categorization was based on the direction that the jets move from the spinneret to the collector. This allows for the definition of three new categories: ES moving uphill, ES moving downhill, and ES moving sideways.

9.3.3 Multi-Jet Electrospinning

Multiple-jets are employed in this electrospinning technique to create the nanofibers. When compared to needle spinning, nanofiber output increased. When there are several jets, a uniform web of nanofiber cannot be formed because of the repletion effect between the jets [72]. Using several jets will enable an upscale electrospinning process. However, the jet deviation causes instability issues thereby facilitating the need for auxiliary plate setup and extra electrodes [73].

9.3.4 Centrifugal Electrospinning

The spinneret is whirled in a polymer solution during centrifugal electrospinning, moistening the spinneret's surface. Higher voltages are then supplied to the spinneret, causing polymer jets to form on its surface and the production of nanofibers, which are then deposited on the collector. The centrifugal electrospinning employs electrical and centrifugal forces that produce discontinuous fibers [74].

9.3.5 Portable Electrospinning

Portable electrospinning process was developed to directly electrospin fibers on the spot. In-situ electrospun nanofibers are more efficient than ones that are kept when used for wound dressing in the medical field [75]. The interest in portable electrospinning is increasing day by day owing to the need for economical and instantaneous production of in-situ electrospun nanofibers.

9.4 MATERIALS USED IN ELECTROSPINNING

The capacity of electrospinning to operate with a range of polymers to produce nanofibers for a number of uses is by far its most prominent feature. Many types of polymers, including synthetic, natural, and mixed polymers, are employed to create nanofibers [76]. Naturally polymer-based electrospun nanofibers usually mimic the physicochemical features of the extracellular matrix. Lipids, proteins, nucleic acids, polysaccharides, and other elements make up natural polymers. Natural polymers outperformed synthetic polymers in terms of biocompatibility and immunogenicity in the medical field. Natural polymers with particular protein arrangements, such as glycine and aspartic acid, are excellent candidates for tissue engineering technology because of their inborn preference for close contact to cells. In descriptions of electrospinning, natural polymers including silk fibroin, collagen, elastin, casein, cellulose acetate, gelatin, chitosan, chitin, fibrinogen, and others are commonly cited. In clinical settings, tissue scaffolds made of natural polymers perform better. However because synthetic polymers break down more quickly and have better mechanical properties like strength and viscoelasticity, they are preferable to natural polymers. As hydrophobic biodegradable polyesters, synthetic polymers including polylactide, polyglycolide, and poly (E-caprolactone) are widely employed in medicinal applications [77]. Moreover, using composite polymers made of natural and artificial polymers in the electrospinning process has advantages over using only one type of polymer [78].

9.5 PROCESS PARAMETERS IN ELECTROSPINNING

The following critical elements influence how the polymer solution is turned into nanofibers during the electrospinning process.

- Viscosity, conductivity, surface tension, the polymer's molecular weight, and its concentration are among the factors that are associated with solutions.
- The needle tip's composition, collector form, collector absorption rate, field strength, flow rate, distance from tip to collector, applied voltage, placement, and design.
- Characteristics connected to the pressure, temperature, and humidity that influence the environment.

Any adjustment to one of these sets of parameters will undoubtedly alter the electrospinning process's progress, which will alter the nanofibers' distinctive properties. It is crucial to consider the set that includes these criteria and assess the appropriate impact on the process. The factors impacting the electrospinning method may be divided into two categories: (i) those connected to the polymer utilized in the procedure, and (ii) those linked to the electrospinning equipment. Several polymer-related factors are as follows.

9.5.1 Parameters Related to Polymer Used in Electrospinning Process

Figure 9.4 shows a schematic of the factors associated with the properties of the polymers employed in the electrospinning process.

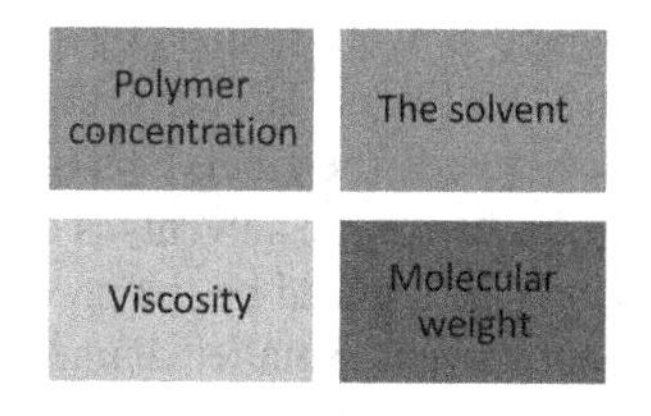

FIGURE 9.4 Schematic of the parameters related to the characteristics of the polymers.

9.5.1.1 Polymer Concentration

The electrospinning approach involves the ideal level of polymer concentration, which is critical for the occurrence of chain entanglement. Surface tension, viscosity, and fiber length all influence polymer concentration selection. The polymer solution acts as a precursor for the development of globules rather than fibers during the electrospinning process [79]. Once the polymer solution becomes overly concentrated, it interferes with the feed and comes to a halt at the capillary tip during the electrospinning process.

According to studies on the electrospinning of polyethylene oxide/aqueous solutions with different percentages of polyethylene oxide, droplets can form when the viscosity is less than 800 centipoise. Nevertheless, at high viscosity values more than 4000 centipoise, the solution becomes excessively thick.

According to a different electrospinning study, when fiber diameter grows, the pore size distribution is reduced by increasing the polystyrene concentration in tetrahydrofuran. Larger fiber diameters are produced by the electrospinning process's ideal polymer concentration range [80].

9.5.1.2 Solvent

Choosing the correct solvent to produce the polymer solution is one of the most important procedures in the electrospinning process. The chosen solvent should have the right evaporation rate, boiling temperature, vapor pressure, and other parameters. The molecular interactions in binary solvents are entirely determined by the species of the solvent. Before solid polymer fibers are produced, phase separation and solvent evaporation occur simultaneously as a polymer jet flows from the tip to the collector. After the solvent evaporates, nanofiber electrospinning begins. The velocity of evaporation of the solvent has a considerable effect on this process [81]. The rate of solvent evaporation is greatly controlled by the solvent's vapor pressure. Many solvents are utilized in the electrospinning of synthetic polymers, including ethanol, hexafluoro-isopropanol, chloroform, dichloromethane, and dimethylformamide. The outcomes demonstrate that the solvent used affects the physical characteristics of polymers. The ecosystem is harmed by and relatively expensive natural solvent. Additionally, the traces of organic matter from the solvents are still present in electrospun materials, posing a risk for biological applications such as tissue engineering and wound healing. Further washing or drying is thus necessary to prepare for application [82]. Electrospinning polymerization is reliant on the solvent composition and surface tension rather than the polymer concentration. For electrospinning, a low solvent surface tension is not necessarily advantageous [83]. The rate of solvent evaporation has an impact on fiber shape as well. Instead of the more common circular fibers, fast evaporation can create flat fibers. They actually form flat fibers when a tiny amount of solvent is caught between the fibers. As the solvent evaporates, the fibers become flatter [84]. The volatility of the solvents is crucial for the selection of the solvent, even if it is significantly influenced by the process circumstances. Fiber diameter is affected by the chosen solvent. Smaller diameter electrospun polyvinyl chloride fibers are produced by dimethylformamide solution or a combination of tetrahydrofuran and dimethylformamide solution. The kind of solvent has an impact on fiber porosity as well. When 100% tetrahydrofuran (more volatile) solutions were electrospun, a significant density of holes was found; whether or not this resulted in an increase in fiber surface area of 20–40% or more depends on the fiber diameter. The electrospinning technique produces smooth fibers with nearly no tiny imperfections because to the solvent concentration (low volatility). The pore density fell between these two limits when pore size and pore depth decreased [85] more.

9.5.1.3 Viscosity

The vast majority of studies demonstrate that viscosity is important for understanding fiber structure, diameter, and molecular weight. The synthesis of polymer fibers increases as the polymer concentration rises. Greater diameters and fibers produce an appearance that is more uniform while decreasing the possibility of beading. There is no chain tangling and no continuous fiber production at very low viscosities. Droplets form when the fiber jet disintegrates. Nevertheless, shooting a jet out of the solution is problematic if the viscosity of the polymer solution is high enough to block the flow of polymer from the needle tip. The concentration of the polymer and the viscosity of the solution have a strong relationship. Polylactic-co-glycolic acid, polyethylene oxide, polyvinyl alcohol, polymethylmethacrylate, and poly-l-lactic acid were used as case studies to evaluate the viscosity, concentration, and shape of electrospun fibers. In the electrospinning process, viscous liquids result in longer stress relaxation durations. Furthermore, larger, more uniform fibers are created by thickening and condensing the solution. When examining the concentration range over which continuous fibers can be formed, viscosity is crucial. An effective way to tell if beaded fibers are being formed is to look for a solution with low viscosity and surface tension. The concentration of the solution had an effect on the morphology, and continuous Nano fiber structures were generated when the concentration was larger than it should have been [86].

9.5.1.4 Molecular Weight

Another major factor that determines the form of polymer nanofibers is their molecular weight. The molecular weight of a material has a considerable impact on rheological characteristics such as viscosity, surface tension, conductivity, and dielectric strength [87].

Because they provide the proper viscosity, large molecular weight polymers are therefore commonly used in the manufacture of fibers. Moreover, Casper et al. discovered that the polymer's increased molecular weight was what produced the change in fiber shape. As a result, fibers with more constant biophysical characteristics can now be produced [88,89]. Instead of fibers, low molecular weight solutions often form globules. In low molecular weight fluids, electrospray rather than electrospinning occurs. Nonetheless, the average fiber diameter rises as the molecular weight does [90]. Despite the low polymer concentration, chain entanglement in the electrospinning method is necessary when utilizing high molecular weight poly-L lactic acid, for example. Evidently, the ideal molecular weight offers enough chain entanglement and viscosity to create homogeneous fibers. Moreover, it reduces surface tension's negative effects, which are crucial for bead formation. Several sources claim that when it comes to producing dependable nanofibers, chain entanglement is more crucial than high molecular weight. The development of nanofibers is said to occur when interactions between polymer molecules are strong enough to qualify for chain entanglement in this article. For instance, a polymer with a significant capacity for quadruple hydrogen bonding will behave during electrospinning similarly to an inert, high-molecular-weight polymer. Similar principles are used to create a non-woven membrane by electrospinning phospholipids from a lecithin solution. Although chain entanglement is important for the polymers used to create fibers, solution viscosity is a far more prevalent feature in ceramic substrates because they may be electrospun while having low molecular weight [91].

9.5.2 Parameters Related to the Electrospinning Equipment

Figure 9.5 shows a schematic of the settings for the electrospinning apparatus.

9.5.2.1 Feed Rate

The flow rate of the polymer within the syringe has an impact on both the speed and intensity of the jet, which both affect the diameter and form of electrospun polymer fibers. Gravity, pumping pressure, and electric field strength are other factors that influence the polymer solution flow during electrospinning. A lower flow rate of the polymer solution may result in finer fibers with ample time for the solvent to evaporate, whereas a faster flow rate results in fibers with a bigger diameter. Because the solvent does not have enough time to evaporate before reaching the collecting plate, higher flow rates might result in the creation of beaded fibers. Both fiber form and porosity are impacted by flow rate. Also, it was discovered that increasing the flow velocity of tetrahydrofuran and polystyrene increased fiber diameter and pore size. High flow rates can partially dry fibers before they reach the collector, which can lead to significant bead flaws. Moreover, a lack of drying might produce fibers that are flat or ribbon-like rather than fibers having a circular cross-section [92].

9.5.2.2 Applied Voltage

One of the key factors affecting the electrospinning of polymer nanofibers is the voltage that is applied. The influence of voltage on the size and structure of electrospun polymer nanofibers has been extensively discussed. While some experts assert that thinner fibers are created as a consequence of electrostatic repulsion forces on the polymer jet, others assert that thicker fibers are produced as a result of greater voltage as there are more polymers leaving the polymer jet. Higher enforced electric forces in the jet, however, typically cause faster solvent evaporation and smaller fiber diameter. This raises the possibility of bead generation at high voltages as well [93–96].

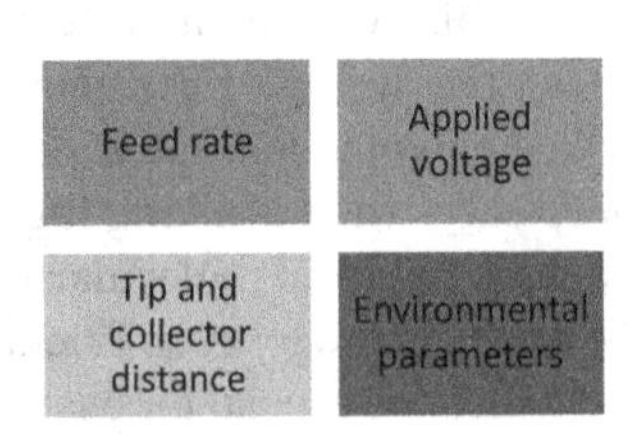

FIGURE 9.5 Schematic of the parameters related to the electrospinning equipment.

9.5.2.3 Tip and Collector Distance

The fibers' shape and diameter are impacted by the separation of the spinneret and collector. In order to avoid producing beads too close to or too far from the collector, fibers must have enough time to dry before being put to it. The shape of the fibers is influenced by the distance seen between spinneret and collector, which also dictates as to if electrospray or electrospinning will occur. The distance from the Taylor cone reduces fibre diameter regardless of the concentration of the polymer solution; shorter distances result in flat or beaded fibres. Many variables, such as the solvent's composition, have an impact on the optimum separation. Water is a low volatility solvent that requires longer distances, but high volatility solvents like acetone demand shorter distances. To produce homogenous, bead-free fibers, there must be a minimum gap of 8 to 15 cm for sufficient evaporation [97].

9.5.3 Environmental Parameters

Environmental variables such as temperature, air flow rate, and humidity, as well as the electrospinning equipment and substrate properties, all have an impact on the quality of electrospun nanofibers. Viscosity drops, resulting in smaller fiber diameters, when temperature increases from 25 to 60 degrees Celsius. The level of humidity can affect the growth of tiny, spherical pores on the surface of the fibers, which eventually connect to one another and affect fiber quality. Due to enhanced evaporation through convection caused by air movement above the injection needle, high humidity can result in electrospun solution discharge and larger fiber diameters, whereas low humidity can result in the solvent evaporating too rapidly. These results show that in order to get the best electrospinning results, environmental factors must be carefully taken into account [98].

9.6 CHARACTERISTICS OF POLYMER NANOFIBERS

9.6.1 Mechanical Characteristics

Several industries, such as the biomedical, textile, aviation, reinforcing components in composites, filter media, structure, and nano-sensors, use electrospun nanofibers. Electrospun polymer nanofibers must meet the necessary mechanical specifications for these applications in order to perform as intended. Examples of their mechanical performance include interactions within the polymer nano fibrous network, friction, and complete deformation of nanofibers under static and dynamic responses. It is required to look at the features of a single fiber in order to determine the mechanical characteristics of electrospun nanofibers. Due to their fragility and thinness, single electrospun nanofibers are commonly grouped on a nonwoven bunch during the electrospinning process. As shown in the schematic block design in Figure 9.6, characterizing the mechanical behavior of a single electrospun nanofiber is difficult and necessitates advanced tools and procedures.

The use of these sophisticated instruments (Figure 9.6) provides more accurate and reliable mechanical measurements of individual electrospun nanofibers. For example, the nanoindentation technique uses a sharp probe to indent the surface

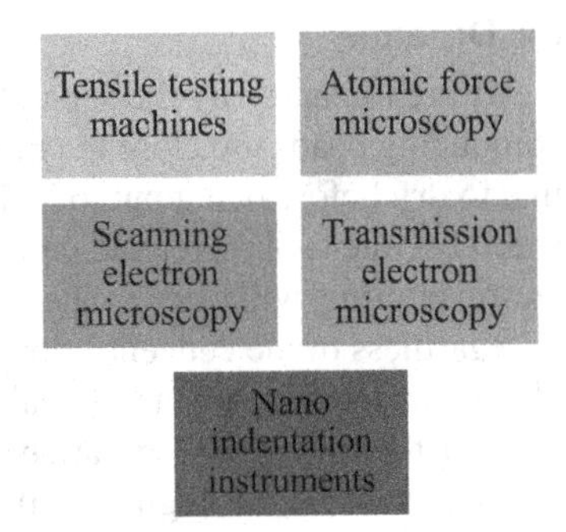

FIGURE 9.6 Schematic of the sophisticated instruments used for mechanical characterization of electrospun nanofibers.

of the nanofiber, measuring the resulting deformation and calculating the Young's modulus. Bending and stretching tests involve applying controlled loads to the nanofiber while it is suspended between two supports, measuring the resulting deflection and strain. By applying a modulated shear force to the nanofiber and detecting the oscillation that results, shear modulation frequency techniques can learn more about the viscoelastic characteristics of the fiber. These methods, which can help in the creation of nanofiber-based materials for a number of purposes, typically make it possible to more accurately understand the mechanical characteristics of specific electrospun nanofibers [99].

9.6.2 Tensile Characterization

The nanofiber is put on a micro-gripper or a specifically made nanofiber clamp, which grips the nanofiber without harming it, during the tensile test. The micro-gripper is then attached to the load cell of a micro/nano-tensile testing machine, which is capable of applying a controlled load to the nanofiber while simultaneously measuring the elongation of the fiber [100]. The force is applied to the nanofiber in a controlled manner until the fiber fractures. Young's modulus, a measurement of the fiber's stiffness, and tensile strength, a measurement of the maximum load that the fiber can withstand before breaking, may both be calculated using the data obtained from the tensile test. It is also possible to determine the strain at fracture, which measures the distortion of the fiber at the moment of failure [101]. The tensile test is a useful technique for determining the mechanical characteristics of electrospun nanofibers, and it is frequently employed in the study and creation of innovative materials for a variety of applications, such as tissue engineering, therapeutic agents, and sensors.

9.6.3 Characterization by Atomic Force Microscopy

This AFM-based method requires that one end of the fiber be connected to a substrate while the opposite side be put on an AFM tip in order to physically characterize polymer nanofibers. The movement of the AFM tip, which exerts a little tensile force on the fiber, records the stress-strain characteristics along the fiber axis using an optical or scanning electron microscope [102]. The mechanical performance of specific nanofibers may be studied using this method. An AFM-based Nano indentation device was created, enabling the fiber to be extended by the stepper motor of the AFM system, for tensile testing of micro- or nanoscaled fiber bundles [103].

9.6.4 Microelectromechanical System Characterization

The load displacement relations for a single polymer nanofiber are investigated using MEMS-based apparatus that includes a leaf-spring load cell and piezoelectric actuators. You may use an optical or scanning electron microscope to see how this device puts the nanofiber in between the gripping elements. The MEMS force sensor is accurate, but its use requires in situ SEM or requires attaching the sensors to the fiber after putting it on a MEMS testing ground [104].

Three nominal strain rates were applied to electrospun PAN nanofibers using MEMS-based apparatus, and their behavior was observed under an optical microscope. They looked at PAN nanofibers with lengths of 12 m and diameters varying between 300 to 600 nm. Optical pictures were used to concurrently monitor the movement and load cell deformation after the nanofibers were applied to the grips to use a micromanipulator and epoxy glue. Optical images were used to calculate the load cell's deflection and the displacement of the fiber grip simultaneously [105].

9.6.5 Characterization by Nano Indentation

Finding the elastic modulus of nanofibers is done using a technique called nano indentation. Using an AFM tip, a light indentation is made on the nanofiber's surface by applying a normal force. The elastic modulus is then determined by probing the surface curve created on the fiber. Yet, there are a lot of variables at play in the procedure, and risks must be considered. Axial stretching failure is anticipated because this approach does not reveal the prevalent deformation mode [106].

9.6.6 Characterization by Bending Test

The mechanical behaviour of a specific electrospun nanofiber is investigated using the AFM method, which employs both two- and three-point flexural tests. The nanofiber is tethered at both ends and is bent in two different directions during two-point bending studies. The force is delivered with the aid of the AFM tip, and the nanofiber's deflection is then measured. With the use of this method, the nanofiber's bending stiffness and Young's modulus may be calculated [107]. In three-point bending studies, the nanofiber is anchored at both ends and is forced at a third point in the middle. By analyzing the deflection of the nanofiber, the Young's modulus and the flexural rigidity may be calculated [108].

With the help of these techniques, it is possible to analyze the mechanical characteristics of single-electrospun nanofibers and learn vital details about their elastic and plastic characteristics. But it's vital to remember that elements like the nanofiber's shape and the interaction in between sample and its tip might impact how accurate the measurements are.

9.6.7 Chemical Characterization

Chemical characterization, which is essential for fully comprehending the physical as well as chemical characteristics of nanofibers and is required for their effective use in a variety of industries, is one of the most significant characterization methodologies. A technique known as small angle X-ray scattering (SAXS) is employed to look at the nanoscale structure of materials. For studying the lamellar structure of semi-crystalline polymers, it is especially helpful. An X-ray beam is used to expose the sample in SAXS, and the dispersed X-rays are therefore gathered and analyzed to reveal more about the sample's makeup. Calculations about the size, shape, and spacing between the material's particles can be made using the scattering pattern that results. In the domains of physics, biology, and materials research, a non-destructive technique known as SAXS is widely used to analyze materials in their natural state. Scientists use a method known as differential scanning calorimetry to examine the thermal behavior of materials (DSC). When a sample is heated or cooled at a controlled rate using DSC, its heat capacity is measured. DSC can be used to detect phase transitions, such as melting, crystallization, and glass transitions, since they affect the material's heat capacity. DSC is a useful technique for analyzing a material's thermal stability and determining its thermal properties, including its melting point, glass transition temperature, and heat of fusion [109].

9.6.8 Thermal Characterization

Among other thermal properties, differential scanning calorimetry is a useful technique for determining the melting and crystallization processes that take place in electrospun nanofibers. For further information, see the website. Among other thermal characteristics, differential scanning calorimetry is a useful technique for determining the melting and crystallization processes that take place in electrospun nanofibers. The heat of fusion of a completely crystalline substance (Hf), the melting enthalpy (Hf), and the enthalpy of crystallization (Hc) obtained from differential scanning calorimetry traces may be used to compute the percent crystallinity (c) of a sample.

9.6.9 Electrical Properties

The nanofibers' electrical characteristics are crucial for describing them. Impedance spectroscopy is another method frequently employed to examine the electrical characteristics of nanofibers in addition to the methods already discussed. This method may be used to calculate the electrical conductivity and dielectric constant of nanofibers [111] and offers information on the dielectric behavior of materials as a function of frequency. Using scanning probe microscopy methods like conductive atomic force microscopy and scanning tunnelling microscopy, it is also possible to study the electrical characteristics of nanofibers [112]. These methods offer highly detailed imaging of the electrical characteristics of specific fibers or fiber networks. In general, electrical characterization methods are crucial for comprehending how nanofibers behave in electronic and energy applications.

9.6.10 Optical Properties

Nagata and colleagues investigated the optical characteristics of poly (2-methoxy-5-(2-ethylhexyloxy)-1,4-phenylenevinylene) electrospun nanofibers using a variety of optical characterization methods, including luminescence spectrofluorometric and ultraviolet visible spectrophotometers. They found that the electrospun nanofibers had a noticeable red shift in comparison to thin film at all concentrations, which was supported by photoluminescence tests at higher polymer concentrations. The optical features of electrospun nanofibers generated from conjugated polymer blends were also studied by Babel et al. They discovered that the binary conjugated polymer blend has a variable composition dependent on its optical properties and might be used in field effect transistors. Similar red shifts in absorbance were seen by Balderas et al. in electrospun fibers produced from a mixture of poly (9-vinylcarbazole) and MEH-PPV [113].

9.7 APPLICATIONS

9.7.1 Biomedical Applications

Many potential uses for polymer nanofibers in the biomedical sector exist, notably in the fields of bioengineering and drug delivery. Nanoscale fibers are useful for imitating biological surroundings and natural extracellular matrices due to their resemblance in size to biological components. High porosity, a large surface area to volume ratio, and the interconnectivity of porous matrices akin to those seen in macromolecules are a few advantages of nanofibrous meshes. These features improve a number of biological processes, including greater cell adhesion, differentiation, and proliferation. Moreover, the pores in these porous structures allow for the interchange of nutrients and waste materials. The following section discusses the two primary study directions in this area [114].

9.7.1.1 Tissue Engineering

Due to their biodegradability, fibrous shape, and capacity to promote cell proliferation, adhesion, and development for a number of tissues, including neuron, bone, blood vessels, collagen, and skin, electrospun polymer nanofibers have become popular as scaffolds in tissue engineering. Researchers are actively attempting to improve these materials' biological compatibility, chemical composition, and mechanical qualities in order to increase their efficiency for tissue regeneration, although there is still room for improvement [115].

9.7.1.2 Wound Healing and Dressing

In combination to natural and synthetic polymers, researchers have looked into employing composite materials for wound dressing. When compared to nanofibers composed of just the polymer, it has been discovered that electrospun nanofibers that combine "silk fibroin" and "poly-lactic co glycolic acid" have enhanced mechanical strength and biocompatibility. These fibers have the ability to support skin cell growth and adhesion. Comparatively to nanofibers manufactured only of chitosan, composite nanofibers constructed of chitosan and polyethylene oxide have shown superior mechanical and antibacterial capabilities. They could therefore be used in applications for skin wound healing. The application of electrospun nanofibrous nanofibers in tissue engineering and wound healing, especially in the development of biomaterials for the regeneration of nerve, bone, blood vessel, cartilage, and skin tissues, has considerable potential despite problems such as the poor mechanical characteristics and biodegradability of natural polymers. Further development is anticipated in the field thanks to improvements in composite materials and scaffold design [116].

9.7.1.3 Drug Delivery Systems

Electrospun nanofibers have been investigated for precise and monitored drug delivery to treat a number of disorders like cancer, cardiovascular issues, and infections in addition to external chemotherapy and biomaterials for wound dressings. For instance, electrospun nanofibers can be functionalized with specific ligands or antibodies that can recognize and bind to specific cells or tissues, leading to enhanced targeting and therapeutic efficiency. In combination to conventional chemotherapy and materials for wound dressings, electrospun nanofibers have been studied for specific and tightly controlled drug delivery to treat a variety of illnesses like cancer, cardiovascular problems, and infections. Overall, electrospun nanofibers hold great promise for advanced drug delivery systems that can improve patient outcomes and reduce side effects [117].

9.7.1.4 Biosensor Applications

Very sensitive and precise detection capabilities are provided by nano-based sensors and biosensors for a variety of applications in industry and medicine. These sensors may be made from electrospun polymer nanofibers because of their unique characteristics, including high porosity, a large surface area, and remarkable mechanical qualities. Fluorescent polymer nanofibers are more sensitive to the detection of ferric, nitro, and mercury ions than film sensors. Due to their excellent electrical characteristics, conductive electrospun polymer nanofibers, such as polyaniline nanowires, make excellent contenders for sensing applications. Ultrathin filters and ultrasensitive sensing applications have been demonstrated to benefit greatly from the use of nano fiber systems with diameters less than 20 nm.

Electrospun polymeric nanofibers are employed as sensing interface agents to sense various gases in gas sensing technologies, where they have also been utilized to detect a variety of gases. As compared to a film-based sensor, an electrode enhanced with an electrospun poly vinyl pyrrolidone/lithium tantalum oxide composite nanofiber demonstrated a quicker reaction time and better sensitivity for hydrogen gas detection [118].

9.7.2 Defense Applications

In the defense industry, polymer nanofibers have shown tremendous promise as intelligent textiles for highly sensitive detection of biological and chemical warfare threats. Its ability to detect pollutants at exceedingly low concentrations, like parts per billion, makes them intriguing candidates for sensing interfaces for warfare chemicals. In one investigation, detoxifying agents that were very effective at capturing chemical warfare toxins were mounted on a PVC nanofiber. The nanofiber was an appropriate filtering medium for this application due to its high porosity and hydrophilicity. Moreover, electrospun nanofibers can have nanoparticles added to them to improve their qualities for application in smart textiles. For instance, these nanoparticles of magnesium oxide, silicon dioxide, titanium dioxide, zinc oxide, and zirconium dioxide can be added to polymers like polyamide, polypropylene, and polyvinyl alcohol to improve their mechanical, thermal, and physical characteristics. Nanometal oxides can also be used as a matrix to improve properties including UV protection, anti-flammability, and antibacterial activity. These developments could significantly increase the usefulness of smart fabrics used in the defense sector [119].

9.7.3 Energy Devices

Particularly in energy storage, harvesting, and conversion, nanofibers have shown to offer considerable potential in applications for renewable energy. Nanofibers are desirable materials for application in energy technologies including fuel cells, solar cells, and lithium-ion batteries because of their high surface area to volume ratio, which enhances conductivity and for efficient charge transfer [120].

Owing to their large porosity and significant specific surface area, which effectively promote effective charge transfer and separation, nanofibers have showed remarkable photoelectric conversion efficiency in solar cells. Furthermore, it has been demonstrated that nanofiber-based electrodes in solar cells exhibit excellent cycle stability and specific capacity [121].

It is also promising to use nanofibers in lithium-ion batteries, which are becoming more and more important for energy storage. Due to their increased mechanical strength, specific surface area, and enhanced electrochemical performance, nanofibers can extend the life and increase the energy density of lithium-ion batteries [122].

The application of electrospun polymer nanofibers as photocatalytic in the water splitting method for generating hydrogen is another illustration of the potential for further research and development in this field. Overall, the incorporation of nanofibers into technologies for renewable energy could offer a long-term response to the world's rising energy needs while lowering dependency on fossil fuels [123].

9.8 CONCLUSIONS AND SCOPE FOR FUTURE WORK

Electrospun polymer nanofibers have developed as a promising material with numerous uses in a variety of disciplines. They are a popular choice for use in creating a variety of technologies due to their special qualities and simplicity of production. This review included information on EPNF's mechanical and other qualities, key electrospinning process parameters, and production methods. The chapter also examined the growing uses of EPNF in numerous disciplines, including biomedical, sensing, and biosensing applications, as well as applications in defense and energy. It also covered mathematical models. To increase the effectiveness and precision of the electrospinning processes and to broaden and enhance the use of EPNF in a variety of applications, however, there is still considerable work to be done in the experimental, computational, and theoretical domains. As this field of study advances, it is projected that EPNF will have a bigger impact on a range of businesses. Since a range of polymers with natural and synthetic origins can be used to make nanofibers, electrospinning is regarded as a well-researched process for doing so. These fibers have qualities that make them desirable for application in biomedical materials, including solubility, biocompatibility, and degradability. Due to their distinctive characteristics, electrospun fibers have found numerous uses in tissue engineering, targeted therapy, wound dressing, and medication administration. They are simple to combine with nanomaterials and biopharmaceuticals that are responsive to magnetic, chemical, and photo-catalytic modalities. This chapter has given a summary of current developments and forecast what electrospun nanofibers will be like in the future as intelligent systems for biomedical applications.

REFERENCES

[1] Frenot, A.; Chronakis, I.S. Polymer nanofibers assembled by electrospinning. Curr. Opin. Colloid Interf. Sci. 2003, 8, 64–75.

[2] Yang, B.; Wang, L.; Zhang, M.; Luo, J.; Lu, Z.; Ding, X. Fabrication, applications, and prospects of aramid nanofiber. Adv. Funct. Mater. 2020, 30, 2000186.

[3] Zahmatkeshan, M.; Adel, M.; Bahrami, S.; Esmaeili, F.; Rezayat, S.M.; Saeedi, Y.; Mehravi, B.; Jameie, S.B.; Ashtari, K. Polymer based nanofibers: Preparation, fabrication, and applications. In Handbook of Nanofibers; Springer: Berlin/Heidelberg, Germany, 2019; pp. 215–261.

[4] Ramesh Kumar, P.; Khan, N.; Vivekanandhan, S.; Satyanarayana, N.; Mohanty, A.; Misra, M. Nanofibers: Effective generation by electrospinning and their applications. J. Nanosci. Nanotechnol. 2012, 12, 25.

[5] Kai, D.; Liow, S.S.; Loh, X.J. Biodegradable polymers for electrospinning: Towards biomedical applications. Mater. Sci. Eng. C 2014, 45, 659–670.

[6] Teo, W.E.; Ramakrishna, S. A review on electrospinning design and nanofibre assemblies. Nanotechnology 2006, 17, R89.

[7] Bhardwaj, N.; Kundu, S.C. Electrospinning: A fascinating fiber fabrication technique. Biotechnol. Adv. 2010, 28, 325–347.

[8] Huang, Z.-M.; Zhang, Y.-Z.; Kotaki, M.; Ramakrishna, S. A review on polymer nanofibers by electrospinning and their applications in nanocomposites. Compos. Sci. Technol. 2003, 63, 2223–2253.

[9] Meghana, B.; Umesh, D.; Abhay, S.; Vilasrao, K. Electrospinning nanotechnology-A robust method for preparation of nanofibers for medicinal and pharmaceutical application. Asian J. Pharm. Res. Dev. 2020, 8, 176–184.

[10] Liu, C.; Tan, Y.; Liu, Y.; Shen, K.; Peng, B.; Niu, X.; Ran, F. Microporous carbon nanofibers prepared by combining electrospinning and phase separation methods for supercapacitor. J. Energy Chem. 2016, 25, 587–593.

[11] DeFrates, K.; Markiewicz, T.; Xue, Y.; Callaway, K.; Gough, C.; Moore, R.; Bessette, K.; Mou, X.; Hu, X. Air-jet spinning corn zein protein nanofibers for drug delivery: Effect of biomaterial structure and shape on release properties. Mater. Sci. Eng. C 2021, 118, 111419.

[12] Nain, A.S.; Phillippi, J.A.; Sitti, M.; MacKrell, J.; Campbell, P.G.; Amon, C. Control of cell behavior by aligned micro/nanofibrous biomaterial scaffolds fabricated by spinneret-based tunable engineered parameters (STEP) technique. Small 2008, 4, 1153–1159.

[13] Yin, Z.; Wu, F.; Zheng, Z.; Kaplan, D.L.; Kundu, S.C.; Lu, S. Self-assembling silk-based nanofibers with hierarchical structures. ACS Biomater. Sci. Eng. 2017, 3, 2617–2627.
[14] Cheng, K.C.; Bedolla-Pantoja, M.A.; Kim, Y.-K.; Gregory, J.V.; Xie, F.; De France, A.; Hussal, C.; Sun, K.; Abbott, N.L.; Lahann, J. Templated nanofiber synthesis via chemical vapor polymerization into liquid crystalline films. Science 2018, 362, 804–808.
[15] Wu, S.; Zhang, F.; Yu, Y.; Li, P.; Yang, X.; Lu, J.; Ryu, S. Preparation of PAN-based carbon nanofibers by hot-stretching. Compos. Interf. 2008, 15, 671–677.
[16] Bera, B. Literature review on electrospinning process (a fascinating fiber fabrication technique). Imper. J. Interdiscip. Res. 2016, 2, 972–984.
[17] Wang, G.; Yu, D.; Kelkar, A.D.; Zhang, L. Electrospun nanofiber: Emerging reinforcing filler in polymer matrix composite materials. Progr. Polymer Sci. 2017, 75, 73–107.
[18] Agarwal, S.; Greiner, A.; Wendorff, J.H. Functional materials by electrospinning of polymers. Progr. Polymer Sci. 2013, 38, 963–991.
[19] Ingavle, G.C.; Leach, J.K. Advancements in electrospinning of polymeric nanofibrous scaffolds for tissue engineering. Tiss. Eng. Part B Rev. 2014, 20, 277–293.
[20] Baji, A.; Mai, Y.-W.; Wong, S.-C.; Abtahi, M.; Chen, P. Electrospinning of polymer nanofibers: Effects on oriented morphology, structures and tensile properties. Compos. Sci. Technol. 2010, 70, 703–718.
[21] Chronakis, I.S. Novel nanocomposites and nanoceramics based on polymer nanofibers using electrospinning process—A review. J. Mater. Proc. Technol. 2005, 167, 283–293.
[22] Bhattacharyya, D.; Fakirov, S. Synthetic Polymer-Polymer Composites; Carl Hanser Verlag GmbH Co KG: Munich, Germany, 2012.
[23] Hwang, K.Y.; Kim, S.-D.; Kim, Y.-W.; Yu, W.-R. Mechanical characterization of nanofibers using a nanomanipulator and atomic force microscope cantilever in a scanning electron microscope. Polym. Test. 2010, 29, 375–380.
[24] Yarin, A.L.; Koombhongse, S.; Reneker, D.H. Bending instability in electrospinning of nanofibers. J. Appl. Phys. 2001, 89, 3018–3026.
[25] Reneker, D.H.; Yarin, A.L.; Fong, H.; Koombhongse, S. Bending instability of electrically charged liquid jets of polymer solutions in electrospinning. J. Appl. Phys. 2000, 87, 4531–4547.
[26] Ji, Y.; Li, B.; Ge, S.; Sokolov, J.C.; Rafailovich, M.H. Structure and nanomechanical characterization of electrospun PS/clay nanocomposite fibers. Langmuir 2006, 22, 1321–1328.
[27] Teo, W.-E.; Ramakrishna, S. Electrospun nanofibers as a platform for multifunctional, hierarchically organized nanocomposite. Compos. Sci. Technol. 2009, 69, 1804–1817.
[28] Zhang, Y.; Venugopal, J.; Huang, Z.-M.; Lim, C.T.; Ramakrishna, S. Crosslinking of the electrospun gelatin nanofibers. Polymer 2006, 47, 2911–2917.
[29] Kulkarni, A.; Bambole, V.; Mahanwar, P. Electrospinning of polymers, their modeling and applications. Polym. Plastics Technol. Eng. 2010, 49, 427–441.
[30] Hu, X.; Liu, S.; Zhou, G.; Huang, Y.; Xie, Z.; Jing, X. Electrospinning of polymeric nanofibers for drug delivery applications. J. Control. Release 2014, 185, 12–21.
[31] Dzenis, Y. Spinning continuous fibers for nanotechnology. Science 2004, 304, 1917–1919.
[32] Taylor, S.R. Abundance of chemical elements in the continental crust: A new table. Geochim. Cosmochim. Acta 1964, 28, 1273–1285.
[33] Reneker, D.; Yarin, A.; Zussman, E.; Xu, H. Electrospinning of nanofibers from polymer solutions and melts. Adv. Appl. Mechan. 2007, 41, 43–346.
[34] Matarrese, S.; Pantano, O.; Saez, D. General relativistic dynamics of irrotational dust: Cosmological implications. Phys. Rev. Lett. 1994, 72, 320.
[35] Brown, T.D.; Dalton, P.D.; Hutmacher, D.W. Melt electrospinning today: An opportune time for an emerging polymer process. Progr. Polymer Sci. 2016, 56, 116–166.

[36] Yu, M.; Dong, R.H.; Yan, X.; Yu, G.F.; You, M.H.; Ning, X.; Long, Y.Z. Recent advances in needleless electrospinning of ultrathin fibers: From academia to industrial production. Macromol. Mater. Eng. 2017, 302, 1700002.
[37] El-Sayed, H.; Vineis, C.; Varesano, A.; Mowafi, S.; Carletto, R.A.; Tonetti, C.; Abou Taleb, M. A critique on multi-jet electrospinning: State of the art and future outlook. Nanotechnol. Rev. 2019, 8, 236–245.
[38] Kim, Y.; Ahn, K.; Sung, Y. A Manufacturing Device and the Method of Preparing for the Nanofibers via Electro-Blown Spinning Process. KR Patent Registration No. 1005434890000, 23 January 2006.
[39] Wang, L.; Ahmad, Z.; Huang, J.; Li, J.-S.; Chang, M.-W. Multi-compartment centrifugal electrospinning based composite fibers. Chem. Eng. J. 2017, 330, 541–549.
[40] He, X.-X.; Zheng, J.; Yu, G.-F.; You, M.-H.; Yu, M.; Ning, X.; Long, Y.-Z. Near-field electrospinning: Progress and applications. J. Phys. Chem. C 2017, 121, 8663–8678.
[41] Han, D.; Steckl, A.J. Coaxial electrospinning formation of complex polymer fibers and their applications. ChemPlusChem, 2019, 84, 1453–1497.
[42] Buzgo, M.; Mickova, A.; Rampichova, M.; Doupnik, M. Blend electrospinning, coaxial electrospinning, and emulsion electrospinning techniques. In Core-Shell Nanostructures for Drug Delivery and Theranostics; Woodhead Publishing Series in Biomaterials; Elsevier: London, UK, 2018; pp. 325–347.
[43] Al-Abduljabbar, A.; Farooq, I. Electrospun polymer nanofibers: Processing, properties, and applications. Polymers, 2023, 15, 65. https://doi.org/10.3390/polym15010065
[44] Zheng, Y.; Gong, R.H.; Zeng, Y. Multijet motion and deviation in electrospinning. RSC Adv. 2015, 5, 48533–48540.
[45] Wang, X.; Wang, X.; Lin, T. Electric field analysis of spinneret design for needleless electrospinning of nanofibers. J. Mater. Res. 2012, 27, 3013–3019.
[46] Niu, H.; Lin, T.; Wang, X. Needleless electrospinning. I. A comparison of cylinder and disk nozzles. J. Appl. Polym. Sci. 2009, 114, 3524–3530.
[47] Niu, H.; Zhou, H.; Yan, G.; Wang, H.; Fu, S.; Zhao, X.; Shao, H.; Lin, T. Enhancement of coil electrospinning using two-level coil structure. Indust. Eng. Chem. Res. 2018, 57, 15473–15478.
[48] Thoppey, N.M.; Bochinski, J.R.; Clarke, L.I.; Gorga, R.E. Edge electrospinning for high throughput production of quality nanofibers. Nanotechnology 2011, 22, 345301.
[49] Jiang, G.; Zhang, S.; Qin, X. High throughput of quality nanofibers via one stepped pyramid-shaped spinneret. Mater. Lett. 2013, 106, 56–58.
[50] Wei, L.; Sun, R.; Liu, C.; Xiong, J.; Qin, X. Mass production of nanofibers from needleless electrospinning by a novel annular spinneret. Mater. Design 2019, 179, 107885.
[51] Xiong, J.; Liu, Y.; Li, A.; Wei, L.; Wang, L.; Qin, X.; Yu, J. Mass production of high-quality nanofibers via constructing pre-Taylor cones with high curvature on needleless electrospinning. Mater. Design 2021, 197, 109247.
[52] Yan, X.; Yu, M.; Ramakrishna, S.; Russell, S.J.; Long, Y.-Z. Advances in portable electrospinning devices for in situ delivery of personalized wound care. Nanoscale 2019, 11, 19166–19178.
[53] Hekmati, A.H.; Rashidi, A.; Ghazisaeidi, R.; Drean, J.-Y. Effect of needle length, electrospinning distance, and solution concentration on morphological properties of polyamide-6 electrospun nanowebs. Textile Res. J. 2013, 83, 1452–1466.
[54] De, B.; Banerjee, S.; Verma, K.D.; Pal, T.; Manna, P.; Kar, K.K. Carbon nanofiber as electrode materials for supercapacitors. In Handbook of Nanocomposite Supercapacitor Materials II; Series in Materials Science, Springer Nature Switzerland Springer; World Scientific Publishing Co. Pte. Ltd.: Singapore, 2020; pp. 149–181, 179–200.
[55] Aleisa, R. Electrospinning. In Handbook of Synthetic Methodologies and Protocols of Nanomaterials: Volume 3: Unconventional Methods for Nanostructure Fabrication; World Scientific Publishing Co. Pte. Ltd.: Singapore, 2020; pp. 149–181.

[56] Gu, S.; Ren, J.; Vancso, G. Process optimization and empirical modeling for electrospun polyacrylonitrile (PAN) nanofiber precursor of carbon nanofibers. Eur. Polymer J. 2005, 41, 2559–2568.
[57] Gu, X.; Li, N.; Luo, J.; Xia, X.; Gu, H.; Xiong, J. Electrospun polyurethane microporous membranes for waterproof and breathable application: The effects of solvent properties on membrane performance. Polymer Bull. 2018, 75, 3539–3553.
[58] Liu, F.; Li, M.; Shao, W.; Yue, W.; Hu, B.; Weng, K.; Chen, Y.; Liao, X.; He, J. Preparation of a polyurethane electret nanofiber membrane and its air-filtration performance. J. Colloid Interf. Sci. 2019, 557, 318–327.
[59] Ding, B.; Kim, H.Y.; Lee, S.C.; Shao, C.L.; Lee, D.R.; Park, S.J.; Kwag, G.B.; Choi, K.J. Preparation and characterization of a nanoscale poly (vinyl alcohol) fiber aggregate produced by an electrospinning method. J. Polymer Sci. Part Polymer Phys. 2002, 40, 1261–1268.
[60] Maslakci, N.N.; Ulusoy, S.; Uygun, E.; Çevikba¸s, H.; Oksuz, L.; Can, H.K.; Uygun Oksuz, A. Ibuprofen and acetylsalicylic acid loaded electrospun PVP-dextran nanofiber mats for biomedical applications. Polymer Bull. 2017, 74, 3283–3299.
[61] Karakas, H. Electrospinning of nanofibers and their applications. J. Algebr. Statistic 2015, 13, 1447–1454.69. Maurmann, N.; Sperling, L.-E.; Pranke, P. Electrospun and electrosprayed scaffolds for tissue engineering. Cut. Edge Enabling Technol. Regen. Med. 2018, 10, 79–100.
[62] Shin, M.K.; Kim, Y.J.; Kim, S.I.; Kim, S.-K.; Lee, H.; Spinks, G.M.; Kim, S.J. Enhanced conductivity of aligned PANi/PEO/MWNT nanofibers by electrospinning. Sens. Actuat. Chem. 2008, 134, 122–126.
[63] Aussawasathien, D.; Sahasithiwat, S.; Menbangpung, L. Electrospun camphorsulfonic acid doped poly (o-toluidine)–polystyrene composite fibers: Chemical vapor sensing. Synt. Met. 2008, 158, 259–263.
[64] Janani, G.; Kumar, M.; Chouhan, D.; Moses, J.C.; Gangrade, A.; Bhattacharjee, S.; Mandal, B.B. Insight into silk-based biomaterials: From physicochemical attributes to recent biomedical applications. ACS Appl. Bio Mater. 2019, 2, 5460–5491.
[65] Bognitzki, M.; Czado, W.; Frese, T.; Schaper, A.; Hellwig, M.; Steinhart, M.; Greiner, A.; Wendorff, J.H. Nanostructured fibers via electrospinning. Adv. Mater. 2001, 13, 70–72.
[66] Guibo, Y.; Qing, Z.; Yahong, Z.; Yin, Y.; Yumin, Y. The electrospun polyamide 6 nanofiber membranes used as high efficiency filter materials: Filtration potential, thermal treatment, and their continuous production. J. Appl. Polym. Sci. 2013, 128, 1061–1069.
[67] Jin, H.-J.; Fridrikh, S.V.; Rutledge, G.C.; Kaplan, D.L. Electrospinning Bombyx mori silk with poly (ethylene oxide). Biomacro-molecules 2002, 3, 1233–1239.
[68] Yao, C.; Li, X.; Neoh, K.; Shi, Z.; Kang, E. Surface modification and antibacterial activity of electrospun polyurethane fibrous membranes with quaternary ammonium moieties. J. Membran. Sci. 2008, 320, 259–267.
[69] Wei, N.; Sun, C.; Wang, J.; Huang, L.Q. Research on electrospinning of cellulose acetate prepared by acetone/DMAc solvent. In Proceedings of Applied Mechanics and Materials; Trans Tech Publications: Bäch, Switzerland, 2014; Volume 469, pp. 126–129.
[70] Behtaj, S.; Karamali, F.; Masaeli, E.; Anissimov, Y.G.; Rybachuk, M. Electrospun PGS/PCL, PLLA/PCL, PLGA/PCL and pure PCL scaffolds for retinal progenitor cell cultivation. BioChem. Eng. J. 2021, 166, 107846.
[71] Dong, B.; Arnoult, O.; Smith, M.E.; Wnek, G.E. Electrospinning of collagen nanofiber scaffolds from benign solvents. Macromol. Rapid Commun. 2009, 30, 539–542.
[72] Koombhongse, S.; Liu, W.; Reneker, D.H. Flat polymer ribbons and other shapes by electrospinning. J. Polymer Sci. Part Polymer Phys. 2001, 39, 2598–2606.
[73] Jayaraman, K.; Kotaki, M.; Zhang, Y.; Mo, X.; Ramakrishna, S. Recent advances in polymer nanofibers. J. Nanosci. Nanotechnol. 2004, 4, 52–65.

[74] Matysiak, W.; Ta ′nski, T.; Smok, W.; Gołombek, K.; Schab-Balcerzak, E. Effect of conductive polymers on the optical properties of electrospun polyacrylonitryle nanofibers filled by polypyrrole, polythiophene and polyaniline. Appl. Surf. Sci. 2020, 509, 145068.

[75] Joseph, B.; Augustine, R.; Kalarikkal, N.; Thomas, S.; Seantier, B.; Grohens, Y. Recent advances in electrospun polycaprolactone based scaffolds for wound healing and skin bioengineering applications. Mater. Today Commun. 2019, 19, 319–335.

[76] Ashammakhi, N.; Wimpenny, I.; Nikkola, L.; Yang, Y. Electrospinning: Methods and development of biodegradable nanofibres for drug release. J. Biomed. Nanotechnol. 2009, 5, 19.

[77] Sill, T.J.; Von Recum, H.A. Electrospinning: Applications in drug delivery and tissue engineering. Biomaterials 2008, 29, 1989–2006.

[78] Stojanovska, E.; Canbay, E.; Pampal, E.S.; Calisir, M.D.; Agma, O.; Polat, Y.; Simsek, R.; Gundogdu, N.S.; Akgul, Y.; Kilic, A. A review on non-electro nanofibre spinning techniques. RSC Adv. 2016, 6, 83783–83801.

[79] Doshi, J.; Reneker, D.H. Electrospinning process and applications of electrospun fibers. J. Electrost. 1995, 35, 151–160.

[80] Bae, H.-S.; Haider, A.; Selim, K.; Kang, D.-Y.; Kim, E.-J.; Kang, I.-K. Fabrication of highly porous PMMA electrospun fibers and their application in the removal of phenol and iodine. J. Polym. Res. 2013, 20, 1–7.

[81] Qian, Y.-F.; Su, Y.; Li, X.-Q.; Wang, H.; He, C.L. Electrospinning of polymethyl methacrylate nanofibres in different solvents. Iran. Polym. J. 2010, 19, 79–87.

[82] Haider, A.H.; Kang, S. A comprehensive review summarizing the effect of electrospinning parameters and potential applications of nanofibers in biomedical and biotechnology. Arab. J. Chem. 2015, 10, 24.

[83] Agarwal, S.; Wendorff, J.H.; Greiner, A. Use of electrospinning technique for biomedical applications. Polymer 2008, 49, 5603–5621.

[84] Majumder, S.; Matin, M.A.; Sharif, A.; Arafat, M.T. Understanding solubility, spinnability and electrospinning behaviour of cellulose acetate using different solvent systems. Bull. Mater. Sci. 2019, 42, 9.

[85] Abbasi, N.; Soudi, S.; Hayati-Roodbari, N.; Dodel, M.; Soleimani, M. The effects of plasma treated electrospun nanofibrous poly (ε-caprolactone) scaffolds with different orientations on mouse embryonic stem cell proliferation. Cell J. 2014, 16, 245.

[86] Hodge, J.; Quint, C. The improvement of cell infiltration in an electrospun scaffold with multiple synthetic biodegradable polymers using sacrificial PEO microparticles. J. Biomed. Mater. Res. Part 2019, 107, 1954–1964.

[87] Santoro, M.; Shah, S.R.; Walker, J.L.; Mikos, A.G. Poly (lactic acid) nanofibrous scaffolds for tissue engineering. Adv. Drug Deliv. Rev. 2016, 107, 206–212.

[88] Zafar, M.; Najeeb, S.; Khurshid, Z.; Vazirzadeh, M.; Zohaib, S.; Najeeb, B.; Sefat, F. Potential of electrospun nanofibers for biomedical and dental applications. Materials 2016, 9, 73.

[89] Nagam Hanumantharao, S.; Rao, S. Multi-functional electrospun nanofibers from polymer blends for scaffold tissue engineering. Fibers 2019, 7, 66.

[90] Tan, E.; Lim, C. Mechanical characterization of nanofibers—A review. Compos. Sci. Technol. 2006, 66, 1102–1111.

[91] Bazbouz, M.B.; Stylios, G.K. The tensile properties of electrospun nylon 6 single nanofibers part B polymer physics. J. Polymer Science Part Polym. Phys. 2010, 48, 1719–1731.

[92] Zhou, X.; Ding, C.; Cheng, C.; Liu, S.; Duan, G.; Xu, W.; Liu, K.; Hou, H. Mechanical and thermal properties of electrospun polyimide/rGO composite nanofibers via in-situ polymerization and in-situ thermal conversion. Eur. Polymer J. 2020, 141, 110083.

[93] Bauchau, O.A.; Craig, J.I. Structural Analysis: With Applications to Aerospace Structures; Springer Science & Business Media: Dordrecht, The Netherlands; Heidelberg, Germany; London, UK; New York, NY, USA, 2009; Volume 163.

[94] Tan, E.; Ng, S.; Lim, C. Tensile testing of a single ultrafine polymeric fiber. Biomaterials 2005, 26, 1453–1456.
[95] Hua, L.; Hu, C.; Li, J.; Huang, B.; Du, J. Mechanical characterization of TiO2 nanowires flexible scaffold by nanoindentation/scratch. J. Mechan. Behav. Biomed. Mater. 2022, 126, 105069.
[96] Janković, B.; Pelipenko, J.; Škarabot, M.; Muševič, I.; Kristl, J. The design trend in tissue-engineering scaffolds based on nanomechanical properties of individual electrospun nanofibers. Int. J. Pharm. 2013, 455, 338–347.
[97] Ding, Y.; Zhang, P.; Jiang, Y.; Xu, F.; Yin, J.; Zuo, Y. Mechanical properties of nylon-6/SiO2 nanofibers prepared by electrospinning. Mater. Lett. 2009, 63, 34–36.
[98] Liao, C.-C.; Wang, C.-C.; Chen, C.-Y.; Lai, W.-J. Stretching-induced orientation of polyacrylonitrile nanofibers by an electrically rotating viscoelastic jet for improving the mechanical properties. Polymer 2011, 52, 2263–2275.
[99] Wu, N.; Chen, L.; Wei, Q.; Liu, Q.; Li, J. Nanoscale three-point bending of single polymer/inorganic composite nanofiber. J. Textile Inst. 2012, 103, 154–158.
[100] Guhados, G.; Wan, W.; Hutter, J.L. Measurement of the elastic modulus of single bacterial cellulose fibers using atomic force microscopy. Langmuir 2005, 21, 6642–6646.
[101] Croisier, F.; Duwez, A.-S.; Jérôme, C.; Léonard, A.; Van Der Werf, K.; Dijkstra, P.J.; Bennink, M.L. Mechanical testing of electrospun PCL fibers. Acta Biomater. 2012, 8, 218–224.
[102] Carlisle, C.R.; Coulais, C.; Namboothiry, M.; Carroll, D.L.; Hantgan, R.R.; Guthold, M. The mechanical properties of individual, electrospun fibrinogen fibers. Biomaterials 2009, 30, 1205–1213.
[103] Baker, S.; Sigley, J.; Helms, C.C.; Stitzel, J.; Berry, J.; Bonin, K.; Guthold, M. The mechanical properties of dry, electrospun fibrinogen fibers. Mater. Sci. Eng. C 2012, 32, 215–221.
[104] Carlisle, C.R.; Coulais, C.; Guthold, M. The mechanical stress–strain properties of single electrospun collagen type I nanofibers. Acta Biomater. 2010, 6, 2997–3003.
[105] Sadrjahani, M.; Hoseini, S.; Mottaghitalab, V.; Haghi, A. Development and characterization of highly oriented pan nanofiber. Braz. J. Chem. Eng. 2010, 27, 583–589.
[106] Ratner, B.; Chilkoti, A.; Castner, D. Contemporary methods for characterizing complex biomaterial surfaces. In Biologically Modified Polymeric Biomaterial Surfaces; Springer: Dordrecht, The Netherlands, 1992; pp. 25–36.
[107] Zhang, Y.; Huang, Z.-M.; Xu, X.; Lim, C.T.; Ramakrishna, S. Preparation of core− shell structured PCL-r-gelatin bi-component nanofibers by coaxial electrospinning. Chem. Mater. 2004, 16, 3406–3409.
[108] Li, H.; Ke, Y.; Hu, Y. Polymer nanofibers prepared by template melt extrusion. J. Appl. Polym. Sci. 2006, 99, 1018–1023.
[109] Unser, A.M.; Xie, Y. Electrospinning of nanofibers. The Nanobiotechnology Handbook; CRC Press: Boca Raton, FL, 2012; pp. 293–320.
[110] Ma, P.X.; Zhang, R. Synthetic nano-scale fibrous extracellular matrix. J. Biomed. Mater. Res. Off. J. Soc. Biomater. Japan. Soc. Biomater. Aust. Soc. Biomater. 1999, 46, 60–72.
[111] Kim, S.H.; Nam, Y.S.; Lee, T.S.; Park, W.H. Silk fibroin nanofiber. Electrospinning, properties, and structure. Polym. J. 2003, 35, 185–190.
[112] Peresin, M.S.; Habibi, Y.; Zoppe, J.O.; Pawlak, J.J.; Rojas, O.J. Nanofiber composites of polyvinyl alcohol and cellulose nanocrystals: Manufacture and characterization. Biomacromolecules 2010, 11, 674–681.
[113] Luzio, A.; Canesi, E.V.; Bertarelli, C.; Caironi, M. Electrospun polymer fibers for electronic applications. Materials 2014, 7, 906–947.
[114] Srivastava, Y.; Marquez, M.; Thorsen, T. Multijet electrospinning of conducting nanofibers from microfluidic manifolds. J. Appl. Polym. Sci. 2007, 106, 3171–3178.
[115] Prabhakaran, M.P.; Ghasemi-Mobarakeh, L.; Jin, G.; Ramakrishna, S. Electrospun conducting polymer nanofibers and electrical stimulation of nerve stem cells. J. Biosci. Bioeng. 2011, 112, 501–507.

[116] Nune, M.; Kumaraswamy, P.; Maheswari Krishnan, U.; Sethuraman, S. Self-assembling peptide nanofibrous scaffolds for tissue engineering: Novel approaches and strategies for effective functional regeneration. Curr. Protein Peptide Sci. 2013, 14, 70–84.
[117] Mohammadzadehmoghadam, S.; Dong, Y.; Barbhuiya, S.; Guo, L.; Liu, D.; Umer, R.; Qi, X.; Tang, Y. Electrospinning: Current status and future trends. Nano-size Polym. 2016, 10, 89–154.
[118] Zhang, Y.; Rutledge, G.C. Electrical conductivity of electrospun polyaniline and polyaniline-blend fibers and mats. Macromolecules, 2012, 45, 4238–4246.
[119] McCullen, S.D.; Stevens, D.R.; Roberts, W.A.; Ojha, S.S.; Clarke, L.I.; Gorga, R.E. Morphological, electrical, and mechanical characterization of electrospun nanofiber mats containing multiwalled carbon nanotubes. Macromolecules 2007, 40, 997–1003.
[120] Chronakis, I.S.; Grapenson, S.; Jakob, A. Conductive polypyrrole nanofibers via electrospinning: Electrical and morphological properties. Polymer 2006, 47, 1597–1603.
[121] Babel, A.; Li, D.; Xia, Y.; Jenekhe, S.A. Electrospun nanofibers of blends of conjugated polymers: Morphology, optical properties, and field-effect transistors. Macromolecules 2005, 38, 4705–4711.
[122] Wong, S.-C.; Baji, A.; Leng, S. Effect of fiber diameter on tensile properties of electrospun poly (ε-caprolactone). Polymer 2008, 49, 4713–4722.
[123] Yuan, B.; Wang, J.; Han, R.P. Capturing tensile size-dependency in polymer nanofiber elasticity. J. Mechan. Behav. Biomed. Mater. 2015, 42, 26–31.

10 Suitability of Electrospun Nanofibers for Textile Applications

Sedat Kumartasli and Ozan Avinc

10.1 INTRODUCTION

Nanofibers are materials with nano-sized fiber diameters, very small pore diameters, and high surface area, which exhibit unusual and significantly enhanced physical, chemical, and biological properties [1]. Although 100 nm and below is a generally accepted criterion in nanotechnology, this limit can be exceeded in commercial applications [2]. For example, in textile applications, fibers smaller than 1 μm or 0.5 μm in diameter are expressed as nanofibers. Although nanofibers are nano-sized in diameter, they can reach kilometers in length, thus establishing a link between nanoscale and macro-scale [3].

Nanofibers have many special and unique properties that are summarized in the following part. Nanofibers offer the ability to absorb liquids effectively owing to their high specific surface area. The total porosity of nanofibers can reach up to 95%, and the pore diameters can vary between 50 and 200 nm [3]. Porosity is the transport of fluids and molecules important for filtration applications through the nanofiber mat. For biological scaffolds, it provides the transport of gases, nutrients, and drugs. Porosity also affects the wettability of nanofibers, allowing control of water retention and diffusion, allowing the creation of superhydrophobic or self-cleaning surfaces. Due to the enhanced placement of polymer chains along the length axis of the nanofiber and the increment in strength in filament-shaped materials with the decrease in fiber diameter, polymer nanofibers are more developed in terms of mechanical properties compared to the film or bulk form of the polymer. Since the decrease in fiber diameter reduces the outer surface per unit length of the fiber, less crack formation and therefore less surface defects occur in the material, which provides a more durable structure [4].

In this chapter, the production of electrospun nanofibers and their use in the textile field are examined. In the next section, information about nanofiber manufacturing techniques is given.

10.2 MANUFACTURING TECHNIQUES OF NANOFIBERS

The production of polymeric nanofibers is done by various methods based upon physical, chemical, thermal, and electrostatic production techniques [5]. The most commonly used methods are bicomponent extrusion, phase separation, template

DOI: 10.1201/9781003333814-10

synthesis, drawing, meltblowing, electrospinning, and centrifugal spinning. The advantages and disadvantages of these methods are summarized [2, 3, 5].

10.2.1 Bicomponent Extrusion: Island in the Sea

Nanofiber production using bicomponent production method is promising day by day [6]. Bicomponent fibers could be described as extruding two polymers together in the same fiber from the same spinneret [7]. One of the techniques used to solve the performance problems of the existing nanofiber technology is the bicomponent method. Considering the conventional method, this production method has advantages. In this fiber production method, complex structures such as solvents or compressed air channels are not used [8]. The island in the sea type is formed by spinning fibers using the bicomponent method of two components that cannot be mixed with each other. While forming a sea of polymer, the other polymer is fed into it [9]. Islands-in-the-sea form fibers can also be named matrix-filament fibers since in cross-section, these fibers be seen as one polymer that was incorporated into a matrix of a second polymer. Islands-in-the-sea fibers might possess a uniform or nonuniform diameter of the island portion. Fundamentally, those filaments were spun from the mixture of two polymers in the necessary proportion, where one polymer was suspended as drops in the melt of the seconds. Rapid cooling of the fiber underneath the spinneret holes was a significant element in fibers manufacture. One of the fiber components could be eliminated with the usage of heat, a solvent, or a chemical; or utilizing mechanical tools [10, 11]. In the bicomponent extrusion, two polymers were sent to a plain spinneret hole, split by a blade edge or septum that feeds the two segments into side-by-side arrangements [10, 12]. Polymer streams consisting of two components determine the thickness of the obtained filaments. Nylon, polypropylene, and polyethylene are other polymers used in island components [13, 14]. This method is also used in fine filters and cleaning cloths. The number of islands in the very fine multifilament yarn depends on the nozzle design. The ratio of the island component to the marine component is determined by the gravity of each component [13, 14]. The pipe in pipe technique is one of most utilized procedures to produce bicomponent fibers where one of the streams constituents covers the other stream component at the end of the tube organized continuous PET nanofibers with a 39 nm diameter by sea-island type from the flow-drawn fiber with further drawing and exclusion of the sea component [15]. Although the sea and island method requires additional costs, it is successfully applied to obtain suede type synthetic leather products [9, 14].

10.2.2 Phase Separation

In this method, which is based on phase separation with temperature differences, the polymer solution can be separated into two polymer-poor and polymer-rich phases to form an interconnected and highly porous three-dimensional structure [4, 6]. Since phase separation cannot be controlled, this method is the thermodynamic separation of two different liquid phases. One of the phases is the solvent, and the other is the polymeric part that remains after the solvent is removed from the medium [15]. The phase separation process takes place in five key stages: polymer dissolution, gelation, solvent extraction, freezing, and freeze drying [3,4,16]. First of all, a polymer solution is prepared in a certain solvent, this solution is transferred to a small container and cooled rapidly in the cooler, and gelation is achieved by phase separation. Then, this container is immersed in water and solvent extraction is performed. The gel obtained is removed from the water, frozen, and finally freeze-drying under vacuum and solid polymer is obtained. In this method, the critical step in controlling the pore structure of nanofibers is the gelation stage [17]. The gelation time is greatly reliant on the composition of the solution, its concentration, the solvent chosen, and the temperature [4]. While nanofibrous structure is formed at low temperature, platelet structure is formed at high gelation temperature. In addition, the cooling rate is a crucial parameter for obtaining uniform nanofibers. It is very difficult to control fiber diameters with this technique. The method is limited to certain polymers only. Nanofibers could be manufactured in the diameter range of 50–500 nm [18].

Self-assembly is a method that enables the creation of well-defined nanostructures from polymers, peptides, and macromolecules starting at the molecular level [15]. Self-regulation is the spontaneous transition of physical systems to a more thermodynamically stable structure [4]. Molecules form a more stable structure by rearrangement by intermolecular interactions such as Van der Waals, electrostatic, hydrogen bonding, and hydrophobic forces [4, 19]. With this method, a broad range of structures such as fibers, films, nanoparticles, tubes, and capsules can be formed [20].

10.2.2.1 Drawing

Very long and single nanofibers can be produced by the drawing process carried out together with the solidification process [21]. In the drawing method, this method is also considered as the dry spinning process at the molecular level, since the solidified material turns into a fiber by evaporation of the solvent [2, 21].

With the evaporation of the solvent over time, the viscosity of the solution at the corner of the drop increases and the fibers break due to cohesion forces. Therefore, viscoelastic materials that can withstand strong deformations during the drawing process are suitable for this process [2].

10.2.2.2 Meltblown Technique

Meltblown technique includes a single step manufacturing of fibers by a polymer melt extruding through with an orifice die and drawing down the extrudate along with a hot air, normally at like temperature as the melted polymer. Here, the air exerts the drag force to attenuate the melt extrudate into fibers that were then assembled in the formation of a nonwoven mat [22].

Commercially, nanofibers were produced in pilot machines using various thermoplastic polymers such as polypropylene (PP), polyethylene terephthalate (PET), polybutylene terephthalate (PBT), and PLA using the melt blown technique [23]. In a study, the diameters of submicron fibers produced using modular meltblown dies ranged from 50 nm to 1,000 nm, with an average diameter of 400–600 nm. Therefore, this modified meltblowing process will be an exclusive and novel approach to manufacture submicron fibers from thermoplastic polymers on an industrial scale. It is also possible under controlled process conditions in other techniques such as submicron fiber webs, electrospinning, without ropes, throws, or fiber breakage. It could be manufactured consistently and successfully at production speeds several times faster than normal [24].

10.2.2.3 Template Synthesis

This method, in which it is possible to synthesize various polymer nanofibers and nanostructures using an outer mold, is based on the production of membranes with nanopores [4]. In this method, nanofibers are formed by applying water pressure to the polymer solution on the mold, allowing the polymer to pass through the nanopores and interact with the solidification solution.

With this method, nanofibers with shorter fiber lengths can be produced. It is achieved in a length of only a few micrometers. The diameters of the fibers vary according to the pore size of the membrane used [25, 26]. Using membranes of different diameters to produce nanofibers of different diameters is one of the advantages of this method.

With this method, polymer nanofibers could be manufactured by using a monomer instead of a polymer solution and polymerizing the pores with various chemical or electrochemical methods [4].

Template synthesis was another frequently utilized methodology primarily to manufacture inorganic nanofibers e.g., carbon nanotubes and nanofibers or conductive polyaniline (PANI), polypyrrole (PPy) etc. [27–29]. Template synthesis entails the utilization of a template or mold to obtain a desired material or structure. Consequently, the casting procedure and DNA replication could be considered as template-based synthesis. For the example of nanofiber generation, the template mentions to a metal oxide membrane with nano-scale diameter thickness pores. The utilization of water pressure together with the porous membrane control results in the extrusion of the polymer that by contacting with a solidifying solution, delivers nanofibers whose diameters were handled by the pores [30].

10.2.2.4 Centrifugal Spinning

Centrifugal spinning was previously used in the glass fiber production industries [25]. Electrospinning was certainly the favored technique for nanofibers manufacture; nonetheless it possesses some downsides, for example, high electric field requirements, solutions with greater dielectric characteristics, low manufacturing ratio, high manufacturing cost, and numerous other safety associated themes. Thus, electrospinning cannot be appropriate for mass manufacturing of specific materials [25, 31–33]. In recent years, their nozzleless cylindrical production has attracted a lot of attention. In the system, there is a channel containing the polymer and its solution on one side of the rotating head, and in this way, fibers are produced with fast rotation [34]. Centrifugal spinning, or force spinning, is a newly created nanofiber formation technique and it attracts huge attention mostly because of its high manufacturing ratio, which is 500 times quicker than conventional electrospinning [35]. Instead of utilizing electrostatic force, centrifugal spinning develops centrifugal force to accomplish the high-ratio nanofiber manufacturing [36]. Centrifugal spinning could be utilized to produce nanofibers by utilizing polymer solutions or polymer melts, without the dielectric constant limitations and the participation of high voltage electrical field. Furthermore, carbon, ceramic, and metal fibers could also be produced by centrifugal spinning [1, 37]. It is significant to notice that the centrifugal spinning procedure was primarily created by Hooper to manufacture artificial silk fiber from viscose by employing centrifugal forces to a viscous material [2]. Consequently, this technique was utilized for fiber manufacturing as it was created by Hooper. The fiber forming procedure of centrifugal spinning bases on the competition between centrifugal force and Laplace force (result from surface curvature) [3].

10.2.2.5 Electrospinning

Electrospinning is an easy and fast procedure used for the manufacturing of nonwoven structures consisting of continuous ultrafine fibers with diameters changing from micrometer to nanometer [20,38,39]. In this method, nanofibers were manufactured from polymer solution or melt with the help of electrostatic forces [38]. The electrospinning system comprises of three main components. These are: a high-voltage power supply, a nozzle and a collector [38].

When high voltage is employed, the droplet suspended in the nozzle is statically charged and the electrostatic force attracts this droplet as a liquid jet. In the jet under the electric field, the solvent evaporates, and solid fibers are generated [38,39]. The electrospinning technique is simple and cost-effective, being suitable for the use of a broad range of synthetic, natural, and blended polymers, and forming highly porous and continuous fibers are the remarkable features of this method [20,40].

10.2.3 Nanofiber Production by Electrospinning Method

Although the utilization of electrospinning technique has become widespread with the developments in nanotechnology after the 1980s, the foundations of this method are much older [41]. The first study of this technique was made by Rayleigh in 1897 and then it was studied in detail by Zeleny in 1914 [40]. The first patent for the conversion of polymer solution into nanofibers by electrostatic forces was received by J. F. Cooley in 1902. Between 1934 and 1944, Anton Formhals further developed the electrospinning method and equipment and obtained new patents [41]. In the 1990s, especially the studies of Reneker and his working group on the production of fine fibers from various organic polymers made a significant contribution to the widespread usage of electrospinning procedure for the production of nanofibers [38].

10.2.3.1 Stages and Types of Nanofiber Production by Electrospinning Method

In the electrospinning procedure, fiber generation takes place as an outcome of the following six consecutive steps [42]: droplet formation, Taylor cone formation, jet formation, drawing section, whipping indecision zone, and solidification as nanofiber.

In the first step, the polymer solution is pumped towards the end of the apparatus with a certain flow rate and a suspended drop is formed at the capillary end. With the application of high voltage (voltage can be positive or negative, but positive voltage is preferred in most cases) charges accumulate on the drop surface [4]. Depending on this load, the (+) and (-) loads are separated in the drop. Charge separation generates a force against the surface tension of the droplet. When the surface tension forces acting inward on the charged droplet exceed the electrostatic forces acting outward, the droplet maintains its stability [42]. When the applied voltage reaches a critical value where the electrostatic forces and the surface tension are in balance, the spherical droplet changes shape and turns into a shape known as a Taylor cone. Any voltage enhancement that will disrupt this balance causes electrostatic forces to exceed the surface tension, resulting in jet formation from the Taylor cone [4, 42]. The reason for jet formation is to create the required additional surface area to increase the deposition of surface charges [42]. The Coulomb thrust forces of the surface charges on the jet are the axial component that elongates the jet along its path towards the collector. The diameter of the resulting jet decreases rapidly due to elongation under the electric field and evaporation of the solvent. The elongation modulus of the rapidly drying jet precludes the onset of capillary instability and creates a stable jet [42]. Thus, the jet, which was initially thin, accelerates under the electric field, elongates, and continues on its way by curling due to instability. This region is called the Whipping instability region [43]. Whipping instability is the main mechanism that causes the nanofiber diameter to decrease during the electrospinning process. In the region of instability, the solvent evaporation rate increases with a significant increase in the jet area [43].

Electrospinning systems are generally horizontal and vertical types according to the geometrical arrangement of the capillary and collecting layer. The vertical type, on the other hand, is divided into shaft type and opposite type [44]. Various modifications have been made to the electrospinning method to increase the control of the procedure or to make the process suitable for special materials and applications. These modifications can be applied to the level, the electric field, or the collector. With the coaxial nozzle design, both the display and protection of the functionalizing agent can be ensured or a material that cannot be electrospun can be spun, and the production speed could be enhanced by the usage of multiple nozzles or the installation of non-nozzle electrospinning. In addition, it is possible to manipulate the electrical field by using auxiliary electrodes to monitor the shape, position, and polarity of the electrospinning jet. In addition, aligned or patterned fibers can be produced by using rotating disk, cylinder, frame, or parallel electrode pair as collector [38].

10.2.3.2 Parameters Affecting Nanofiber Production by Electrospinning Method

The morphologies and diameters of the nanofibers acquired by the electrospinning method are affected by many parameters. As summarized later, these parameters can be grouped under three main groupings: solution parameters, porosity parameters, and ambient parameters [38,44].

10.2.3.2.1 Solution Parameters

The concentration of the polymer solution is a substantial parameter that influences fiber production by the electrospinning method. This effect is explained by the solution concentration being a) too low, b) below the appropriate value, c) at the appropriate value, and d) above the appropriate value. While polymeric micro (nano) particles are formed in very low concentration, a mixed structure with beads and fibers is formed below the appropriate value. While smooth nanofibers are formed at the appropriate concentration value, helix shaped micro strips are formed instead of nanofibers at concentrations above the appropriate value [46]. Generally, fiber diameter increases with increasing concentration. Therefore, the optimum concentration should be determined in order to avoid beading at low concentrations and to prevent large diameter fiber formation at high concentrations [40].

The molecular weight of the polymer is directly related to the solution viscosity as it is an indicator of polymer chain entanglement. Even if the concentration is low, solutions with sufficient chain entanglement provide the necessary viscosity for the electrospinning process [1]. While beaded structure formation is observed in low molecular weight solutions at the same concentrations, smooth fibers are formed as the molecular weight increases; however, it was determined that micro-stripes were formed with further increase in molecular weight [46].

The viscosity of the polymer solution varies contingent upon the concentration, the molecular weight of the polymer, the ambient temperature, and the impurities in the solution [45]. The viscosity ranges at which electrospinning can be performed for different polymer solutions are different [40]. Viscosity is a critical parameter which influences the electrospinning procedure and fiber morphology. While continuous fiber production cannot be achieved at very low viscosities, jet formation from solution becomes difficult at high viscosities [37,46]. At very low viscosities, polymer particles are generated instead of fibers, while beaded fibers are generated at low viscosities [47]. The increase in viscosity generally leads to the formation of larger and more evenly dispersed fibers [40]. Thus, it is very significant to ascertain the optimum viscosity value.

The conductivity of the solution varies contingent upon the type of polymer and solvent utilized and the salt in the environment. Therefore, conductivity could be adjusted by altering the polymer/solvent pair and solution concentration or by adding ionic salts such as NaCl, KH_2PO_4 to the solution [45,46]. The dielectric constant indicates how much electrical charge a material can store under an electric field [42]. Generally, the use of materials with high dielectric constants creates a bead structure and reduces the fiber diameter. It also improves nanofiber quality and efficiency during the electrospinning process [42,46]. To augment the dielectric property of the solution and improve the fiber morphology, solvents with high dielectric constant such as dimethylformamide (DMF) can be added to the solution. In an ideal electrospinning process, all solvent is required to evaporate before reaching the collector. If the solvent does not evaporate completely, the wet fibers may coalesce and coalesce. When a highly volatile solvent is used, the solvent may dry out in the capillary and block the flow. For these reasons, the choice of solvent with suitable volatility is important. In addition, very fast drying may inhibit the formation of small diameter nanofibers, since proper jet elongation is required for small diameter fiber formation [42]. Therefore, shorter spinning distances should be used for highly volatile solvents [45].

10.2.3.2.2 Process Parameters

Electrospinning process parameters are explained below.

10.2.3.2.2.1 Voltage

In order for the electrospinning process to start, a voltage must be applied at which the electrostatic forces will exceed the surface tension forces [2]. Although the influence of implemented voltage on fiber diameter is variable, high voltage is likely to form beaded structure [40,49]. With increasing voltage, fiber diameter and bead formation increase [47].

10.2.3.2.2.2 Solution Flow Rate

In the electrospinning technique, the solution flow ratio must be at a level that can meet the amount of solution moving away from the capillary end [42]. For this reason, enough solution must be fed to maintain the Taylor cone under constant voltage. When the solution flow rate is low, the production is intermittent. At high flow rates, more solution is drawn from the tip, so the time required for the solvent to evaporate cannot be provided, and as an outcome, large diameter and beaded fibers are generated [2,42].

10.2.3.2.2.3 Spinning Distance

In the electrospinning technique, the solution jet must get rid of its solvent before it reaches the collector. When the spinning distance is reduced, the jet path will be shortened and the effect of voltage will increase, so sufficient time may not be provided for the solvent to be removed, and as a result, beaded fibers may form [2]. If the distance is too short, the solvent cannot move away from the solution and an interconnected structure is formed as a result of incomplete solidification [2,45]. Keeping the other parameters constant, increasing the spinning distance generally reduces the fiber diameter, but sometimes depending on other parameters, the fiber diameter increases, or it can stop the electrospinning process [2,42]. The fiber diameter may be reduced as long distance will provide more elongation time to the solution, but it is also possible to enhance the fiber diameter as the fiber elongation will diminish due to the decrease of the electrostatic effect at long distance [2].

10.2.3.2.3 Environment Parameters

10.2.3.2.3.1 Humidity Humidity causes effects that vary according to the solvent used and the hydrophilic properties of the solute. In aqueous solutions, as the evaporation rate decreases with increasing humidity, the solidification time increases and therefore both the formation of beaded structure and the fiber diameter can increase. However, the increased moisture reduces the fiber diameter in the time it takes for the jet to completely move away from the water before it reaches the collector. In non-aqueous solutions, high humidity sometimes causes finer fiber production due to long solidification time, while sometimes incomplete solidification can be observed. In addition, non-aqueous solutions can absorb ambient moisture, and this can increase the fiber diameter by reducing the evaporation rate and cause the formation of a porous structure on the fiber surface due to solvent concentration differences [44]. In a study using polyvinylpyrrolidone (PVP) and cellulose acetate (CA), it was observed that water absorbed into the solution at different humidity levels decreased the PVP fiber diameter and increased the CA fiber diameter. As a result, it was determined that the moisture increase up to a certain value decreases the nanofiber diameter, but when this value is exceeded, the nanofibers combine and increase the fiber diameter [48].

If the ambient humidity is too low, the solvent in the solution evaporates rapidly, causing the capillary tip to become clogged [40]. Ambient temperature is a parameter that influences the evaporation rate, surface tension, and viscosity of the solvent. As the solvent evaporation ratio enhances with increasing temperature, viscosity and surface tension decrease [1,45]. In a study in the literature, it was determined that polymer drops were formed when working at low temperature, and the fibrous structure was obtained as a result of the decrease in surface tension with increasing temperature. It was also monitored that the fiber diameter decreased due to the decreasing viscosity with the increase in temperature [1].

10.2.4 Usage Areas of Nanofibers and Textile Applications

The largest usage area of polymer nanofibers is medical applications with a ratio of approximately 2/3; half of the remaining share is filtration applications, and the other half is protective clothing, sensors, and various industrial applications. Various usage areas of polymeric nanofibers have been mentioned [51].

10.2.4.1 Medical Applications

In tissue engineering, a three-dimensional matrix is created that will provide cellular development, proliferation, and new tissue formation for the regeneration of destroyed tissues. Electrospinning method has become the most widely used method in tissue engineering for forming scaffolds in tissue engineering, as it is a simpler and more economical method to create a loosely bonded 3D porous structure with high surface area compared to other fiber production methods. In order for scaffolds to be replaced by regenerated tissues, they must degrade over time. Various natural and artificial polymers or mixtures of natural and artificial polymers are used to create biocompatible, biodegradable scaffolds. Since electrospinning has the flexibility of material selection and at the same time provides the control of the properties of the scaffold, the scaffolds produced by this method are suitable for usage in numerous tissues and organs for instance bone, cartilage, skin, tendon, bone connective tissues, arterial blood vessels, heart, and nerves [40].

Polymeric nanofibers could also be utilized in the treatment of wounds and burns, as well as in haemostatic devices [41]. Wound healing is prolonged, especially in severe wounds, and the risk of scar formation is high. In addition, in cases with strong discharge, standard wound bandages are short-lived and not economical as they are rapidly saturated with body fluid. Nanofibers, on the other hand, are good absorbers for liquids because they have high porosity and high surface area. Nanofibers also prevent the wound from drying out by creating a moist environment; enable fluid and gas exchange with the environment thanks to its pores; and contribute to the wound healing process by supporting tissue adhesion, spreading, and proliferation with its high surface areas. Various natural and artificial biodegradable polymers can be used on their own to create a nanofiber dressing material, or they can be used with components that add various properties such as providing a biocidal effect [3]. By using polymeric biocompatible delivery matrices and biodegradable polymers, controlled drug release can be achieved at a certain rate over a certain time period [40]. With the decrease in the size of the drug and drug-containing coating material, drugs are better absorbed into the human body. In drug release with polymeric nanofibers, the dissolution ratio of the drug is enhanced with the increment of the surface area of the drug and the carrier [41]. Contingent upon the polymer carrier utilized, the release of the drug dosage could be designed as fast, instantaneous, delayed, or modified dissolution [40].

The usage of electrospun nanofibers in its medical application has attracted great interest because it is like the natural tissue structure. The natural polymers of chitin are surprising materials with many benefits for use in medical dressing applications [53]. These materials are both biocompatible and biodegradable and are abundant in nature as a renewable resource. The addition of those materials to nanofibers provides advantages in biomedical applications because of the large surface area that gives them functional characteristics [53,54].

Stellenbosch Nanofiber Company (SNC) Ltd., headquartered in Cape Town, South Africa, is a company working to commercialize nanofiber materials for wound dressings, drug delivery materials and cell culture scaffolds, and the biomedical industry [55]. According to recent research on cell response to sterilized electrospun poly(E-caprolactone) scaffolds to assist in vivo tendon regeneration utilizing PCL nanofibers produced from SNC Ltd., sterilized electrospun PCL scaffold appeared to perform likewise to the gold standard autograft. Fabrication of electrospun PCL fibers had no influence on the performance of the scaffold, allowing the fabrication procedure to be simply outsourced and scaled up for commercial translation. It has been understood that longer-term in vivo studies are needed for the electrospun PCL scaffold to be an alternate intervention in patients necessitating tendon repair [56].

10.2.4.2 Filtration

Filters are used to remove substances from air or liquid media in domestic and industrial applications [39]. Fibrous materials have benefits such as high-level filtration effectiveness and low air resistance. Filtration efficiency is enhanced with diminishing fiber diameter. In general, nanofiber filters can easily hold particles smaller than 0.5 μm with their high surface area/volume rate and high surface cohesion [41]. Nanofiber-based air filters produced by a company called Donaldson by electrospinning method are used industrially [39].

10.2.4.3 Protective Clothing and Defense Garment

Protective clothing should be light, breathable, insoluble in any solvent, permeable to air and water vapor, and resistant to toxic chemical. Nanofiber membranes obtained by the electrospinning method have the feature of neutralizing chemical substances and do not resist the passage of air and water vapor thanks to their high surface area. In addition, their high porosity offers the opportunity to create a light-weight material that is breathable and prevents the penetration of harmful chemical substances in aerosol form [40, 41]. The use of nanofibers in defensive and protective clothing and equipment is beneficial to enhance the protection and survival of people operating in severe conditions and dangerous situations. For such situations, the goal is to protect from injury from environmental hazards such as toxic gas and hazardous environments. Improving fabric performance and functions will be of great help in professions such as defense forces and emergency response services [57]. Today, ballistic characteristics have been improved with the adding of carbon nanotubes. In this way, nanofibers that are super strong, have functional properties, and can be combined with fabrics have been produced. These new nanofibers are bonded to fabrics, making the body armor suit as light as a cotton shirt, and bulletproof.

It is possible to make commercially available bulletproof vests that are 4 times harder than spider silk and 17 times harder than aramid fibers [58]. In a study, nanofibers were produced from polyvinylidene fluoride (PVDF) and its copolymer, polyvinylidene fluoride trifluoroethylene (PVDF-TrFE). These fibers were then spun into yarn and compared with the same material used in conventional cable making. When the mechanical properties were compared, it was observed that nanofiber yarns exhibited higher mechanical properties. It has also been observed to increase toughness [59]. In a different study, different properties were gained by adding functional materials to electrospun nanofibers. In the study, which was designed as a material that can provide protection against toxic substances, protection against both biological and chemical substances was provided. After adding magnesium oxide (MgO) nanoparticles into the Nylon 6 polymer, nanofibers were produced. As a result of the tests, it provides protection against Gram-negative *E. coli* and Gram-positive *S. aureus* pathogens, while also improving flame resistance properties. In this way, it has been seen that it can be used as protective clothing for soldiers [60]. Fabric formation by electrospinning is one of the most difficult processes. A company known as Electroloom has demonstrated that it designs wearables using electrospinning. After removing the mold of the garment using the 3D molding method, electrospun fibers are sprayed into the mold in a layered manner. Then the fabrics removed from the mold become wearable products. In this technique, the production speed is low, and it is difficult to produce colored fabrics [61]. Fabrics manufactured from electrospun nanofiber structures are generally white because of the translucency characteristics of the polymers used. Therefore, nanofiber-based fabrics are not preferred for wearable fashion because of their color. In one study, the polymer solution was colored before electrospinning. The nanofiber membrane made from colored poly(vinyl butyral) (PVB)/cationic dyestuff displayed nice color fastness [62]. The major concern for fabrics made with electrospun fibers is maintaining their durability and functionality as they are subjected to repetitive washing and use. In a study, the fabrics produced preserved their structural integrity following 1,500 cycles of repeated compression and torsion after being passed under the hot press [63].

In another study, permeability tests of a triple nanofiber/spunbond laminated nanofiber/spunbond layer and a military laundry cycle with detergent applied on nanofibers were performed. The results showed a decrease in aerosol retention effectiveness up to 66% and 86% of the initial yield, respectively [64]. When the studies on electrospun nanofibers are examined, the commercialization of nanofibers in apparel appears to be in its infancy. It is desired that the sensors should be small in size; low in production costs; multifunctional; and high in sensitivity, selectivity, and reliability. In order for the sensors to be highly sensitive and respond quickly, they must possess a large surface area and a porous structure. For this reason, nanofibers are materials suitable for the desired properties for sensor production. In order to impart sensing ability to nanofibers, it is necessary to use a polymeric sensing material, to add sensing molecules into the nanofibers or to cover the nanofiber surface with a sensing material by coating/grafting technique [40]. Conductive nanofibers can be utilized for the manufacture of small electronic devices or machines. As the ratio of electrochemical reactions is proportional to the electrode surface area, conductive nanofiber membranes are very appropriate for usage as porous electrodes in the production of high-performance batteries [39]. Nanofibers obtained by electrospinning of high modulus polymers such as polybenzimidazole (PBI) can be used for composite material reinforcement [51].

10.3 CONCLUSIONS

Owing to their small diameter and high surface area, nanofibers hold huge potential in engineering and many industrial applications. Electrospinning is still one of the most broadly utilized procedures in nanofiber manufacturing. At this point, current advances in production scalability in electrospinning technologies have increased. The commercialization of electrospun nanofiber materials and electrospun nanofiber-based industrial products plays a very significant role. The potential in the commercialization and application of electrospun nanofiber-based materials increases the value of electrospinning products economically. Thanks to electrospinning, strong and durable functional nanofiber-based fabrics are used in filtration, defense and protection clothing, medical applications, home furnishings, and a range of consumer products. Electrospinning methods have facilitated the production and application of nonwovens analogues. Therefore, there is no doubt that nanofiber production technology will be used in all areas of the functional textile sector for more advanced product applications in the coming years. Still, there are concerns about the current technology, which are important to both manufacturers and researchers. The problems are the hazardous waste emitted during production and after use when creating nanofiber webs that must be recycled or disposed of in an eco-friendly way. Solving this problem will require equipment and cost. Another problem is the shelf life of electrospun nanofiber fabrics before washing and restoration. The fabric must be strong enough to withstand both the appropriate mechanical properties and the physical and chemical stresses of washing. It is seen that the limitations that arise during the production and development of electrospun nanofibers will be solved in the future. Therefore, the usage of electrospun nanofibers in various textile applications is expected to be enhanced day by day.

REFERENCES

[1] Wei, Q., Tao, D., and Xu, Y. Nanofibers: principles and manufacture. Functional nanofibers and their applications. Woodhead Publishing, 2012. 3–21

[2] Ramakrishna, S. An introduction to electrospinning and nanofibers. World Scientific, 2005

[3] Van der Schueren, L., and De Clerck, K. Nanofibrous textiles in medical applications. Handbook of medical textiles. Woodhead Publishing, 2011. 547–566.

[4] Pisignano, D. Polymer nanofibers: building blocks for nanotechnology. *Royal Society of Chemistry* 29 (2013).
[5] Beachley, V., and Wen, X. Polymer nanofibrous structures: Fabrication, biofunctionalization, and cell interactions. Progress in Polymer Science 35.7 (2010): 868–892.
[6] Tian, L., et al. Synergistic effect of topography, surface chemistry and conductivity of the electrospun nanofibrous scaffold on cellular response of PC12 cells. Colloids and Surfaces B: Biointerfaces 145 (2016): 420–429.
[7] Hegde, R. R., Dahiya, A., and Kamath, M. G. Bicomponent fibers, May (2006). http://www.engr.utk.edu/mse/pages/Textiles/Bicomponent% 20fibers.htm
[8] Ponting, M., Hiltner, A., and Baer, E. Polymer nanostructures by forced assembly: process, structure, and properties. Macromolecular Symposia 294.1 (2010).
[9] Mukhopadhyay, S., and Ramakrishnan, G. Microfibres. Textile Progress 40.1 (2008): 1–86.
[10] Lewin, M., and Sello, S. B. Handbook of Fiber Science and Technology. M. Dekker, 1985.
[11] Almetwally, A. A., et al. Technology of nano-fibers: Production techniques and properties-Critical review. Journal of the Textile Association 78.1 (2017): 5–14.
[12] Fitzgerald, W. E., and Knudsen, J. P. Mixed-streamspinning of bicomponent fibers. Textile Research Journal, 37.6 (1967): 447–453
[13] Nakajima, T., Kajiwara, K., and McIntyre, J. E. eds. Advanced fiber spinning technology. Woodhead Publishing, 1994
[14] Gün, A., Demiröz, B. D., and Şevkan, A. Mikroliflerin üretim yöntemleri, özellikleri ve kullanim alanlari. Tekstil ve mühendis 18.83 (2011): 38–46.
[15] Nakata, K., Fujii, K., Ohkoshi, Y., Gotoh, Y., Nagura, M., Numata, M., and Kamiyama, M. Poly (ethylene terephthalate) nanofibers made by sea-islandtype conjugated melt spinning and laser-heated flow drawing. Macromolecular Rapid Communications, 28.6 (2007): 792–795.
[16] Kurimoto, R., Niiyama, E., and Ebara, M. Fibrous materials. Biomaterials Nanoarchitectonics. William Andrew Publishing, 2016. 267–278.
[17] Liu, H., and Webster, T. J. Nanomedicine for implants: a review of studies and necessary experimental tools. Biomaterials 28.2 (2007): 354–369.
[18] Katta, P., Alessandro, M., Ramsier, R.D., and Chase, G.G. Continuous electrospinning of aligned polymer nanofibers onto a wire drum collector. Nano Letters, 4.11 (2004): 2215–2218
[19] Ma, Peter X., and Ruiyun Zhang. Synthetic nano-scale fibrous extracellular matrix. Journal of Biomedical Materials Research: An Official Journal of The Society for Biomaterials, The Japanese Society for Biomaterials, and The Australian Society for Biomaterials 46.1 (1999): 60–72
[20] Eatemadi, A., et al. Nanofiber: synthesis and biomedical applications. Artificial Cells, Nanomedicine, and Biotechnology 44.1 (2016): 111–121
[21] Ondarçuhu, T., and Joachim, C. Drawing a single nanofibre over hundreds of microns. EPL (Europhysics Letters) 42.2 (1998): 215.
[22] Hassan, M. A., et al. Fabrication of nanofiber meltblown membranes and their filtration properties. Journal of membrane science 427 (2013): 336–344.
[23] Uppal, R., et al. Meltblown nanofiber media for enhanced quality factor. Fibers and Polymers 14.4 (2013): 660–668
[24] Han, W., Wang, X., and Bhat, G. S. Structure and air permeability of melt blown nanofiber webs. The Journal of Nanomaterials & Molecular Nanotechnology 2.3 (2013): 1–5.
[25] Zhang, X., and Lu, Y. Centrifugal spinning: an alternative approach to fabricate nanofibers at high speed and low cost. Polymer Reviews 54.4 (2014): 677–701.
[26] Kumar, P. Effect of colletor on electrospinning to fabricate aligned nano fiber, Thesis (BTech), Department of Biotechnology & Medical Engineering National Institute of Technology, Rourkela, India, 2012.

[27] Yang, H, et al. A simple melt impregnation method to synthesize ordered mesoporous carbon and carbon nanofiber bundles with graphitized structure from pitches. The Journal of Physical Chemistry B 108.45 (2004): 17320–17328.
[28] Li, X., et al. One-step route to the fabrication of highly porous polyaniline nanofiber films by using PS-b-PVP diblock copolymers as templates. Langmuir 21.21 (2005): 9393–9397.
[29] Feng, J., et al. Synthesis of polypyrrole nano-fibers with hierarchical structure and its adsorption property of Acid Red G from aqueous solution. Synthetic Metals 191 (2014): 66–73.
[30] Feng, L., et al. Super-hydrophobic surface of aligned polyacrylonitrile nanofibers. Angewandte Chemie International Edition 41.7 (2002): 1221–1223.
[31] Souza, M. A., et al. Controlled release of linalool using nanofibrous membranes of poly (lactic acid) obtained by electrospinning and solution blow spinning: A comparative study. Journal of Nanoscience and Nanotechnology 15.8 (2015): 5628–5636.
[32] Balogh, A., et al. Melt-blown and electrospun drug-loaded polymer fiber mats for dissolution enhancement: a comparative study. Journal of Pharmaceutical Sciences 104.5 (2015): 1767–1776.
[33] Chen, W, et al. Improved performance of PVdF-HFP/PI nanofiber membrane for lithium ion battery separator prepared by a bicomponent cross-electrospinning method. Materials Letters 133 (2014): 67–70.
[34] Ding, B., Wang, X., and Yu, J., eds. Electrospinning: Nanofabrication and Applications. William Andrew, 2018.
[35] Ren, L., Ozisik, R., and Kotha, S. P. Rapid and efficient fabrication of multilevel structured silica micro-/nanofibers by centrifugal jet spinning. Journal of Colloid and Interface Science 425 (2014): 136–142.
[36] Sarkar, K., et al. Electrospinning to forcespinning™. Materials Today 13.11 (2010): 12–14.
[37] Padron, S., et al. Experimental study of nanofiber production through forcespinning. Journal of Applied Physics 113.2 (2013): 024318.
[38] Wang, L., and A. J. Ryan. Introduction to electrospinning. Electrospinning for Tissue Regeneration. Woodhead Publishing, 2011. 3–33.
[39] Fang, J., et al. Applications of electrospun nanofibers. Chinese Science Bulletin 53.15 (2008): 2265–2286.
[40] Bhardwaj, N., and Kundu, S. C. Electrospinning: a fascinating fiber fabrication technique. Biotechnology Advances 28.3 (2010): 325–347.
[41] Huang, Z.-M., et al. A review on polymer nanofibers by electrospinning and their applications in nanocomposites. Composites Science and Technology 63.15 (2003): 2223–2253.
[42] Andrady, A. L. Science and Technology of Polymer Nanofibers. John Wiley & Sons, 2008.
[43] Subbiah, T., et al. Electrospinning of nanofibers. Journal of Applied Polymer Science 96.2 (2005): 557–569.
[44] Yang, C., et al. Comparisons of fibers properties between vertical and horizontal type electrospinning systems. 2009 IEEE Conference on Electrical Insulation and Dielectric Phenomena. IEEE, 2009.
[45] Robb, B., and Lennox, B. The electrospinning process, conditions and control. Electrospinning for Tissue Regeneration. Woodhead Publishing, 2011. 51–66.
[46] Li, Z., and Wang, C. One-dimensional Nanostructures: Electrospinning Technique and Unique Nanofibers. Springer Berlin Heidelberg, 2013.
[47] Valizadeh, A., and Farkhani, S. M. Electrospinning and electrospun nanofibres. IET Nanobiotechnology 8.2 (2014): 83–92.
[48] Baji, A., et al. Electrospinning of polymer nanofibers: Effects on oriented morphology, structures and tensile properties. Composites Science and Technology 70.5 (2010): 703–718.

[49] Angammana, C. J., and Jayaram, S. H. Fundamentals of electrospinning and processing technologies. Particulate Science and Technology 34.1 (2016): 72–82.

[50] De Vrieze, S., et al. The effect of temperature and humidity on electrospinning. Journal of Materials Science 44.5 (2009): 1357–1362.

[51] Burger, C., Hsiao, B. S., and Chu, B. Nanofibrous materials and their applications. Annual Review of Materials Research 36 (2006): 333–368.

[52] Liu, H., et al. A functional chitosan-based hydrogel as a wound dressing and drug delivery system in the treatment of wound healing. RSC Advance 8.14 (2018): 7533–7549.

[53] Lemma, M., Frédéric Bossard, S., and Rinaudo, M. Preparation of pure and stable chitosan nanofibers by electrospinning in the presence of poly (ethylene oxide). International Journal of Molecular Sciences 17.11 (2016): 1790.

[54] Abraham, A., Soloman, P. A., and Rejini, V. O. Preparation of chitosan-polyvinyl alcohol blends and studies on thermal and mechanical properties. Procedia Technology 24 (2016): 741–748.

[55] The Stellenbosch Nanofiber Company (SNC). www.innovus.co.za/spin-out-companies/thestellenbosch-nanofiber-company-snc.html (accessed on 28 August 2022).

[56] Bhaskar, P., et al. Cell response to sterilized electrospun poly (ε-caprolactone) scaffolds to aid tendon regeneration in vivo. Journal of Biomedical Materials Research Part A 105.2 (2017): 389–397.

[57] Sinha, M. K., et al. Exploration of nanofibrous coated webs for chemical and biological protection. Zaštita materijala 59.2 (2018): 189–198.

[58] Yao, J., Bastiaansen, C. W. M., and Peijs, T. High strength and high modulus electrospun nanofibers. Fibers 2.2 (2014): 158–187.

[59] Baniasadi, M., et al. High-performance coils and yarns of polymeric piezoelectric nanofibers. ACS Applied Materials & Interfaces 7.9 (2015): 5358–5366.

[60] Dhineshbabu, N. R., et al. Electrospun MgO/Nylon 6 hybrid nanofibers for protective clothing. Nano-Micro Letters 6.1 (2014): 46–54.

[61] White, J., Foley, M., and Rowley, A. A novel approach to 3D-printed fabrics and garments. 3D Printing and Additive Manufacturing 2.3 (2015): 145–149.

[62] Yan, X., et al. Colorful hydrophobic poly (vinyl butyral)/cationic dye fibrous membranes via a colored solution electrospinning process. Nanoscale Research Letters 11.1 (2016): 1–9.

[63] Faccini, M., C. Vaquero, and D. Amantia. Development of protective clothing against nanoparticle based on electrospun nanofibers. Journal of Nanomaterials 2012 (2012).

[64] Graham, K., Gogins, M., and Schreuder-Gibson, H. Incorporation of electrospun nanofibers into functional structures. International Nonwovens Journal 2 (2004): 1558925 004os–1300209.

11 Electrospinning for Food Packaging and Antibacterial Applications

P. Sankarganesh, A. Surendra Babu, A. O. Adeyeye Samuel, and N. Guruprasad

11.1 INTRODUCTION

Electrospinning is a process that produces ultrafine nanofibers. Though several technologies have been explored for nanofibers production, electrospinning is the most extensively utilized nanofiber generation technology, because it is simple to execute and allows for exact control over the shape and quality of the fibers. So, it is employed in mass production at industrial level. In the presence of a high voltage electric field, it charges and ejects a polymer melt or solution through a spinneret. Electrospun nanofiber films are increasingly being investigated in the field of food packaging due to their huge specific surface area and desirable features such as strong biocompatibility, high porosity, effective antioxidant and antibacterial capabilities, and great water permeability.

Electrospinning has allowed the creation of ultrathin fiber mats which can be used in food packing such as wrapper, sachet, and bag. In addition, electrospinning has been used to produce nanofibers from various biopolymers, both natural and synthetic. Synthetic polymers are generally stronger and more versatile than natural polymers. There are several methods of electrospinning methods used to prepare food packaging (Table 11.1).

11.2 BLEND ELECTROSPINNING

Blend electrospinning is a technique that involves combining two or more polymers with an active substance in a solvent system, consecutively working as an electrospinning solution. The blending of synthetic polymers such as polyvinylidene fluoride, polyethene, polystyrene, polylactic acid, etc., with natural polymers such as starch, chitosan, polyvinyl alcohol, etc., to improve biocompatibility [32]. Further, it enables the attainment of both polymers' required properties, resulting in a novel nanofiber with desirable properties for food packaging, including improved mechanical properties, high hydrophobicity, spinnability, better stability, and biological activity [33,34]. Zhang et al. fabricated gluten/zein nanofiber film using a blend of xylose [1]. The obtained

DOI: 10.1201/9781003333814-11

TABLE 11.1
Comparison of Various Types of Electrospinning Methods Developed for Food Packaging Applications

Type of electrospinning	Type of polymer	Polymer concentration (%)	Bioactive compound encapsulated
Blend electrospinning	Gluten and zein [1]	30 (1:0, 3:1, 1:1)	–
	Polylactic acid [2]	6	–
	Carboxymethyl cellulose (CMC), polyvinyl alcohol (PVA), and polyvinylpyrrolidone (PVP) [3]	16.6%, 33.4%, and 50% (CMC), 6% (PVA), 12% (PVP)	–
	Gum arabic and pullulan [4]	20	–
	Chitosan (Ch) and polyvinyl alcohol [5]	30:70, 40:60 50:50, 60:40 70:30	–
	Hordein and chitosan [6]	11 and 0.4	–
	Poly(lactic acid) (PLA) and cellulose nanocrystals (CNCs) [7]	10 (PLA) and 1, 2 4% (CNCs)	–
	Whey protein isolate (WPI) and guar gum (GG) [8]	7–12 (WPI) and 0.7–0.9 (GG)	–
	PLA and tea polyphenol [9]	5:1, 4:1, 3:1, and 2:1	–
	Casein and pullulan [10]	2:1 to 1:4	–
Emulsion Electrospinning	Gelatin [11]	25%, 35%, 45%	β-carotene (5%)
	Soy protein isolate (SPI) and polyvinyl alcohol (PVA) [12]	11% (SPI and PVA)	–
	Polycaprolactone (PCL), poly-L-lactic acid (PLLA), and polyvinyl alcohol (PVA) [13]	12% PCL and 1.4% NaCl solution and 10mL of 13% PLLA and 1.4% NaCl	Bioactive compound: phycocyanin
	Gelatin[14]	25%	Bioactive compound: Corn oil (0.2, 0.4, 0.6, and 0.8 v/v)
	Polyvinyl alcohol PVA [15]	7.5, 9 and 10.5%	Bioactive compound: Fish oil (5, 7.5, 10%)
	Polyvinyl alcohol (PVA) [16]	5–12%	Bioactive compound: (R)-(+)-limonene (1:4, 1:6, 1:8) (dispersed phase:continuous phase, w/w)
	Pullulan and β-cyclodextrin [17]	20 and 25%	Bioactive compound: (R)-(+)-limonene (90 wt %)

(Continued)

TABLE 11.1 (Continued)

Type of electrospinning	Type of polymer	Polymer concentration (%)	Bioactive compound encapsulated
Suspension Electrospinning	Polyvinylpyrrolidone (PVP) [18]	10%	–
	PLA [19]	10%	–
	PVA [20]	10%	–
	Polyacrylonitrile (PAN) [21]	10%	–
	Poly(lactic acid) (PLA) [22]	13%	–
	PVP and Cellulose Nanocrystal [23]	Different concentrations	–
Coaxial electrospinning	PLA (Shell) [24]	9% (shell) 4% (core)	Cinnamaldehyde, tea polyphenol (core)
	Poly lactide-co-glycolide (Shell) [25]	12% (shell) and 75% (core)	Thymol (core)
	Gelatin (Shell) [26]	20%	Sour cherry concentrate (core)
	Zein prolamine (Shell) [27]	25, 30, 35%	Orange essential oil (core)
	PVP (Shell) [28]	27–31% (shell) 25% Zein (core)	Fish oil (30%)
Triaxial electrospinning	CA (Shell) PCL (Intermediate) PVP (Core0) [29]	12% (shell), 11% (intermediate), 19% (core)	Nisin (0.1%)
	Gelatin (Shell) Poly(ε-caprolactone) (Intermediate) Gelatin (Core) [30]	17% (shell), 11% (intermediate), 10% (core)	–
	PCL (Shell) PCL (Intermediate) PVP (core) [31]	10% (shell), 10% (intermediate), 15% (core)	–

nanofilm exhibited excellent water stability and elasticity because of the enhanced intermolecular cross-linkages and interactions between the polymers. The study proved that blending electrospinning of natural polymers could produce desirable properties for intelligent food packaging. In further study, Mahdian-Dehkordi et al. developed a chitosan/polyvinyl alcohol (PVA) blend scaffold by the electrospinning method [5]. Borax was utilized as a novel crosslinking agent. The resultant nanofiber showed improved structural stability, tensile strength, and antimicrobial activity against gram +ve and gram -ve bacteria. Electrospinning of polymers such as arabic gum and pullulan was reported by Ma et al. [4]. Application of pullulan to the gum arabic solution improved the electrospinnability and strength. The fabricated nanofiber membrane has high thermal stability and is found to be an ideal carrier for probiotic microorganisms with good viability.

11.3 EMULSION ELECTROSPINNING

Another innovative technique called emulsion electrospinning is used to develop core-shell nanofibers that can encapsulate bio-active materials (e.g., antioxidants, flavonoids, essential oils, enzymes, etc.). In this method, electrospinning can be done using an O/W (oil-in-water) or W/O (water-in-oil) emulsion [14] [35]. An emulsifier is used to stabilize the original emulsions. Electrospinning involves the rapid solidification by solvent evaporation of oil in water dispersion, forming a solid phase in the nanofiber that results in an encapsulated matrix. Natural bioactive compounds are lipophilic; these substances can be entrapped in nanofibers with greater encapsulation efficiency [14]. For example, fish oil and vitamin C were successfully encapsulated in a gelatin-based core-shell electrospun nanofibrous membrane [36]. Adding vitamin C did not affect the core cell structure but increased nanofiber's fragility. Further, vitamin C incorporation prevented lipid oxidation. The study results confirmed that the electrospun nanofibrous membrane could be used as an edible film to effectively deliver vitamin C and fish oil. Animal protein hydrolysate and whey protein isolate might be used in emulsion electrospinning. In a study, fish oil was encapsulated into polyvinyl alcohol (PVA) based electrospun nanofiber membranes using emulsion electrospinning. Plant protein, WPI (whey protein isolate), and animal protein (fish protein hydrolysate) were used as emulsifiers in the electrospinning solution. PVA/WPI emulsions with smaller particle sizes exhibited higher emulsifying properties, providing superior physical stability for PVA/WPI emulsions. In addition, the electrospun fiber showed better encapsulation efficiency of omega-3 fish oil [15].

11.4 SUSPENSION ELECTROSPINNING

Suspension is a process of electrospinning polymer solution dispersed with nanoparticles and incorporating nanoparticles in the interior or top of the nanofiber matrix. The suspension electrospinning uses AgNPs, TiO2, ZnO, SiO2 MgO, and polytetrafluoroethylene for efficient nano-effects in drug delivery, water treatment, and wound dressing [37,38]. This technique also finds application in food packaging by incorporating nano-materials to impart antioxidant and antimicrobial properties [39]. Zhu et al. have designed a polyvinylpyrrolidone based composite film dispersed in silver nanoparticles using ultrasound-assisted electrospinning [18], while Zhang et al. (2020) fabricated a ZnO nanoparticle dispersed antimicrobial fibrous membrane with polylactic acid using suspension electrospinning [19]. As per their results, the barrier and mechanical characters of the PLA membranes were improved by incorporation with ZnO nanoparticle as nanofiller with robust antimicrobial activity against *Escherichia coli* and *Staphylococcus aureus*. In a similar study, Wang et al. developed a biodegradable PLA based nanocomposite film entangled with cellulose nanocrystal (CNC)-ZnO [40]. Embedding CNC-ZnO in PLA nanofilm improved the crystallization and increased the degradation rate by 28% compared to the 8% pure PLA film. In another study, a nanofilm was prepared by embedding TiO2 nanoparticles in a polyacrylonitrile polymer matrix and tested for its efficiency in delaying ripening of banana fruit. The resultant nanofilm efficiently reduced the color alternations during the storage period and the fruit's softening. The results stated the PLA/TiO2 nanofilm's role in packaging material to delay the ripening of fruits [41].

11.5 COAXIAL ELECTROSPINNING

Currently, coaxial electrospinning technology is the key focus of research. It finds application in drug release, food additives, microelectronic devices, etc. [42,43]. In the coaxial electrospinning technique, the core and sheath fluids are ejected from the spinneret, forming double layers. After volatilization of the final solvent, the core-sheath structure of the nanofiber is produced [44]. This method helps regulate the location of active material and polymer in nanofiber for food packaging applications. The active material may be embedded in the core liquid to disperse in the core layer of nanofiber, which leads to the sudden release of the active compounds from the nanofiber surface [45]. Han et al. encapsulated cinnamaldehyde and tea polyphenols in polylactic acid nanofiber film by coaxial electrospinning [24]. The developed film showed decreased particle diameter with uniform distribution, high thermal stability, and mechanical properties. In addition, the nanofiber film presented effective antimicrobial efficiency against *S. putrefaciens*. Fish oil was embedded in a zein/polyvinylpyrrolidone (PVP) composite nanofiber using co-axial electrospinning. The co-axial electrospinning technology effectively improved the thermal and oxidative stable property of the composite film. Further, the shelf life of fish oil was enhanced by applying composite nanofiber [46]. Similarly, an innovative antimicrobial nanocomposite membrane based on zein prolamine was developed using the coaxial electrospinning method. The orange essential oil was encapsulated in the core matrix. Results showed a potential antibacterial activity over *E. coli*, suggesting its application in active food packaging systems [28].

11.6 TRIAXIAL ELECTROSPINNING

Triaxial electrospinning is an excellent technique for delivering various functional compounds. Triaxial fibers are of great interest since the middle layer can be utilized as an encapsulation layer amid the core and outer shell. This enables the embedding of hydrophilic substances such as proteins in the outer shell, thereby depriving the premature release of encapsulated active materials from the nanofiber core [47]. Triaxial electrospinning comprises a spinneret with three needles aligned concentrically. Three solutions will be delivered from the pump via tip of the spinneret. The solution undergoes deformation under the influence of the electrostatic field, and the triaxial jet arises during the surface tension of the solution replaces the electrostatic field. Further, the polymer solution is subjected to whipping motion, vaporization, and deposition as dry fibers [48,49]. Active packaging involves a regulated release of active ingredients. Food packaging encapsulated with active ingredients at various diffusion rates can be developed through triaxial electrospinning. However, the application of the triaxial electrospinning role in the food packaging industry is still limited and needs extensive research. Han et al. encapsulated nisin in the core of three-layered nanofiber by coaxial electrospinning [31]. The core layer was made of polyvinyl pyrrolidone embedded with nisin, while the middle layer is hydrophobic poly(ε-caprolactone) and hydrophilic cellulose acetate remains as an outer protective shell. The nisin encapsulated triaxial fiber exhibited potential biocidal and antimicrobial efficiency. Similarly, a novel biodegradable triaxial nanofiber structure was fabricated by Liu et al. using gelatin as the shell and core material while hydrophobic poly(ε-caprolactone) was used as an intermediate layer [50]. These studies confirm that multilayered nanofiber can find promising applications in the food industry for packaging.

11.7 ELECTROSPINNING IN FOOD PACKAGING

Food packaging has a crucial function in the storage and transportation of all types of food, preserving it from external environmental damage and transferring it from producers to consumers. As a result, resolving this issue correctly will be cost-effective and convenient for both producers and customers. Electrospun nanofibers are in the development stage and are attracting increasing attention in nearly 10 years of research on their application in food packaging. Antimicrobial electrospun nanofibers for food packaging are frequently made with biodegradable natural and synthetic polymers. Natural polymers, including proteins and polysaccharides such as soy, egg albumin, zein, wheat protein, gelatin, whey protein, starch, cyclodextrin, cellulose acetate, cellulose, and chitosan, are already used for the purpose of drug release [51,52].

There is a growing demand for health-promoting foods. Poor stability of functional foods such as prebiotics, probiotics under packing, and storage conditions has been reported. This results in low bioavailability in the digestive system. Bioactive substances encapsulated in nanofilms were created using the electrospinning process. The chemical stability, encapsulation efficiency, and oral bioavailability of the encapsulated nanofilm are all improved [53].

Aytac et al. developed food packaging materials including citric acid, thyme oil, and the nisin-based zein fibers that effectively inhibited *Escherichia coli* and *Listeria innocua*. In order to make a thymol-encapsulated coaxial electrospun core-shell nanofiber film [54], Zhang et al. employed poly (lactic-co-glycolic acid) (PLGA) as the sheath solution and thymol as the core solution [55]. The headspace of nanofiber film could be gently released with thymol to limit the growth of bacteria on food surfaces.

Ma et al. demonstrated that adding pullulan polysaccharide to arabic gum (GA) improved not only its electrospinnability but also prevented the nanofiber membrane from interfering with GA's probiotic carrier function [56]. Cinnamaldehyde (CMA) and tea polyphenol (TP) were encapsulated in a PLA nanofiber membrane via physical mixing [57]. CMA/TP synergistic antibacterial performance had the best effect on *S. putrefaciens*, making it a great way to combine food preservatives and coaxial electrospinning technology. Coaxial electrospinning was utilized to encapsulate the nisin in the core of multi-layered fibers by Han et al. [58]. Polyvinyl pyrrolidone (PVP) loaded with nisin, hydrophobic PCL as the intermediate layer, and cellulose acetate (CA) as the outer protective layer, were the three layers that comprised the innermost structure. Recently, different electrospinning strategies adopted for the preparation of anti-microbials blended double-layered nanofibers for food packaging applications as shown in Figure 11.1 [59].

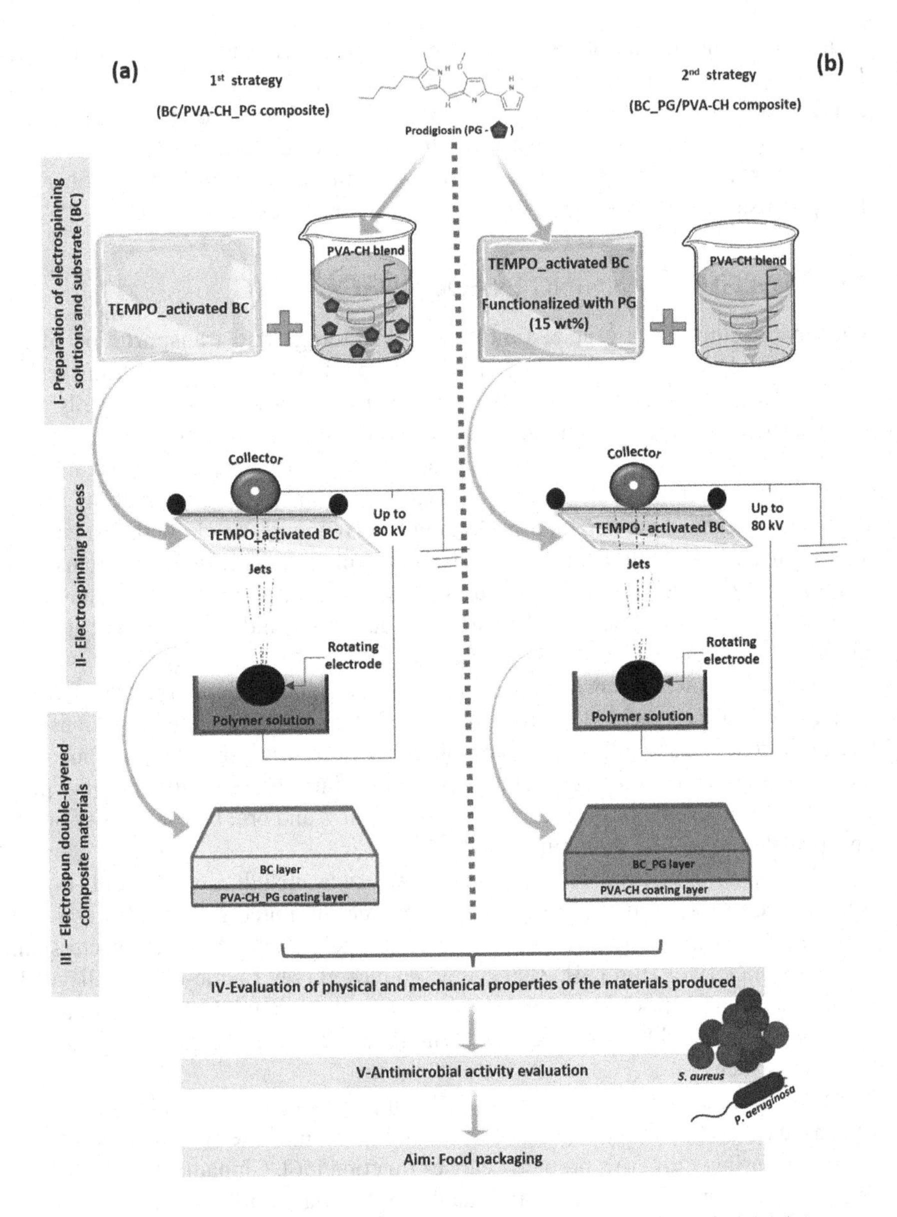

Abbreviations: BC bacterial cellulose; CH chitosan; *P. aeruginosa Pseudomonas aeruginosa;* PG prodigiosin; PVA poly(vinyl alcohol); *S. aureus Staphylococcus aureus;* TEMPO 2,2,6,6-tetramethylpiperidine-1-oxyl radical.

FIGURE 11.1 Method of electrospinning for food package applications.

11.8 FOOD SAFETY

Food safety has always been a topic of widespread concern, as it relates not only to people's daily requirements, but also to their desire for a more luxurious lifestyle. Food packaging is an important technology that serves as the foundation and is crucial to ensuring food safety. The main methods for keeping food fresh are as follows: confining oxygen and restricting food from oxidation; sterilization and antibacterial treatment to prevent microorganisms from growing on food; delaying food water loss; minimizing fruit and vegetable respiration; and reducing nutrient loss in food. There are a number of important considerations, including thermal stability, surface hydrophobicity, water resistance, water vapor barrier properties, and mechanical strength. Significant losses in items such as fruits, vegetables, meats, dairy, and baked goods are caused not just by transportation and processing, but also by oxidation and microbial degradation caused by packaging. This limits the applicability of functional foods and nutraceuticals. Electrospun antimicrobial hybrid mats can circumvent these restrictions by incorporating natural antimicrobials into nanofibers created by electrospinning [60].

11.9 ANTIMICROBIALS IN FOOD PACKAGING

Antimicrobials in food packaging systems help in the prevention of microbial contamination during food transit and storage [61]. For instance, plant-based essential oils, silver nanoparticles thymol, phlorotannin, nisin, curcumin, N-calamine, ZnO, eugenol, chitosan, carvacrol, polyphenol, allyl isothiocyanate, antimicrobial peptides, and flavonoids were all found to be effective at killing pathogens such as *Listeria monocytogens, Staphylococcus aureus, Escherichia coli 0157:H7, Salmonella typhimurium, Bacillus cereus, Campylobacter jejunii, Pseudomonas aeruginosa, Staphylococcus epidermidis 1878*, and *Clostridium perfringens* [62,63].

11.10 ENCAPSULATION OF ANTIMICROBIALS

Foods that are infected by microorganisms have a shorter shelf life and a higher risk of foodborne illness. In order to inhibit the proliferation of foodborne pathogens and microbial deterioration, efficient control mechanisms must be developed. It is possible to improve food safety, quality, and shelf life by allowing antimicrobial chemicals to slowly migrate through a polymer film into the product. Both the antibacterial chemicals and the polymers play a role in antimicrobial packaging materials' effectiveness. Antimicrobials are naturally occurring compounds with low solubility, high volatility, and a high degree of chemical instability. During storage, they've been shown to interact with food components and seep out of packing materials, although further research is needed.

There must be a regulated delivery rate of antimicrobial agents to ensure their long-term action and minimize undesirable impacts on the flavor of packaged food goods. It is therefore necessary to control the dispersion and release of antimicrobial chemicals from polymeric packaging materials, as well as their transfer into and through food, in order to build efficient antimicrobial packaging.

Encapsulation is the technique of encasing an active component (antimicrobials) in a matrix (wall material) to generate an active-loaded particle [64]. Using encapsulation to address these issues is a feasible solution to the problem. Antimicrobial agents' handling, dispersibility, stability, and release properties can all be improved by encapsulation technology. Antimicrobials are commonly encapsulated in food-grade polymers to protect them from harmful environmental conditions and to regulate their retention and release in various environmental situations. Because of this, encapsulated components are released only when and where they are required.

There are a number of aspects to consider when selecting wall materials for encapsulated products, including the physical and chemical characters of the active substances as well as the intended use. For instance, gums such as arabic, alginate, and polyethylene have been employed, as well as proteins such as gelatin, albumin, and casein; carbohydrates such as cellulose, starch, and chitosan; and lipids such as wax, fatty acids, and paraffin [65].

Chemically unstable or extremely volatile compounds are ideal for encapsulation because of their hydrophobicity present in essential oils. Encapsulation can enhance the activity of active compounds by shielding them from reactions caused by exposure of oxygen, moisture, temperature, and or light. In most cases, formulations and processing processes can be used to adjust the particle size and surface area, enhancing the antibacterial activity.

Thermally labile antimicrobials, like citral and trans-cinnamaldehyde, cannot be included directly in packaging materials made utilizing melt extrusion processes; for example, ethylene-vinyl alcohol copolymer films can include antimicrobials encapsulated in beta-cyclodextrin. It was subsequently shown that these active packaging materials extended beef shelf life and reduced its release rate while being stored in the refrigerator. When the cyclodextrin complexes encapsulated the antimicrobials, its release rate and thermal stability were reduced [64].

The antibacterial effectiveness of cyclodextrin complexes containing coriander oil against a broad range of foodborne pathogenic microbes was demonstrated by F. Silva et al [66]. These complexes could be used in active food packaging films due to their good antimicrobial activity. For the delivery of antimicrobial drugs, electrospinning has proven popular among the numerous encapsulating technologies available. Polyvinyl alcohol nanofibers incorporating cinnamon oil/cyclodextrin complexes have been demonstrated to be effective against *E. coli* and *Staphylococcus aureus*.

In a study, starch food packaging film integrated with nisin, EDTA, and lysozyme was developed [67]. The investigation on antimicrobial effect of the packaging material supported the hypothesis that nisin and EDTA have a significant synergistic effect on antibacterial activities by proving the efficiency of the combined effect of antibacterial agents. Antibacterial films developed by incorporating crude pediocin into polyhydroxybutyrate (PHB) has showed antibacterial efficacy against foodborne microbes [68]. The film has antibacterial efficacy against foodborne microbes.

In a study, zein films blended with partly purified lysozyme [69] showed the release rate of lysozyme ranging between 7 and 29 U/cm^{2}min and the rate of diffusion increased with increasing lysozyme concentration. Murielgalet et al. created a unique antimicrobial film in which lysozyme is covalently attached to EVOH 29 and EVOH 44, ethylene-vinyl alcohol copolymers [70]. The antibacterial properties of the resultant films are comparable to those of the free enzyme against *L. monocytogenes*.

Strawberry shelf life was also extended when nanofibers containing cinnamon oil were employed as packaging materials. It was also demonstrated that electrospun films were more effective in inhibiting microbial growth than casting films. This was ascribed to the mild conditions of the electrospinning process, which reduced

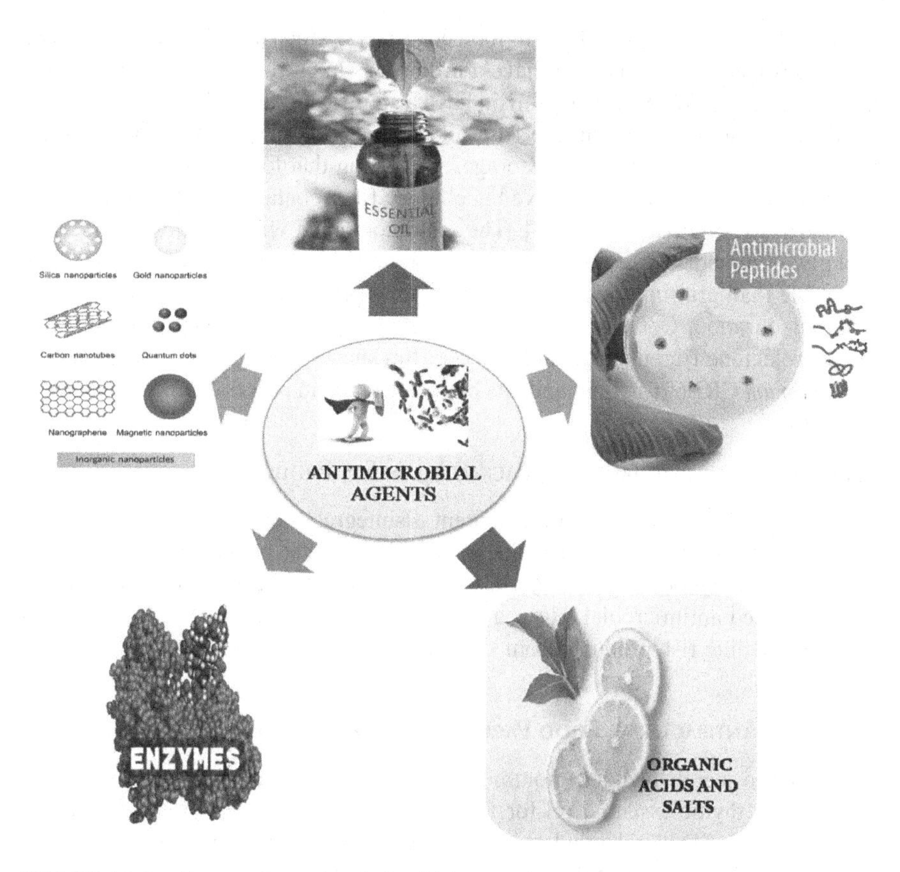

FIGURE 11.2 Commonly used antimicrobial agents in food packaging applications.

the loss of essential oil. A viable possibility for the development of new antimicrobial food packaging materials is shown in Figure 11.2 [71].

11.11 APPLICATIONS

11.11.1 Antibacterial Food Packaging in Meat

Throughout the world, people's diets rely heavily on animal protein and micronutrients from meat and animal products. They are extremely susceptible to microbiological deterioration caused by pathogenic organisms because of the high water, protein, lipid, and other nutritional content of meat and meat products [72]. Seasoned beef is packaged and stored at a low temperature (4°C) and RT (room temperature) for 60 days. As a result, in seasoned beef, nisin promotes the growth of mesophilic microbes slightly more than seasoned beef without nisin, whereas seasoned beef without nisin promotes the growth of these microbes significantly more. That is, even at room temperature, nisin possesses good antibacterial capabilities. Nisin increases the mechanical qualities of food containers, which helps preserve seasoned meat. Packages containing nisin have a minor loss in cutting force while stored at 25°C, whereas packages absent of nisin have a significant decrease.

Nisin was added to a cellulose film for meatpacking [73]. Growth of *L. monocytogenes* slows after 14 days in storage, suggesting that food packaging's antimicrobial qualities have been improved. Cellulose fiber containing nisin was used in antimicrobial food packaging [74]. The nisin combined with cellulose nanofibers strongly inhibited *B. subtilis.* Blending nisin with cellulose nanofibers is more accessible than grafting; nevertheless, the former preserves antibacterial action for a shorter duration. Correa et al. developed a biodegradable polyhydroxybutyrate/polycaprolactone film using nisin [75]. The film successfully inhibits the growth of *L. Plantarum* CRL691 in ham and has high efficiency in processed meatpacking.

11.11.2 Antibacterial Food Packaging in Sea Products

Fish and shellfish with a high lipid content disintegrate rapidly, resulting in a considerable loss of moisture and influencing the chemical breakdown and sensory properties of the product. Active antimicrobial packaging technologies, such as biopolymer-based antimicrobial films, are therefore a potential choice for preserving and safeguarding fish products from surface microbial contamination [76].

11.11.3 Antibacterial Food Packaging in Fruits and Vegetables

Sweet potato starch blended montmorillonite nanoclay nanocomposite film was activated by thyme essential oil for food packaging purposes. Lin et al. encapsulated TTO and -CD in polyethylene oxide to develop an antibacterial nanofiber food packaging. TTO is bactericidal against *E. coli* O157:H7 on beef for seven days following plasma therapy. Plasma treatment at 4 or 12°C has significant antibacterial properties and inhibitory effectiveness of 99.99%. Proteolysis of protein in the proteoliposomes was employed to control the release of CEO from the PEO nanofibers, resulting in a controlled release of CEO [77].

11.11.4 Antibacterial Food Packaging in Dairy Products

Owing to huge food waste, microbial contamination of milk products is a serious worldwide health and sustainability concern [78]. Dairy products are susceptible to microbial contamination because they are a rich source of macro and micronutrients on which bacteria can feed [79]. Lactose- and casein-free cheese, yogurt, and sour cream are known sources of high-nutritional-value dairy products vital to the human diet. However, humidity level, O_2, bacteria, and light sources contribute to undesired changes in dairy products, such as oxidation, microbial contamination, discoloration, and the formation of off-flavors, culminating in their fast degradation [64].

Chitosan, cellulose, and nisin were combined in a composite film and employed as an antimicrobial packaging for ultra-filtered (UF) cheese by Divsalar et al [80]. Antimicrobial capabilities of the pure chitosan-cellulose film were shown to be ineffective against *L. monocytogenes*, but, the incorporation of nisin into the cellulose film effectively inhibited *L. monocytogenes*. Nevertheless, *L. monocytogenes* does not considerably alter ultra-filtered white cheese refrigerated at 4°C for 14 days in a composite film containing nisin.

11.12 CHALLENGES

Electrospinning has a big disadvantage in the food packaging industry, because it is exclusively used in the laboratory. Commercial-scale fiber production will be required to progress toward industrialization and widespread application in food packaging films. Electrospinning-based fibrous mats for food packaging need to be enhanced in several aspects, including stability, barrier properties, usability, and mechanical properties. However, there is not much proof that they have been employed commercially [81]. Additionally, silver nano particles, encapsulated antimicrobials, can be easily absorbed by the skin and inhaled. This damages the immune system, causing brain and liver damage, and can lead to death. To alleviate this bottleneck problem, new packaging innovations that will not come into contact with the food are required [81,82].

Encapsulating antimicrobials in food packaging requires extensive research [71], because the delivery speed at which antimicrobial packaging reaches the food surface affects its efficacy. Antibacterial action is delayed if the release rate is too slow. Alternatively, releasing antimicrobial chemicals too quickly will cause their depletion before their storage life is up, resulting in rapid MIC reach. The rate of antimicrobial release from the food packaging to the surface of the food must be monitored over time to assess efficacy, a tedious and costly process. A heterogenous mixture of materials such as carbohydrates and water, food has many different antimicrobials in development and storage time-temperature combinations to consider. Synthetic antibacterial packaging is more effective than natural antibacterial chemicals at inhibiting bacterial growth. Lowering the toxicity of synthetic fibers is another crucial problem requiring more study [83].

11.13 CONCLUSION

Electrospin technology is better suited to the production of antimicrobial nanofibers in the food packaging industry. It's critical for extending food's shelf life and reducing microorganism decay. The implementation of active packaging materials made from these nanofibers on a wider range of food items under more realistic storage conditions requires more investigation, however. Electrospun nanofibers' antibacterial activity against a wide range of bacteria, viruses, fungi, and other foodborne pathogens must also be supported by substantial scientific evidence.

REFERENCES

[1] Zhang, Y.; Deng, L.; Zhong, H.; Zou, Y.; Qin, Z.; Li, Y.; Zhang, H. Impact of glycation on physical properties of composite gluten/zein nanofibrous films fabricated by blending electrospinning. *Food Chemist.* 2022, *366*, 130586.

[2] Dou, H.; Liu, H.; Wang, F.; Sun, Y. Preparation and Characterization of Electrospun Polylactic Acid Micro/Nanofibers under Different Solvent Conditions. *FDMP-Fluid Dynamic. & Material. Process.* 2021, *17*(3), 629–638.

[3] Hashmi, M.; Ullah, S.; Ullah, A.; Saito, Y.; Haider, M.; Bie, X.; Koasi, W.; Kim, I. S. Carboxymethyl cellulose (CMC) based electrospun composite nanofiber mats for food packaging. *Polymer*. 2021, *13*(2), 302.
[4] Ma, J.; Xu, C.; Yu, H.; Feng, Z.; Yu, W.; Gu, L.; Liu, Z.; Lijun, C.; Zhenmai, J.; Hou, J. Electro-encapsulation of probiotics in gum Arabic-pullulan blend nanofibres using electrospinning technology. *Food Hydrocolloid*. 2021, *111*, 106381.
[5] Mahdian-Dehkordi, M.; Sarrafzadeh-Rezaei, F.; Razi, M.; Mahmoudian, M. Fabrication of chitosan-based electrospun nanofiber scaffold: Amplification of biomechanical properties, structural stability, and seeded cell viability. *Veterinary Research Forum*, 2021, *12*(1), 25–32. doi: 10.30466/vrf.2020.123047.2893
[6] Li, S.; Yan, Y.; Guan, X.; Huang, K. Preparation of a hordein-quercetin-chitosan antioxidant electrospun nanofibre film for food packaging and improvement of the film hydrophobic properties by heat treatment. *Food Packaging and Shelf Life*, 2020, *23*, 100466.
[7] Patel, D. K.; Dutta, S. D.; Hexiu, J.; Ganguly, K.; Lim, K. T. Bioactive electrospun nanocomposite scaffolds of poly (lactic acid)/cellulose nanocrystals for bone tissue engineering. *Internat. J. Biologic. Macromolecul*. 2020, *162*, 1429–1441.
[8] Aman mohammadi, M.; Ramazani, S.; Rostami, M.; Raeisi, M.; Tabibiazar, M.; Ghorbani, M. Fabrication of food-grade nanofibers of whey protein Isolate–Guar gum using the electrospinning method. *Food Hydrocolloid*. 2019, *90*, 99–104.
[9] Liu, Y.; Liang, X.; Wang, S.; Qin, W.; Zhang, Q. Electrospun Antimicrobial Polylactic Acid/Tea Polyphenol Nanofibers for Food-Packaging Applications. *Polymer* 2018, *10*(5), 561. doi: 10.3390/polym10050561
[10] Tomasula, P. M.; Sousa, A. M.; Liou, S. C.; Li, R.; Bonnaillie, L. M.; Liu, L. Electrospinning of casein/pullulan blends for food-grade applications. *J. Dairy Sci*. 2016, *99*(3), 1837–1845.
[11] Liu, Q.; Cheng, J.; Sun, X.; Guo, M. Preparation, characterization, and antioxidant activity of zein nanoparticles stabilized by whey protein nanofibrils. *Internat. J. Biologic. Macromolecule*. 2021, *167*, 862–870.
[12] Bruni, G. P.; de Oliveira, J. P.; Gómez-Mascaraque, L. G.; Fabra, M. J.; Martins, V. G.; da Rosa Zavareze, E.; López-Rubio, A. Electrospun β-carotene–loaded SPI: PVA fiber mats produced by emulsion-electrospinning as bioactive coatings for food packaging. *Food Packag Shelf Life* 2020, *23*, 100426.
[13] Schmatz, D. A.; Costa, J. A. V.; de Morais, M. G. A novel nanocomposite for food packaging developed by electrospinning and electrospraying. *Food Packag. Shelf Life*. 2019, *20*, 100314.
[14] Zhang, C.; Zhang, H. Formation and stability of core–shell nanofibers by electrospinning of gel-like corn oil-in-water emulsions stabilized by gelatin. *J. Agric. Food Chem*. 2018, *66*(44), 11681–11690.
[15] García-Moreno, P. J.; Stephansen, K.; van der Kruijs, J.; Guadix, A.; Guadix, E. M.; Chronakis, I. S.; Jacobsen, C. Encapsulation of fish oil in nanofibers by emulsion electrospinning: Physical characterization and oxidative stability. *J. Food Eng*. 2016, *183*, 39–49.
[16] Camerlo, A.; Vebert-Nardin, C.; Rossi, R. M.; Popa, A. M. Fragrance encapsulation in polymeric matrices by emulsion electrospinning. *Eur. Polym. J*. 2013, *49*(12), 3806–3813.
[17] Fuenmayora, C. A.; Mascheronia, E.; Cosioa, M. S.; Piergiovannia, L.; Benedettia, S.; Ortenzic, M.; . . . Manninoa, S. Encapsulation of R-(+)-limonene in edible electrospun nanofibers. *Chem. Eng*. 2013, *32*, 1771–1776.
[18] Zhu, L.; Zhu, W.; Hu, X.; Lin, Y.; Machmudah, S.; Kanda, H.; Goto, M. PVP/Highly Dispersed AgNPs Nanofibers using ultrasonic-assisted electrospinning. *Polymer* 2022, *14*(3), 599.
[19] Zhang, R.; Lan, W.; Ji, T.; Sameen, D. E.; Ahmed, S.; Qin, W.; Liu, Y. Development of polylactic acid/ZnO composite membranes prepared by ultrasonication and electrospinning for food packaging. *LWT*, 2021, *135*, 110072.

[20] Kowsalya, E.; MosaChristas, K.; Balashanmugam, P.; Rani, J. C. Biocompatible silver nanoparticles/poly (vinyl alcohol) electrospun nanofibers for potential antimicrobial food packaging applications. *Food Packag. Shelf Life*. 2019, *21*, 100379.
[21] Zhu, Z.; Zhang, Y.; Zhang, Y.; Shang, Y.; Zhang, X.; Wen, Y. Preparation of PAN@ TiO2 nanofibers for fruit packaging materials with efficient photocatalytic degradation of ethylene. *Mater*. 2019a, *12*(6), 896.
[22] Rokbani, H.; Ajji, A. Rheological properties of poly (lactic acid) solutions added with metal oxide nanoparticles for electrospinning. *J. Polym. Environ*. 2018, *26*(6), 2555–2565.
[23] Huang, S.; Zhou, L.; Li, M. C.; Wu, Q.; Kojima, Y.; Zhou, D. Preparation and properties of electrospun poly (vinyl pyrrolidone)/cellulose nanocrystal/silver nanoparticle composite fibers. *Mater*. 2016, *9*(7), 523.
[24] Han, Y.; Ding, J.; Zhang, J.; Li, Q.; Yang, H.; Sun, T.; Li, H. Fabrication and characterization of polylactic acid coaxial antibacterial nanofibers embedded with cinnamaldehyde/tea polyphenol with food packaging potential. *Int. J. Biol. Macromol*. 2021, *184*, 739–749.
[25] Zhang, Y.; Zhang, Y.; Zhu, Z.; Jiao, X.; Shang, Y.; Wen, Y. Encapsulation of thymol in biodegradable nanofiber via coaxial electrospinning and applications in fruit preservation. *J. Agric. Food Chem*. 2019, *67*(6), 1736–1741.
[26] Isik, B. S.; Altay, F.; Capanoglu, E. The uniaxial and coaxial encapsulations of sour cherry (Prunus cerasus L.) concentrate by electrospinning and their in vitro bioaccessibility. *Food Chem*. 2018, *265*, 260–273.
[27] Yang, H.; Wen, P.; Feng, K.; Zong, M. H.; Lou, W. Y.; Wu, H. Encapsulation of fish oil in a coaxial electrospun nanofibrous mat and its properties. *RSC adv*. 2018, *7*(24), 14939–14946.
[28] Yao, Z. C.; Chen, S. C.; Ahmad, Z.; Huang, J.; Chang, M. W.; Li, J. S. Essential oil bioactive fibrous membranes prepared via coaxial electrospinning. *J. Food Sci*. 2017, *82*(6), 1412–1422.
[29] Han, D.; Sherman, S.; Filocamo, S.; Steckl, A. J. Long-term antimicrobial effect of nisin released from electrospun triaxial fiber membranes. *Acta biomate*. 2017, *53*, 242–249.
[30] Liu, W.; Ni, C.; Chase, D. B.; Rabolt, J. F. Preparation of multilayer biodegradable nanofibers by triaxial electrospinning. *ACS Macro Lett*. 2013, *2*(6), 466–468.
[31] Han, D.; Steckl, A. J. Triaxial electrospun nanofiber membranes for controlled dual release of functional molecules. *ACS Appl. Mater. Interf*. 2013, *5*(16), 8241–8245.
[32] Costa-Júnior, E. S.; Barbosa-Stancioli, E. F.; Mansur, A. A.; Vasconcelos, W. L.; Mansur, H. S. Preparation and characterization of chitosan/poly (vinyl alcohol) chemically crosslinked blends for biomedical applications. *Carbohyd. Polym*. 2009, *76*(3), 472–481.
[33] Moreira, A.; Lawson, D.; Onyekuru, L.; Dziemidowicz, K.; Angkawinitwong, U.; Costa, P. F.; . . . Williams, G. R. Protein encapsulation by electrospinning and electrospraying. *J. Control Release*. 2021, *329*, 1172–1197.
[34] Wang, X. X.; Yu, G. F.; Zhang, J.; Yu, M.; Ramakrishna, S.; Long, Y. Z. Conductive polymer ultrafine fibers via electrospinning: Preparation, physical properties and applications. *Prog. Mater. Sci*. 2021, *115*, 100704.
[35] Su, S.; Bedir, T.; Kalkandelen, C.; Başar, A. O.; Şaşmazel, H. T.; Ustundag, C. B.; . . . Gunduz, O. Coaxial and emulsion electrospinning of extracted hyaluronic acid and keratin based nanofibers for wound healing applications. *Eur. Polym. J*. 2021, *142*, 110158.
[36] Liu, L.; Tao, L.; Chen, J.; Zhang, T.; Xu, J.; Ding, M.; . . . Zhong, J. Fish oil-gelatin core-shell electrospun nanofibrous membranes as promising edible films for the encapsulation of hydrophobic and hydrophilic nutrients. *LWT*, 2021, *146*, 111500.
[37] Coimbra, P.; Freitas, J. P.; Gonçalves, T.; Gil, M. H.; Figueiredo, M. Preparation of gentamicin sulfate eluting fiber mats by emulsion and by suspension electrospinning. *Mater. Sci. Eng. C*, 2019, *94*, 86–93.
[38] Bruni, G. P.; de Oliveira, J. P.; Gómez-Mascaraque, L. G.; Fabra, M. J., Martins, V. G.; da Rosa Zavareze, E.; López-Rubio, A. Electrospun β-carotene-loaded SPI: PVA fiber mats

produced by emulsion-electrospinning as bioactive coatings for food packaging. *Food Packag. Shelf Life*. 2020, *23*, 100426.

[39] Bardoňová, L.; Mamulová Kutláková, K.; Kotzianová, A.; Kulhánek, J.; Židek, O.; Velebný, V.; Tokarský, J. Electrospinning of fibrous layers containing an antibacterial Chlorhexidine/Kaolinite composite. *ACS Appl. Bio Mater*. 2020, *3*(5), 3028–3038.

[40] Wang, Y. Y.; Yu, H. Y.; Yang, L.; Abdalkarim, S. Y. H.; Chen, W. L. Enhancing long-term biodegradability and UV-shielding performances of transparent polylactic acid nanocomposite films by adding cellulose nanocrystal-zinc oxide hybrids. *Int. J. Biol. Macromol*. 2019, *141*, 893–905.

[41] Zhu, Z.; Zhang, Y.; Shang, Y.; Wen, Y. Electrospun nanofibers containing TiO2 for the photocatalytic degradation of ethylene and delaying postharvest ripening of bananas. *Food Bioproc. Tech*. 2019b, *12*(2), 281–287.

[42] Bhattarai, D. P.; Kim, M. H.; Park, H.; Park, W. H.; Kim, B. S.; Kim, C. S. Coaxially fabricated polylactic acid electrospun nanofibrous scaffold for sequential release of tauroursodeoxycholic acid and bone morphogenic protein2 to stimulate angiogenesis and bone regeneration. *Chem. Eng. J*. 2020, *389*, 123470.

[43] Qiu, Q.; Chen, S.; Li, Y.; Yang, Y.; Zhang, H.; Quan, Z.; . . . Yu, J. Functional nanofibers embedded into textiles for durable antibacterial properties. *Chem. Eng. J*. 2020, *384*, 123241.

[44] Ding, T.; Li, T.; Li, J. Preparation of coaxial polylactic acid–propyl gallate electrospun fibers and the effect of their coating on salmon slices during chilled storage. *ACS Appl. Mater. Interfaces*. 2019, *11*(6), 6463–6474.

[45] Chen, H.; Wang, N.; Di, J.; Zhao, Y.; Song, Y.; Jiang, L. Nanowire-in-microtube structured core/shell fibers via multifluidic coaxial electrospinning. *Langmuir* 2010, *26*(13), 11291–11296.

[46] Yang, H.; Wen, P.; Feng, K.; Zong, M. H.; Lou, W. Y.; Wu, H. Encapsulation of fish oil in a coaxial electrospun nanofibrous mat and its properties. *RSC Adv*. 2017, *7*(24), 14939–14946.

[47] Qiu, L. Y.; Bae, Y. H. Self-assembled polyethylenimine-graft-poly (ε-caprolactone) micelles as potential dual carriers of genes and anticancer drugs. *Biomaterials* 2007, *28*(28), 4132–4142.

[48] Greiner, A.; Wendorff, J. H. Electrospinning: a fascinating method for the preparation of ultrathin fibers. *Angew Chemie Int Ed* 2007, *46*(30), 5670–5703

[49] Lallave, M.; Bedia, J.; Ruiz-Rosas, R.; Rodríguez-Mirasol, J.; Cordero, T.; Otero, J. C.; . . . Loscertales, I. G. Filled and hollow carbon nanofibers by coaxial electrospinning of alcell lignin without binder polymers. *Adv. Mater*. 2007, *19*(23), 4292–4296.

[50] Liu, W.; Ni, C.; Chase, D. B.; Rabolt, J. F. Preparation of multilayer biodegradable nanofibers by triaxial electrospinning. *ACS Macro Letters*. 2013, *2*(6), 466–468.

[51] Senthil Muthu Kumar, T.; Senthil Kumar, K.; Rajini, N.; Suchart S.; Nadir, A; Varada Rajulu, A. A comprehensive review of electrospun nanofibers: Food and packaging perspective. *Comp Part B* 2019, *175* (2019) 107074.

[52] Yibin, W.; Haixia, Xu.; Mian, Wu.; Deng-Guang, Yu. Nanofibers-based food packaging. *ES Food Agroforestr*. 2022. doi: 10.30919/esfaf598.

[53] Peng, W.; Min-Hua, Z.; Robert, J. L.; Kun, F.; Hong, W. Electrospinning: A novel nano-encapsulation approach for bioactive compounds. *Trend Food Sci. & Technolog*. 2017, *79*, 56–68.

[54] Zeynep, A.; Runze, H.; Nachiket, V.; Tao, X.; et al. Development of biodegradable and antimicrobial electrospun zein fibers for food packaging. *ACS Sustain Chem. Eng*. 2020, *8*, 40, 15354–15365 doi: 10.1021/acssuschemeng.0c05917.

[55] Zhang, Y.; Zhang, Y.; Zhu, Z.; Jiao, X.; Shang, Y.; Wen, Y.; Encapsulation of Thymol in Biodegradable Nanofiber via Coaxial Eletrospinning and Applications in Fruit Preservation. *J. Agr. Food. Chem*. 2019, *67*, 1736–1741, doi: 10.1021/acs.jafc.8b06362.

[56] Ma, J.; Xu, C.; Yu, H.; Feng, Z.; Yu, W.; Gu, L.; Liu, Z.; Chen, L.; Jiang, Z. Electro-encapsulation of probiotics in gum Arabic-pullulan blend nanofibres using electro-

spinning technology. *J. Hou. Food Hydrocolloid.*, 2021, *111*, 106381, doi: 10.1016/j.foodhyd.2020.106381.

[57] Han, Y.; Ding, J.; Zhang, J.; Li, Q.; Yang, H.; Sun, T.; Li, H. Fabrication and characterization of polylactic acid coaxial antibacterial nanofibers embedded with cinnamaldehyde/tea polyphenol with food packaging potential. *Int. J. Biol. Macromol.* 2021, *184*, 739–749, doi:10.1016/j.ijbiomac.2021.06.143.

[58] Han, D.; Sherman, S.; Filocamo, S.; Steckl, A. J. Long-term antimicrobial effect of nisin released from electrospun triaxial fiber membranes. *Acta Biomater.* 2017, *53*, 242–249, doi: 10.1016/j.actbio.2017.02.029.

[59] Amorim, L. F. A.; Mouro, C.; Riool, M.; Gouveia, I. C. Antimicrobial food packaging based on prodigiosin-incorporated double-layered bacterial cellulose and chitosan composites. *Polymer.* 2022, *14*, 315. doi: 10.3390/polym14020315.

[60] Luying, Z.; Gaigai, D.; Guoying, Z.; Haoqi, Y.; Shuijian, H.; Shaohua, J. Electrospun functional materials toward food packaging applications: a review. *Nanomaterial.* 2020, *10,150.* doi:10.3390/nano10010150.

[61] Han, J. H. Edible films and coatings: a review. *Innov. Food Packag.* 2005, 239–262. doi: 10.1016/B978-012311632-1/50047-4.

[62] Topuz, F.; Uyar, T. Antioxidant, antibacterial and antifungal electrospun nanofibers for food packaging applications. *Food Res Inter.* 2019, *130*,108927. doi: 10.1016/j.foodres.2019.108927.

[63] Fatemeh, H.; Akbar, B.; Afshin, F. E.; Hedayat, H.; David, J. M.; Leonard, W. Electrospun antimicrobial materials: advanced packaging materials for food applications. *Trends Food Sci Technol.* 2021, *111*, 520–533.

[64] Chen, H.; Wang, J.; Cheng, Y.; Wang, C.; Liu, H.; Bian, H. et al. Application of protein-based films and coatings for food packaging: a review. *Polymer*, 2019, *11*(12), 2039. doi: 10.3390/polym11122039.

[65] Melo, D.; Ribeiro-santos, R.; Andrade, M. Use of Essential Oils in Active Food Packaging: Recent Advances and Future Trends. *Trends Food Sci. Technol.* 2017, *61*, 132–140, doi: 10.1016/j.tifs.2016.11.021.

[66] Silva, F.; Caldera, F.; Trotta, F.; Nerín, C.; Domingues, F. C. Encapsulation of coriander essential oil in cyclodextrin nanosponges: A new strategy to promote its use in controlled-release active packaging. *Innov. Food Sci. Emerg. Technol.* 2019, *56*, 102177. doi: 10.1016/j.ifset.2019.102177.

[67] Bhatia, S.; Bharti, A. Evaluating the antimicrobial activity of Nisin, Lysozyme and Ethylenediaminetetraacetate incorporated in starch based active food packaging film. *J. Food Sci. Technol.* 2015, *52*, 3504–3512.

[68] Narayanan, A.; Neera M.; Ramana, K. V. Synergized antimicrobial activity of eugenol incorporated polyhydroxybutyrate films against food spoilage microorganisms in conjunction with pediocin. *Appl. Biochem. Biotechnol.* 2013, *170*, 1379–1388.

[69] Mecitoglu, C.; Yemenicioglu, A.; Arslanoglu, A.; Elmaci, Z. S.; Korel, F.; Cetin, A. E. Incorporation of partially purified hen egg white lysozyme into zein films for antimicrobial food packaging. *Food Res. Int.* 2006, *39*, 12–21.

[70] Murielgalet, V.; Talbert, J. N.; Hernandezmunoz, P.; Gavara, R.; Goddard, J. M. Covalent immobilization of lysozyme on ethylene vinyl alcohol films for nonmigrating antimicrobial packaging applications. *J. Agric. Food Chem.* 2013, *61*, 6720–6727.

[71] Rekha Chawla; Sivakumar, S.; Harsimran, K. Antimicrobial edible films in food packaging: current scenario and recent nanotechnological advancements- a review. *Carb. Poly. Technologie. and Application.* 2021, *2*, 100024: 1–19.

[72] Surendhiran, D.; Li, C.; Cui, H.; Lin, L. Fabrication of high stability active nanofibers encapsulated with pomegranate peel extract using chitosan/PEO for meat preservation. *Food Packag. Shelf Life.* 2020, 23, 100439.

[73] Nguyen, V. T.; Gidley, M. J.; Dykes, G. A. Potential of a nisin-containing bacterial cellulose film to inhibit Listeria monocytogenes on processed meats. *Food Microbiol.* 2008, *25*, 471–478.

[74] Saini, S.; Sillard, C.; Belgacem, M. N.; Bras, J. Nisin anchored cellulose nanofibers for long term antimicrobial active food packaging. *RSC Adv*. 2016, *6*, 12437–12445.

[75] Pablo Correa, J.; Molina, V.; Sanchez, M.; Kainz, C.; Eisenberg, P.; Blanco Massani, M. Improving ham shelf life with a polyhydroxybutyrate/polycaprolactone biodegradable film activated with nisin. *Food. Packaging. Shelf Life*. 2017, *11*, 31–39.

[76] Valdés, A.; Ramos, M.; Beltrán, A.; Jiménez, A.; Garrigós, M. C. State of the art of antimicrobial edible coatings for food packaging applications. Coatings. 2017, *7* (4), 56. doi: 10.3390/coatings7040056.

[77] Lin, L.; Gu, Y.; Cui, H. Novel electrospun gelatin-glycerin- ε -Poly-lysine nanofibers for controlling Listeria monocytogenes on beef. *Food Packaging and Shelf Life*. 2018, *18*, 21–30. doi: 10.1016/j.fpsl.2018.08.004.

[78] Ibarra-Sánchez, L. A.; El-Haddad, N.; Mahmoud, D.; Miller, M. J.; Karam, L. Invited review: Advances in nisin use for preservation of dairy products. *J Dairy Sci*. 2020, *103*(3), 2041–2052. doi: 10.3168/jds.2019-17498.

[79] Lin, L.; Dai, Y.; Cui, H. Antibacterial poly(ethylene oxide) electrospun nanofibers containing cinnamon essential oil/beta-cyclodextrin proteoliposomes. *Carb. Polymer*. 2017, *178*, 131–40. doi: 10.1016/j.carbpol.2017.09.043.

[80] Divsalar, E.; Tajik, H.; Moradi, M.; Forough, M.; Lotfi, M.; Kuswandi, B. Characterization of cellulosic paper coated with chitosan-zinc oxide nanocomposite containing nisin and its application in packaging of UF cheese. *Int. J. Biol. Macromol*. 2018, *109*, 1311–1318.

[81] Zhang, C.; Y. Li; Wang, P.; Zhang, H. Electrospinning of nanofibers: potentials and perspectives for active food packaging. *Compreh. Rev. Food Sci. Food Saf.* 2020, *19* (2): 479–502. doi: 10.1111/1541-4337.12536.

[82] Neo, Y. P.; Ray, S.; Perera, C. O. Fabrication of functional electrospun nanostructures for food applications, In Handbook of Food Bioengineering, Role of Materials Science in Food Bioengineering, Academic Press, 2018, Pages 109–146. doi:10.1016/B978-0-12-811448-3.00004 –8.

[83] Tianqi, H.; Yusheng, Q.; Jia, W.; Chuncai, Z. Polymeric antimicrobial food packaging and its applications. *Polymer*. 2019, *11*, 560; doi:10.3390/polym11030560.

12 Applications of Electrospun Nanofibers in the Drug Delivery and Biosensing

V. Parthasarathy, S. Mahalakshmi, Annie Aureen Albert and A. Saravanan

12.1 INTRODUCTION

The electrospinning process involves the fabrication of nanofibers with a fiber diameter of up to several hundred nanometers from the polymer solution or melt by the application of an electric field [1, 2]. The electrospinning technique had been employed by many research groups to produce nanofibers in the 1990s. Since then, a wide range of polymers has been made as nanofibers by the electrospinning technique. The applications of nanofibers are innumerable in the bioengineering and biomedical sectors such as wound dressing [3], drug delivery systems [4], tissue engineering [5], biosensors [6], enzyme immobilization [7], medical implant and antimicrobial agents [8] since most of the human organs and tissues are in the nanofibrous form. The voltage-driven electrospinning method is a widely employed technique for fabricating long and continuous fibers. The morphology and the diameter of the electrospun fibers are regulated by various parameters. The diameter of the electrospun fibers is in the range of 10 nm to 1,000 nm. Thus electrospun nanofibers exhibit high porosity, excellent surface-to-volume ratio, good mechanical properties and excellent surface adhesion, which make them promising candidates in many fields [9–11]. The porosity of nanofibers admits adequate nutrient transport, excellent cell attachment, large amount of drug loading, free drug diffusion and rapid waste removal [12, 13].

The controlled drug release, drug solubility, degradability and bioavailability are the major challenges in drug delivery applications. Nanofibers are loaded with low-solubility drugs to achieve controlled drug release or to improve bioavailability. An improvement in cell binding, drug loading, mass transfer and proliferation processes is possible with nanofibers owing to their high surface-to-volume ratio [14]. The drug administration by eternal routes is in the form of capsules, granules and tablets, whereas the parenteral routes of drug administration are intramuscular, intra-arterial and intravenous. The drug administration routes have some disadvantages such as pain or discomfort and first-pass metabolism. These issues can be addressed by delivering drugs directly to the buccal cavity. The bioactive substances

DOI: 10.1201/9781003333814-12

are incorporated into nanofibers to deliver the drugs directly into the buccal mucosa. The drug release mechanism depends on polymeric interaction, drug loading method, swelling rate of polymer and degradation [15]. The electrospun nanofibers are highly suitable for sensing applications owing to their large surface area and high porosity because these properties are responsible for surface and molecular forces such as capillary, steric and Vander walls. The electrospun nanofibers can also be used in biosensors for the detection of neurotransmitters, drugs and glucose.

12.2 APPLICATIONS OF ELECTROSPUN NANOFIBERS IN DRUG DELIVERY

Chronic diseases like hypertension, diabetes, arthritis and cardiovascular disease have been controlled by referring drugs either orally or by systematic injection to the patients. The prescribed drug is not only delivered to the target site, but it also spreads to the healthy sites through blood circulation, which leads to adverse side effects in patients. Hence, it is essential to develop controlled drug delivery systems that will improve the effectiveness of the drugs by minimizing the excess use of drugs. The blend electrospinning method has been employed to fabricate drug-loaded electrospun nanofibers by blending the drug with polymer solutions for drug delivery applications. Thus prepared electrospun nanofiber suffers from denaturation and clearance of drug molecules along with the burst release of hydrophilic drugs [16]. Emulsion electrospinning is an ideal technique to fabricate electrospun nanofibers with controlled drug release. In the emulsion electrospinning process, the hydrophobic polymer is dissolved in the oil phase whereas the drug is dissolved in the water phase. The encapsulation of hydrophilic drug occurs in the fibers since the oil phase quickly evaporates during the solidification process [17]. Therefore, the burst release is restricted considerably in the emulsion electrospinning nanofibers, and this supports the controlled drug release. Hu et al. [18] fabricated two different drug delivery systems by encapsulating metformin hydrochloride (MH)/ metoprolol tartrate (MPT) into poly(e-caprolactone) (PCL) and poly(3-hydroxybutyric acid-co-3-hydroxyvaluric acid) by emulsion electrospinning technique. MH and MPT drugs have been recommended for treating chronic cardiovascular diseases. The oral intake of these hydrophilic drugs is limited due to their short half-life time and low bioavailability. Hence, it is required to develop a promising drug delivery system to deliver these drugs in a controlled manner. The effect of emulsion composition (polymer/drug/span 80) on the drug release behavior of hydrophilic drug-loaded PCL and PHBV nanofibers was studied to optimize their drug release potential. In vitro drug release study of MPT and MH from PHBV and PCL was carried out in phosphate buffer solution (PBS) to conclude the produced MH and MPT-loaded nanofibers as potential drug delivery systems. The MH/MPT drug-loaded emulsion electrospun nanofibers showed a lower burst release in comparison with the blend electrospun nanofibers. However, emulsion electrospinning drug-PHBV nanofibers exhibited a higher burst release as compared to emulsion electrospinning drug-PCL nanofibers. The amount of released drug was 34.82% for emulsion electrospun MPT/PCL and 67.77% for blend electrospun MPT/PCL nanofibers within 12 hrs. The cumulative drug release was 82% for MPT/PCL and 55% for electrospun

MPT/PCL nanofibers. The cumulative percentage of MPT release over 21 days was 75% for blend electrospun nanofiber. The reason for higher burst release from emulsion electrospun drug-PHBV NFs as compared to drug-PCL NFs was due to the different inherent properties of PCL and PHBV. The semicrystalline PHBV possesses high melting point (160°C) and crystallinity (60–80%). The hydrophilic drug molecules were located on the surface of NFs upon the electrospinning process due to their high crystallinity which led to the fast release of both MPT and MH. It was also concluded that PHBV was more hydrophilic than PCL based on water contact angle measurement, which enabled rapid diffusion of drug molecules through PHBV NFs. Therefore, PHBV may not be a suitable matrix for drug delivery systems. PCL was reported as a promising matrix for drug delivery application due to its low melting (60°C) and crystallinity (45–60%).

12.2.1 Absorption of Drugs through the Mucoadhesive System

The adsorption of administrated substance occurs through the mucosal membrane in different ways which depends on the physicochemical and anatomical properties of the mucosal layer as well as the chemical nature of the molecule. The lipid percentage, epithelium percentage and keratinization grade are also considered as crucial factors in determining the adsorption pathway [19]. Passive diffusion is the reliable route for the drug molecules to penetrate through the mucosal membrane, and this process generally happens in two ways such as paracellular and transcellular. The hydrophilic molecules use the paracellular way, while the lipophilic molecules employ the transcellular way. However, the drug molecule with amphoteric properties can access both ways (Figure 12.1)

The electrospun polymer scaffold has been employed for drug delivery applications in various mucosal sites such as ocular, oral, vaginal, gastro-enteric and nasal. The electrospinning parameters such as polymer concentration, electric field, solution conductivity, needle-collector distance, electrospinning setup, environmental parameters, viscosity and flow rate play a major role in fabricating efficient and reproducible mucoadhesive polymer nanofibers. The performance of the mucoadhesive electrospun polymer NFs can be improved by controlling the mucosal performance time and adhesion strength as well as the design of the multilayer systems and unidirectional release for treating several pathologies to ensure the concurrent release of drug in the target tissue and organ. Mucoadhesive polymers have a lot

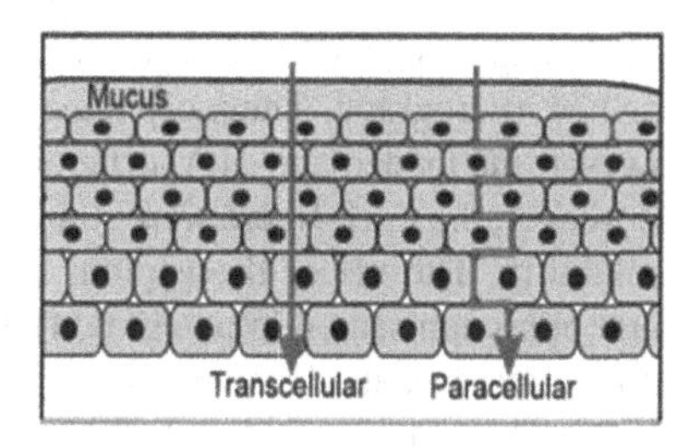

FIGURE 12.1 Schematic illustration of paracellular and transcellular pathways [20].

TABLE 12.1
Adhesive Forces of Various Polymers [20]

Sl. No	Polymers	Adhesive force in percentage
1	Poly (hydroxyethyl methacrylate)	88.4
2	Poly (vinylic alcohol)	94.8
3	Poly (ethylene glycol)	96
4	Poly (vinyl pyrrolidone)	97.6
5	Pectin	100
6	Gelatin	115.8
7	Methyl ethyl cellulose	117.4
8	Hydroxypropyl methylcellulose	125.2
9	Methylcellulose	128
10	Poly (methyl vinyl ether)	147
11	Poly (acrylic acid)	185

of hydrophilic groups such as carbonyl, hydroxyl, sulphate and amide groups, and these groups can attach to the cell membrane or mucins by making strong interaction. The adhesive forces of various bioadhesive polymers are listed in Table 12.1. The adhesive force of the bioadhesive polymers is related to their molecular weight since crosslinking and interpenetration of polymers are favored only with low molecular weight. The types of functional groups present in polymer also determine the mucoadhesion property. The presence of mucins on the mucus surface can establish a strong bond with a cationic polymer. However, the polycationic polymer has less effect in an acidic environment. Therefore, anionic polymers are considered as promising candidates for the mucoadhesive system.

12.2.2 Mucoadhesive Drug Delivery Systems

Many people have been infected by human immunodeficiency virus (HIV) through transfusion of contaminated blood and sexual contact. Prevention better than cure is the only solution to get rid of this deadly disease since the HIV virus presents genetic diversity which makes vaccines ineffective. It is highly required to find an alternative solution to prevent HIV from spreading. The cellulose acetate phthalate (CAP) microfibers were loaded with an anti-HIV drug to prevent HIV transmission by Huang et al. [21]. The diameter of CAP fibers was in the range of 500 nm to 800 nm according to the SEM micrograph. The morphology of the CAP fibers was reported as aberrant. However, the loaded anti-HIV drug did not alter the morphology of the CAP fibers. Moreover, the CAP fibers were stable in vaginal fluid owing to their pH-dependent solubility. The pH of the vaginal flora is below 4.5 while introducing a small amount of human semen into vaginal flora, the CAP fibers dissolved immediately due to the increase in pH from 4.5 to 7.4 resulting in the release of encapsulated anti-HIV drugs. It was concluded that the CAP fibers are nontoxic against vaginal Lactobacilli and vaginal epithelial cells even after being dissolved.

The study on the semen-sensitive (intravaginal) drug release potential of polyurethane (PU)/cellulose acetate phthalate (CAP) coaxial fibers loaded with rhodamine B (RhB) was also investigated by Hua and his co-workers [22]. The rhodamine B-loaded PU/CAP composite fiber was fabricated by the co-axial electrospinning method. The electrospinning was carried out with a flow rate of 3 ml/h for CAP and 0.5 ml/h for PU by maintaining a distance of 12 cm between the metal plate and syringe needle. The collected PU/CAP composite fibers on a metal plate were dried at 40°C for 24 hours. There was an improvement in the tensile strength of PU/CAP composite fibers (13.27±2.32 MPa) as compared to that of pristine CAP fibers (0.2±0.3 MPa). The release of RhB from the PU/CAP fibers was studied under two different pH conditions. The PBS solution with pH 4.2 served as stimulated vaginal fluid (SVF), whereas the solution with pH 7.4 was used as stimulated human semen (SHS). The RhB release study revealed that PU/CAP composite fibers retained RhB in SVF (pH 4.2) even after 3 hours without releasing the loaded RhB. However, RhB release took place in SHS (pH 7.4) within 1 minute, which inferred that PU/CAP composite fibers showed pH-dependent release of RhB. Therefore, the PU/CAP composite fibers were recommended as a promising candidate for intravaginal drug delivery.

Illangakoon et al. [23] fabricated a fast-dissolving drug delivery system by loading caffeine (CAF) and paracetamol (PCM) into polyvinylpyrrolidone (PVP) by electrospinning method. The intermolecular interaction of drugs and PVP was concluded by molecular modelling. The disintegration of the CAF/PCM drugs-loaded PVA mat took place within 0.5 s in the stimulated saliva, and the dissolution study witnessed the simultaneous release of the loaded CAF and PCM drugs in less than 150 s. The CAF/PCM drugs-loaded PVA mat was reported as a suitable candidate for a fast-dissolving drug delivery system, which could be more useful for patients and children with swallowing difficulties.

Oral Lichen Planus (OLP) is an inflammatory disorder that affects the oral cavity with symptoms like patches, swelling and redness. Recurrent aphthous stomatitis (RAS) is known as an oral mucosal disease often characterized by painful ulceration or painful oral lesions. The mouthwashes, ointments or creams containing steroids are usually prescribed to treat OLP and RAS. However, these treatments are ineffective owing to the insufficient contact time of drugs with the lesion. However, oral diseases like RAS and OLP can be effectively treated by using mucoadhesive patches. The oral patches contain a drug-loaded bioadhesive layer for interacting with a mucosal, and impermeable backing layer.

The drug delivery layer of mucoadhesive patches was fabricated by electrospinning the solution mixture of a copolymer of Eudragit RS100 (12.5 wt%), PVP (10 wt%), 20 wt% of polyethylene oxide (PEO) and Clobetasol-17-propionate (steroid) in ethanol[24]. The fabricated drug delivery layer was further coated with an impermeable back layer of polycaprolactone (PCL) by the electrospinning method. It was observed that the mucoadhesive patches loaded with Clobetasol-17-propionate (C-17P) released the drug in a controlled manner in both *ex-vivo* porcine and tissue-engineered oral mucosa, and the drug release profile in an in-vivo animal model and residence time of C-17P-loaded mucoadhesive patches were also evaluated to affirm their prolonged drug release and adhesion. Moreover, the electrospun mucoadhesive patches showed better adhesion without damaging the tissue. Therefore, the research group proposed that the C-17P-loaded mucoadhesive patches can be used for treating oral diseases like RAS and OLP more effectively.

The mucoadhesive NFs have also been used for drug delivery application in the gastroenteric tract because they improve the bioavailability of the drug in the gastroenteric site. The oral strategies exhibit low bioavailability owing to short retaining time and incomplete release of the drug at the absorption zone. However, mucoadhesive NFs have the potential to deliver drugs for a longer time in the gastroenteric tract due to their prolonged contact time with gastric mucosa [25, 26]. The progesterone-loaded carboxymethyl cellulose (CMC) fibers were prepared by Brako and his research team [27] to study the mucoadhesive interaction between the progesterone-loaded CMC fibers and either mucosa membrane or artificial membrane (cellulose acetate) by texture analysis. The relationship between mucoadhesive and interfacial roughness was examined by AFM. The interpenetration of polymer-mucin and filling up of cavities on the mucosa surface led to interfacial roughness resulting in strong mucoadhesion. The adhesion of drug-loaded CMC fibers to lamb oesophageal mucosa was compared with that of artificial cellulose acetate membrane. It was found that the adhesion was 10 times higher for artificial cellulose acetate membrane as compared to lamb oesophageal mucosa. The possible relationship between the interfacial geometry of adhering surface and mucoadhesion was also assessed by mucoadhesive theories.

The developing oral formulations with controlled and sustained drug delivery are essential to delivering drugs slowly into the gastrointestinal tract for a longer period [28]. Gastroretentive drug delivery systems (GRDDS) are superior to conventional drug delivery systems owing to their improved drug solubility, possible reduction of the drug concentration, bioavailability and therapeutic efficacy [29–31]. The electrospun PLA fiber encapsulated with diacerein (DIA) was characterized to understand the physical state of the drug, morphology, release behavior and floating behavior for the gastroretentive drug delivery system [32]. The DIA drug in the PLA fibers was in the amorphous state according to XRD analysis, which improved the drug solubility in the fabricated PLA fibers. PLA fibers witnessed a slow drug release profile by releasing 61.3% of DIA in 30 hours. PLA is an anionic polymer that enabled a slow release of DIA drug at acidic pH. There was a reduction in the mucoadhesive strength of pristine PLA fibers from 38 g/cm^2 to 14 g/cm^2 after encapsulating DIA into PLA nanofibers owing to the increase in fiber density and fiber diameter. The drug release from PLA fibers followed a diffusion pattern by the Peppas model. The Higuchi and Peppas models also showed a uniform and smooth drug release pattern. Therefore, it was recommended as a promising gastroretentive drug delivery system.

12.3 BIOSENSING APPLICATIONS OF ELECTROSPUN NANOFIBERS

The demand for analytical tools like sensors and biosensors has risen due to the advancement in the field of environmental protection and medical diagnostics. The monitoring of biomolecules such as hormones, protein markers and drugs is very

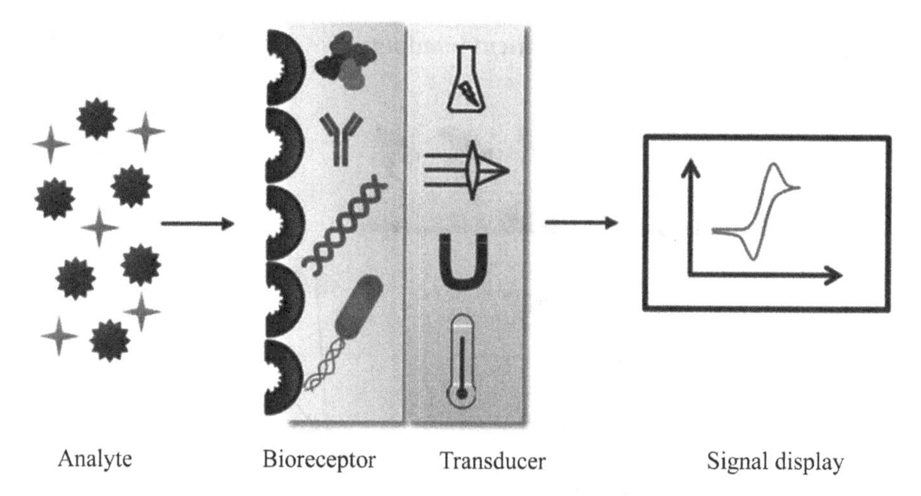

FIGURE 12.2 Schematic operating principle of a biosensor [34].

essential to prevent or control diseases [33]. The sensor consists of two main components like the receptor and transducer. A receptor generally converts the chemical information into a form of energy which is further transformed by the transducers. The recognition of the analyte is the major function of the receptor which is based on physical and chemical principles. However, the recognition of the analyte is associated with biochemical reactions in biosensors [34]. The classification of a biosensor is based on biorecognition. The responsible molecules for the binding of the analyte are nucleic acids, enzymes, cells, bacteria and antibodies [35–37]. The operating principle of a biosensor is illustrated in Figure 12.2. The occurrence of biorecognition can be achieved by the immobilization of biomolecules such as nucleic acids, antibodies and enzymes in the biosensor. The biomolecule is expected to maintain its function, shape, stability and activity while selecting a suitable immobilization method for biorecognition. Various methods like entrapment, adsorption, affinity binding, crosslinking and covalent bonding have been employed for enzyme immobilization (Figure 12.3). Nanomaterials facilitate better immobilization of bioreceptors owing to their high surface-to-volume ratio.

Various nanomaterials such as metal nanoparticles, quantum dots, nanofibers and nanotubes have been employed in sensing devices owing to their size-dependent unique properties which will affect the performance of the sensing platforms. Among the nanomaterials, nanofibers find applications in drug delivery, scaffolds, wound healing and biosensor due to their unique properties like porosity; good mechanical, chemical and physical properties; controllable morphology; flexibility and ease of fabrication [38–40]. There is an improvement in the immobilization of biomolecules with nanofibers which probably enhances the responsibility and sensitivity of the biosensors [41]. The electrospun NFs have been used in biosensors for the detection of microorganisms and biomolecules. The PAN NFs with a diameter of 677 nm were deposited on a gold electrode by Sapountzi et al. [42]. It was further impregnated with Fe(II) p-toulene sulfonate (FeTos) and also coated with conducting polypyrrole for

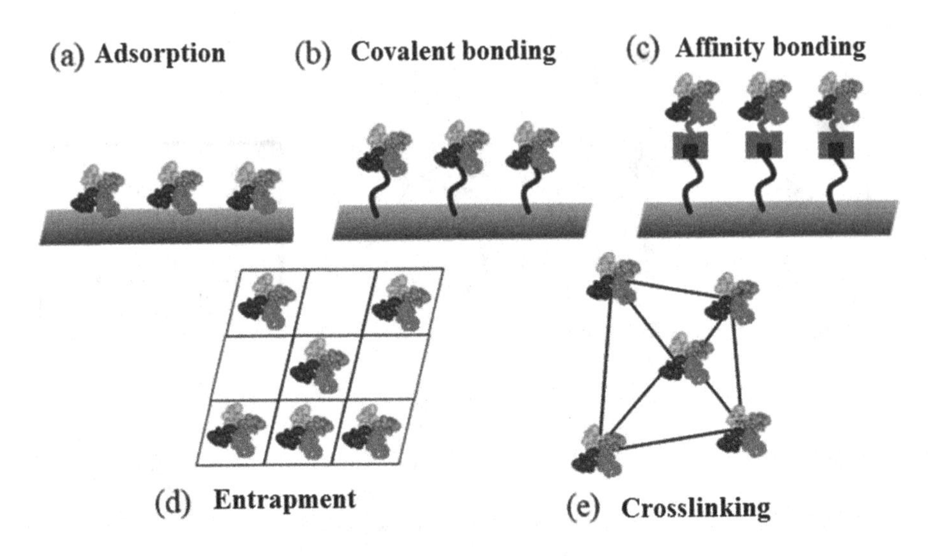

FIGURE 12.3 Enzyme immobilization methods [34].

the detection of glucose. The modified electrode was further covalently immobilized with an enzyme (glucose oxidase) and the glucose detection was carried out by the EIS method. The biosensor showed good linearity in the range of 20 nM–2 uM of glucose concentration, low LOD (2 nm), good selectivity and stability. In another study, the gold chip was coated with graphene oxide (GO) nanofibers via electrospinning and then it was coated with Au NPs that were incorporated with Cu nanoflower-glucose oxidase for glucose detection by Boak and his research team [43]. The Nafion was used as a binding agent for fabricating the surface-modified Au chip with Cu nanoflower@AuNPs-GO nanofibers. The modified Au chip was subjected to electrochemical experiments to assess the analytical performance. The glucose detection ability of the surface-modified Au chip biosensor was assessed by chronoamperometric and cyclic voltammetry analyses. The modified Au chip with Cu nanoflower@Au NPs-GO NFs showed linearity in the range of 0.001–0.1 mM for glucose concentration with a detection limit of 0.018 uM.

The combination of electrospinning and surfactant-assisted adsorption method (anionic) was employed to fabricate Ni/CoO/CNF electrode for the detection of glucose in biofluids quantitatively [44]. The effect of surfactant (sodium dodecyl sulfate) on the intensity of the glucose current response and morphology of Ni/CoO/CNF was investigated to assess the low detection limit and linear range of glucose concentration, and these values were estimated as 0.03 μM and 0.25–600 μM, respectively. The oxidation current of glucose increased linearly with the increasing concentration of glucose from 0.25–600 μM. The Ni/CoO/CNF electrode was kept in the refrigerator for 30 days to detect electrochemical response at regular intervals for the equivalent glucose concentration. The attained response was 86% after 30 days which proved that the stability of the biosensor was remarkably good. The performance of the sensor was also assessed in human serum (real sample) by determining glucose concentration. The i-t curve method was used to detect the glucose content at the applied voltage of +0.15 V in 10 μL of serum. The proposed Ni/CoO/CNF electrode showed a good recovery with variation from 96.37% to 101.20% in human serum. Guo et al. [45] fabricated a graphitic nanofiber/$NiCo_2S_4$ electrode by growing nanowire arrays of $NiCo_2S_4$ on graphitic nanofiber film for biosensing of nonenzymatic glucose. The grown $NiCo_2S_4$ nanowires created a lot of electrochemically active sites owing to their unique core-shell structure. The electrochemical performance of graphitic nanofiber/$NiCo_2S_4$ electrode was analyzed by measuring the peak current (I_p) and the magnitude of I_p was high due to the high electrical conductivity of graphitic nanofiber and the enhanced electrocatalytic activity of $NiCo_2S_4$ nanowires. The behavior of the electrode interface was assessed by the electrochemical impedance spectroscopy (EIS) technique. The term, R_{ct} is associated with the electron transfer ability of the electrode, and the R_{ct} value was found to be high for the fabricated graphitic nanofiber/$NiCo_2S_4$ electrode. Therefore, it was recommended to fabricate sensitive sensors. An increase in oxidation current upon the addition of glucose affirmed the excellent electrocatalytic activity of graphitic nanofiber/$NiCo_2S_4$ electrode towards the oxidation of glucose. While applying graphitic nanofiber/$NiCo_2S_4$ electrode for real sample testing, the glucose determination was affected by the coexisting electrochemical active species. In human serum, the glucose concentration is about 3–8 mM, while the concentration of DA, AA and UA are about 0.1 mM [46]. The graphitic nanofiber/$NiCo_2S_4$ electrode was monitored by the amperometric method to assess the operational stability of the sensor. The sensor exhibited high stability with a current response of 86.2%. The graphitic nanofiber/$NiCo_2S_4$ sensor was stored at room temperature for two weeks to investigate its storage stability. The acceptable storage stability of the fabricated sensor was concluded by the obtained current response of 94.2% for the initial current.

The blend of cellulose acetate (CA) and chitosan was coated onto a glassy carbon electrode (GCE) to fabricate CA-CS nanofiber by Yezer and Demirkol [47]. The GOx was further immobilized onto a modified CA-CS nanofiber electrode by covalent bonds. The modified CA-CS/GOx nanofiber electrode was examined to assess its glucose detention in samples. The electrochemical measurements were performed by carrying out the EIS, differential pulse voltammetry (DPV) and cyclic voltammetry (CV) measurements for the fabricated electrode. The DPV measurement was carried out with the scanning speed of 50 mV/s at the potential rate of –0.4 to +0.8 V in the presence of hexacyanoferrate (5 mM) with KCl as an auxiliary electrolyte. The CV measurement was also performed in the same experimental condition at the potential rate of 0 V to +0.8 V. The frequency range of 0.21×10^{-4}–100 KHz was selected to perform EIS measurement. The fabricated CA-CS/GOx sample was subjected to amperometric measurement to detect glucose levels in artificial samples such as urine, tears, serum and sweat. The recovery of glucose detection was estimated by comparing the measured glucose concentration by CA-CS/GOx electrode with the known glucose concentration in artificial samples. The recovery % was higher towards the glucose detection in artificial urine, tears, serum and sweat. The glucose detection by the various nanofiber-based sensors is listed in Table 12.2.

TABLE 12.2
The Electrochemical Glucose Detection by the Nanofiber-Based Biosensors [34]

Sensor	LOD	Linear range
NiMoO4/CNF	50 nM	0.0003–4.5 mM
PAN/PPy/PPy3COOH	2 nM	20 nM–2 µM
Ni/CoO CNF	0.03 µM	0.25–600 µM
Cu-nanoflower@AuNPs-GO NFs	0.018 µM	0.001–0.1 mM
NiCo2O4/EGF	0.167 µM	0.0005–3.571 mM
CNF@Ni-Co LDH	0.03 µM	1–2000 µM
CA/CS	4.8 µM	5.0 µM–0.75 mM
Ni/CNF	0.57 µM	2 µM–5 mM
PEI/PVA	0.3 mM	2–8 mM, 10–30 mM
CS/GO	0.02 mM	0.05–20 mM
Cellulose/β-CD	9.35×10^{-5} M	0–1 mM
SnO2 NFs	0.05 mM	0.5–5 mM

12.4 CONCLUSION

The efficient and reproducible mucoadhesive polymer NFs are fabricated by controlling the electrospinning parameters such as polymer concentration, electric field, solution conductivity, needle-collector distance, electrospinning setup, environmental parameters, viscosity and flow rate. The mucoadhesion of polymer nanofibers depends on the types of functional groups present in the polymer matrix. The anionic polymers are considered as a potential candidate for mucoadhesive systems because the cationic polymer has less effect in an acidic environment. The mucoadhesive NFs can be used as an efficient drug delivery system in the gastroenteric tract owing to the prolonged contact time with the gastric mucosa. The mucoadhesive patches incorporated with clobetasol-17-propionate are recommended for treating oral diseases like RAS and OLP. The PLA fiber encapsulated with DIA showed a uniform and smooth drug release profile. Therefore, it was recommended as a promising gastroretentive drug delivery system. The glucose detection ability of the biosensor is assessed by chronoamperometric and cyclic voltammetry analyses. The Au chip modified with PAN NFs and Cu nanoflower@Au NPs-GO NFs showed an excellent biosensing property towards the detection of glucose.

REFERENCES

[1] El-Hadi AM, Al-Jabri FY. Influence of electrospinning parameters on fiber diameter and mechanical properties of poly (3-hydroxybutyrate)(PHB) and polyanilines (PANI) blends. Polymers. 2016 Mar 22;8(3):97.

[2] Reneker DH, Chun I. Nanometre diameter fibres of polymer, produced by electrospinning. Nanotechnology. 1996 Sep 1;7(3):21.

[3] Huang ZM, Zhang YZ, Kotaki M, Ramakrishna S. A review on polymer nanofibers by electrospinning and their applications in nanocomposites. Composites science and technology. 2003 Nov 1;63(15):2223–53.
[4] Thenmozhi S, Dharmaraj N, Kadirvelu K, Kim HY. Electrospun nanofibers: New generation materials for advanced applications. Materials Science and Engineering: B. 2017 Mar 1;217:36–48.
[5] Ahadian S, Obregón R, Ramón-Azcón J, Salazar G, Ramalingam M. Clinical/preclinical aspects of nanofiber composites. In Nanofiber composites for biomedical applications 2017 (pp. 507–528). Woodhead Publishing.
[6] Kai D, Liow SS, Loh XJ. Biodegradable polymers for electrospinning: Towards biomedical applications. Materials Science and Engineering: C. 2014 Dec 1;45:659–70.
[7] Kim BC, Nair S, Kim J, Kwak JH, Grate JW, Kim SH, Gu MB. Preparation of biocatalytic nanofibres with high activity and stability via enzyme aggregate coating on polymer nanofibres. Nanotechnology. 2005 Apr 22;16(7):S382.
[8] Ghajarieh A, Habibi S, Talebian A. Biomedical applications of nanofibers. Russian Journal of Applied Chemistry. 2021 Jul;94:847–72.
[9] Talebian A, Mansourian A. Release of Vancomycin from electrospun gelatin/chitosan nanofibers. Materials Today: Proceedings. 2017 Jan 1;4(7):7065–9.
[10] Elhami M, Habibi S. A study on UV-protection property of poly (vinyl alcohol)-montmorillonite composite nanofibers. Journal of Vinyl and Additive Technology. 2021 Feb;27(1):89–96.
[11] Habibi S, Saket M, Nazockdast H, Hajinasrollah K. Fabrication and characterization of exfoliated chitosan–gelatin–montmorillonite nanocomposite nanofibers. The Journal of The Textile Institute. 2019 Nov 2;110(11):1672–7.
[12] Lee JB, Jeong SI, Bae MS, Yang DH, Heo DN, Kim CH, Alsberg E, Kwon IK. Highly porous electrospun nanofibers enhanced by ultrasonication for improved cellular infiltration. Tissue Engineering Part A. 2011 Nov 1;17(21–22):2695–702.
[13] Wang X, Ding B, Li B. Biomimetic electrospun nanofibrous structures for tissue engineering. Materials Today. 2013 Jun 1;16(6):229–41.
[14] Torres-Martínez EJ, Cornejo Bravo JM, Serrano Medina A, Pérez González GL, Villarreal Gómez LJ. A summary of electrospun nanofibers as drug delivery system: Drugs loaded and biopolymers used as matrices. Current Drug Delivery. 2018 Dec 1;15(10):1360–74.
[15] Abdul Hameed MM, Mohamed Khan SA, Thamer BM, Rajkumar N, El-Hamshary H, El-Newehy M. Electrospun nanofibers for drug delivery applications: Methods and mechanism. Polymers for Advanced Technologies. 2023 Jan;34(1):6–23.
[16] Yang Y, Li X, Qi M, Zhou S, Weng J. Release pattern and structural integrity of lysozyme encapsulated in core–sheath structured poly (dl-lactide) ultrafine fibers prepared by emulsion electrospinning. European Journal of Pharmaceutics and Biopharmaceutics. 2008 May 1;69(1):106–16.
[17] Xu X, Zhuang X, Chen X, Wang X, Yang L, Jing X. Preparation of core-sheath composite nanofibers by emulsion electrospinning. Macromolecular Rapid Communications. 2006 Oct 2;27(19):1637–42.
[18] Hu J, Prabhakaran MP, Tian L, Ding X, Ramakrishna S. Drug-loaded emulsion electrospun nanofibers: characterization, drug release and in vitro biocompatibility. RSC Advances. 2015;5(121):100256–67.
[19] Sudhakar Y, Kuotsu K, Bandyopadhyay AK. Buccal bioadhesive drug delivery—A promising option for orally less efficient drugs. J Control Release. 2006;114(1):15–40. doi:10.1016/j.jconrel.2006.04.012.
[20] Pérez-González GL, Villarreal-Gómez LJ, Serrano-Medina A, Torres-Martínez EJ, Cornejo-Bravo JM. Mucoadhesive electrospun nanofibers for drug delivery systems: applications of polymers and the parameters' roles. International Journal of Nanomedicine. 2019 Jul 15:5271–85.

[21] Huang C, Soenen SJ, van Gulck E, Vanham G, Rejman J, Van Calenbergh S, Vervaet C, Coenye T, Verstraelen H, Temmerman M, Demeester J. Electrospun cellulose acetate phthalate fibers for semen induced anti-HIV vaginal drug delivery. Biomaterials. 2012 Jan 1;33(3):962–9.
[22] Hua D, Liu Z, Wang F, Gao B, Chen F, Zhang Q, Xiong R, Han J, Samal SK, De Smedt SC, Huang C. pH responsive polyurethane (core) and cellulose acetate phthalate (shell) electrospun fibers for intravaginal drug delivery. Carbohydrate Polymers. 2016 Oct 20;151:1240–4.
[23] Illangakoon UE, Gill H, Shearman GC, Parhizkar M, Mahalingam S, Chatterton NP, Williams GR. Fast dissolving paracetamol/caffeine nanofibers prepared by electrospinning. International Journal of Pharmaceutics. 2014 Dec 30;477(1–2):369–79.
[24] Colley HE, Said Z, Santocildes-Romero ME, Baker SR, D'Apice K, Hansen J, Madsen LS, Thornhill MH, Hatton PV, Murdoch C. Pre-clinical evaluation of novel mucoadhesive bilayer patches for local delivery of clobetasol-17-propionate to the oral mucosa. Biomaterials. 2018 Sep 1;178:134–46.
[25] Malik R, Garg T, Goyal AK, Rath G. Polymeric nanofibers: targeted gastro-retentive drug delivery systems. Journal of Drug Targeting. 2015 Feb 7;23(2):109–24.
[26] Malik R, Garg T, Goyal AK, Rath G. Diacerein-Loaded novel gastroretentive nanofiber system using PLLA: Development and in vitro characterization. Artificial Cells, Nanomedicine, and Biotechnology. 2016 Apr 2;44(3):928–36.
[27] Brako F, Thorogate R, Mahalingam S, Raimi-Abraham B, Craig DQ, Edirisinghe M. Mucoadhesion of progesterone-loaded drug delivery nanofiber constructs. ACS Applied Materials & Interfaces. 2018 Mar 29;10(16):13381–9.
[28] Chaudhary S, Garg T, Murthy RS, Rath G, Goyal AK. Recent approaches of lipid-based delivery system for lymphatic targeting via oral route. Journal of Drug Targeting. 2014 Dec 1;22(10):871–82.
[29] Goyal AK, Rath G, Garg T. Nanotechnological approaches for genetic immunization. DNA and RNA Nanobiotechnologies in Medicine: Diagnosis and Treatment of Diseases. 2013:67–120.
[30] Goyal G, Garg T, Malik B, Chauhan G, Rath G, Goyal AK. Development and characterization of niosomal gel for topical delivery of benzoyl peroxide. Drug delivery. 2015 Nov 17;22(8):1027–42.
[31] Garg T, Kumar A, Rath G, Goyal A. Gastroretentive drug delivery systems for therapeutic management of peptic ulcer. Critical Reviews™ in Therapeutic Drug Carrier Systems. 2014;31(6).
[32] Malik R, Garg T, Goyal AK, Rath G. Diacerein-Loaded novel gastroretentive nanofiber system using PLLA: Development and in vitro characterization. Artificial Cells, Nanomedicine, and Biotechnology. 2016 Apr 2;44(3):928–36.
[33] Chen L, Hwang E, Zhang J. Fluorescent nanobiosensors for sensing glucose. Sensors. 2018 May 5;18(5):1440.
[34] Halicka K, Cabaj J. Electrospun nanofibers for sensing and biosensing applications—A review. International Journal of Molecular Sciences. 2021 Jun 14;22(12):6357.
[35] Dalirirad S, Steckl AJ. Aptamer-based lateral flow assay for point of care cortisol detection in sweat. Sensors and Actuators B: Chemical. 2019 Mar 15;283:79–86.
[36] Tlili C, Myung NV, Shetty V, Mulchandani A. Label-free, chemiresistor immunosensor for stress biomarker cortisol in saliva. Biosensors and Bioelectronics. 2011 Jul 15;26(11):4382–6.
[37] Ferro Y., Perullini M., Jobbagy M., Bilmes S. A., Durrieu C. Development of a biosensor for environmental monitoring based on microalgae immobilized in silica hydrogels. Sensors. 2012;12:16879–16891. doi: 10.3390/s121216879
[38] Terra I. A. A., Mercante L. A., Andre R. S., Correa D. S. Fluorescent and colorimetric electrospun nanofibers for heavy-metal sensing. Biosensors. 2017;7:61.

[39] Balusamy B., Senthamizhan A., Uyar T. Functionalized electrospun nanofibers as colorimetric sensory probe for mercury detection: A review. Sensors. 2019;19:4763.
[40] Thenmozhi S., Dharmaraj N., Kadirvelu K., Kim H. Y. Electrospun nanofibers: New generation materials for advanced applications. Materials Science and Engineering: B. 2017;217:36–48.
[41] Chen K., Chou W., Liu L., Cui Y., Xue P., Jia M. Electrochemical sensors fabricated by electrospinning technology: An overview. Sensors. 2019;19:3676.
[42] Sapountzi E, Chateaux JF, Lagarde F. Combining electrospinning and vapor-phase polymerization for the production of polyacrylonitrile/polypyrrole core-shell nanofibers and glucose biosensor application. Frontiers in Chemistry. 2020 Aug 4;8:678.
[43] Baek SH, Roh J, Park CY, Kim MW, Shi R, Kailasa SK, Park TJ. Cu-nanoflower decorated gold nanoparticles-graphene oxide nanofiber as electrochemical biosensor for glucose detection. Materials Science and Engineering: C. 2020 Feb 1;107:110273.
[44] Mei Q, Fu R, Ding Y, Wang A, Duan D, Ye D. Electrospinning of highly dispersed Ni/CoO carbon nanofiber and its application in glucose electrochemical sensor. Journal of Electroanalytical Chemistry. 2019 Aug 15;847:113075.
[45] Guo Q, Wu T, Liu L, He Y, Liu D, You T. Hierarchically porous NiCo2S4 nanowires anchored on flexible electrospun graphitic nanofiber for high-performance glucose biosensing. Journal of Alloys and Compounds. 2020 Apr 5;819:153376.
[46] Ci S, Huang T, Wen Z, Cui S, Mao S, Steeber DA, Chen J. Nickel oxide hollow microsphere for non-enzyme glucose detection. Biosensors and Bioelectronics. 2014 Apr 15;54:251–7.
[47] Yezer I, Demirkol DO. Cellulose acetate–chitosan based electrospun nanofibers for bio-functionalized surface design in biosensing. Cellulose. 2020 Nov;27(17):10183–97.

13 Electrospun Nanofibers Based on Microalgae Pigments and Their Applications in Intelligent Food Packaging

Ana Luiza Machado Terra, Suelen Goettems Kuntzler, Jorge Alberto Vieira Costa, Michele Greque de Morais, and Juliana Botelho Moreira

13.1 INTRODUCTION

Packaging is an indispensable component of the food supply chain. It acts as a protection layer or barrier against contamination, prolonging the product's shelf life during storage, transport, and market exposition [1]. As the expiration date is related to the inherent characteristics of the products and environmental factors, it is crucial to monitor in real-time these factors that impact quality, sensory properties, and food safety [1]-[3]. Thus, an innovative packaging system emerges that assigns new functions such as monitoring, detecting, and recording external or internal changes to packaging [4].

Intelligent packaging informs the consumer about biochemical and microbiological changes detected in the food, monitoring the quality of the product. Intelligent systems are categorized into time-pH, time-temperature, and freshness detectors [5]. Intelligent components are added to packaging materials to interact with internal packaging factors and external environmental parameters. As a result, there is a color change on the packaging that informs the viability of the food [4].

Several techniques can manufacture high-efficiency intelligent packaging materials [6]. Among these, the electrospinning process stands out, which has the advantage of being used on a laboratory scale to an industrial scale [7]. Electrospinning is an efficient technique for nanofibers production. Under the action of the electric field, polymeric solutions produce continuous jets that elongate, forming the nanomaterial [8]. Nanofibers have a high surface area, the capacity to incorporate bioactive agents, and high porosity, allowing greater detection sensitivity and reactivity of

DOI: 10.1201978100333814-13

the material. Besides, the absence of remaining solvents from the production process is an advantage for applying these structures in the food industry [9].

Adding intelligent components in nanofibers that identify factors of microbiological deterioration and biochemical changes in food is essential in developing intelligent systems. These components can be dyes that change color to communicate about quality of foods. However, the food industry must control the use of chemical dyes, as they lead to bioaccumulation in the body, causing damage to human health [10]. Therefore, there is a growing search for replacing these dyes with natural pigments, mainly due to their non-toxicity and biocompatibility with human cells and tissues [11]. In this sense, microalgae biomass and its pigments (chlorophylls, carotenoids, and phycocyanin) have been used in producing electrospun nanofibers [2],[3],[12],[13].

The microalgae of the genus *Chlorella* and *Spirulina* contain pigments in their composition that can be used in the food industry without risk to human health due to certification of Generally Recognized as Safe (GRAS). Chlorophyll and carotenoids are natural colorants from *Chlorella* [14]. In addition to these pigments, *Spirulina* is composed of phycocyanin [15]. Chlorophyll is a pigment that contributes to the green coloration in microalgae. The cause of its discoloration is the conversion of chlorophyll into pheophytin due to the influence of pH [16]. The phycocyanin is a blue color characterized as a phycobiliprotein that suffers denaturation on the pH effects, altering the load on the protein and causing changes in coloration [2]. This color-changing ability of these pigments indicates the potential to use these microalgal compounds as intelligent components to detect and monitor food quality. In this context, this review aims to present a new approach to developing intelligent systems for food monitoring from nanofibers produced by electrospinning, microalgal biomass, and microalgae pigments.

13.2 CONVENTIONAL FOOD PACKAGING × INTELLIGENT PACKAGING

Traditional food packaging has the use of preserving the characteristics and increasing the safety of the products. The packaging acts as an inert barrier between the food and the environment (light, moisture, and oxygen), preventing microbiological contamination and chemical changes and allowing wide distribution [4]. In addition to protecting and conserving food, packaging plays a role in communicating with the consumer through marketing, nutritional information, and manufacturing data. In this way, traditional packaging started to conserve and expose the products and attract consumers through their communicative visual aspects [17],[18].

Traditional packaging can be classified according to functionality or production as primary, secondary, or tertiary. The primary packaging comes in direct contact with the food. The secondary one facilitates the handling and presentation of the product and can also protect the primary packaging inside, avoiding excessive shocks and vibrations. Tertiary packaging protects the product during transport and storage [19]. According to production, the packaging is classified into laminate, carton, and multilayer. Laminate packaging is formed by overlapping materials (plastic and metalized films and papers). However, when one of the layers is constituted of cardboard, these are called cartons. When the packaging has more than two layers, it is called a multilayer. This variety of materials used to compose the packaging aims to expand the range of products that can be packaged [20]. However, the increase in demand for safe food by consumers implies producing packaging with additional functions, such as monitoring viability, which has been developed based on science and technology.

Intelligent systems are an extension of the traditional function of packaging. They establish strategies to reduce or eliminate the occurrence of foods unsuitable for consumption. Intelligent packaging is defined as indicator systems, internal or external, capable of detecting traces or signs and constantly monitoring the shelf life of foods (physical-chemical and biochemical parameters). Moreover, these systems monitor the environment and other aspects of packaging integrity during food processing, transportation, and storage [8]. Using these systems can benefit the expansion of the market in the context of globalization, as it assists national and international food safety regulations [21]. Therefore, intelligent packaging can represent advances to avoid food waste and adulteration. Furthermore, it also contributes to improving logistics and product traceability [22].

The intelligent indicator systems are composed of intelligent components, which can be produced from nanotechnology. These compounds interact with internal packaging factors (food components, carbon dioxide (CO_2), oxygen gas (O_2), antimicrobial agents, and volatile compounds) and/or external environmental factors (physical, environmental, and user) [23],[24]. As a result of the interaction between these factors and the indicator, a visual change will occur. This system improves the Hazard Analysis and Critical Control Point (HACCP) and anti-counterfeiting. Moreover, intelligent packaging ensures the effectiveness of food safety and improves convenience and consumer satisfaction [25].

Intelligent systems inform about changes detected in the product or its environment, monitoring foods. Examples of this packaging category include time-pH, time-temperature, and freshness detectors [5]. Indicator systems demand that the consumer interpret the measurement so that, based on this answer, they can make the product purchase decision [4]. The additional functionalities of intelligent food packaging refer to (a) retaining the integrity of the packaging, preventing the loss or tampering of the product; (b) improving the product's attributes (such as appearance, taste, aroma, viscosity, and texture); (c) measure on the packaging the possible variations in the environment (internal and external factors) of the product; (d) communicate product conditions and history to consumers; and (e) ensure the authenticity or counterfeiting of foods [22],[26].

13.3 INTELLIGENT PACKAGING BASED IN ELECTROSPUN NANOFIBERS

Nanofibers present a high surface area to the volume that improves intelligent packaging sensitivity and response time. This property increases material reactivity, accelerates adsorption or release mechanisms, and increases the number of sites for interaction or bonding of reactive materials from sensors and indicators [4]. Incorporating bioactive compounds in nanofibers contributes to preserving the physical and biological characteristics of compounds sensitive to light, temperature, and humidity. That provides improved stability, bioavailability, and controlled release properties of the biomolecule [27]. Some studies encapsulated compounds in nanofibers with the potential to evaluate the deterioration of perishable foods (Table 13.1).

Another important property for developing intelligent packaging is the porosity of the nanofibers. The interstitial pores derive from the interlacing and overlapping of the fibers during the electrospinning process. Pores are considered intrinsic characteristics and are present in individual nanofibers. They can have their size and shape controlled by the electrospinning parameters [37]. The high porosity proportionally reduces the resistance to mass transport of the fluid passing through the membrane pores. It also increases the surface area ratio of the nanofiber. Thus, there is an improvement in the detection performance and response time of intelligent packaging [38]. The possibility of filling this nanostructure with bioactive compounds provides applications in several sensors and indicators.

Drawing, template synthesis, phase separation, self-assembly, and electrospinning are methods to produce nanofibers. In the electrospinning process, electrical charges are applied to the drop of liquid leaving the capillary, deforming the drop interface. These charges generate electrostatic forces inside the droplet (Coulomb

TABLE 13.1

Nanofiber-Based Colorimetric Indicators for Monitoring Food Freshness

Compounds	Polymers	Detection	Application	Reference
Açaí (Euterpe oleracea) extract	Polycaprolactone and polyethylene oxide	pH	–	[28]
Anthocyanins and curcumin	Pullulan and chitin	pH	*Plectorhynchus cinctus* (fish)	[29]
Anthocyanins extract	Polyvinyl alcohol and κ-carrageenan	pH	–	[30]
Curcumin	Gelatin and chitosan	Volatile ammonia	–	[31]
Curcumin	Chitosan and polyethylene oxide	pH and total volatile basic nitrogen	Chicken breast	[32]
Curcumin	Polyvinylpyrrolidone and ethylcellulose/ polyethylene oxide	Volatile amines	–	[33]
Laccase	Zein	Time-temperature	–	[34]
Purple sweet potato extract and carvacrol	Pullulan and zein	pH and total volatile basic nitrogen	Pork	[35]
Red cabbage (*Brassica oleracea L.*) extract	Polyvinyl alcohol	pH	Fruit (Rutab)	[36]

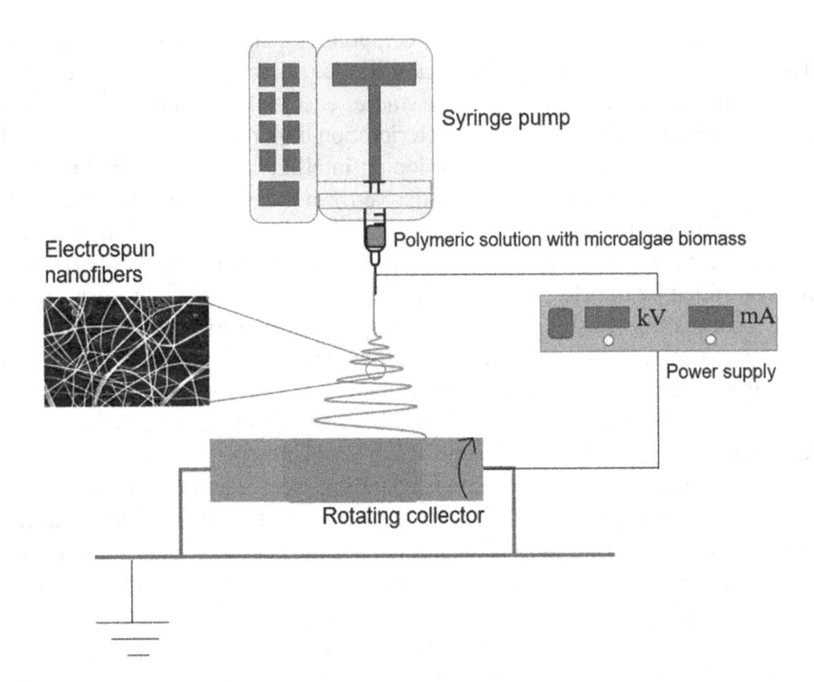

FIGURE 13.1 Schematic diagram of the nanofibers production containing microalgal biomass by the electrospinning.

force), which compete with the droplet's surface tension, forming the Taylor cone posteriorly nanofibers (Figure 13.1). The electrospinning machine included a high-potential electric source, positive detachment pump, capillary, and grounded collector (static or rotary) [4],[8]. Moreover, it uses ambient temperature during the process and does not remain solvent in the product, allowing the incorporation of bioactive and thermosensitive compounds in the nanofibers. The principal advantage of electrospinning over other techniques is the possibility of scale-up and the repeatability of the process [39].

Free-surface electrospinning produces continuous nanofibers on a large scale. This system also uses high electrical potential and is characterized by the ejection of a solution by the conductive wire electrode, forming various polymeric jets [40],[41]. A container with the polymeric solution feeds the conductor wire. This feeding system makes it possible to keep the solution viscosity stable throughout the process [42]. The free-surface electrospinning process is desirable for pilot-scale production due to the higher production yield and the versatility of the injection of the solution [39].

13.4 POTENTIAL APPLICATION OF MICROALGAE AND THEIR PIGMENTS INTO NANOFIBERS

Natural dyes do not harm the environment and are non-toxic and biocompatible with cells and tissues. Moreover, they can be obtained from different sources and are alternatives to synthetic dyes that can potentially cause health risks [43]. Microalgae are key photosynthetic organisms in aquatic ecosystems and are a natural source of biologically active compounds of commercial interest. *Spirulina* and *Chlorella* are the most studied microalgae due to their composition of proteins, lipids, carbohydrates, minerals, and pigments (β-carotene, chlorophyll, phycocyanin) used in several areas [44].

13.4.1 Microalgae Biomass

The microalgae *Chlorella* belongs to the phylum of Chlorophyta, the class Chlorophyceae, and was discovered by researcher Martinus Willem Beijerinck in 1890, which was classified as the first microalgae with a well-defined nucleus [45]. This microorganism is spherical, globular, or ellipsoidal and characterized as eukaryotic, unicellular with a diameter ranging from 2.0 to 10.0 μm [46].

Chlorella is one of the most studied microalgae for its biotechnological importance as a source of human nutrition; their biomass can be sold to consumers or applied in food products [47]. Cells grown under ideal conditions produce biomass rich in proteins, lipids, pigments, vitamins, and minerals [48]. Photosynthetic pigments such as chlorophyll and carotenoids are mainly used for food, pharmaceutical, or cosmetic applications such as natural dyes, food supplements, and sources of bioactive molecules [49].

Spirulina (*Arthrospira*) present a differentiated arrangement of cylindrical multicellular trichomes in an open spiral. The helical shape of the trichomes is characteristic of the genus, although the length and size of the spiral vary with the species [50]. This microalga has high nutritional value, biocompounds, high content of pigments, minerals, and trace elements. The pigment β-carotene, tocopherols, phycocyanin, and phycoerythrin are part of the composition of *Spirulina* [44]. Both microalgae (*Spirulina* and *Chlorella*) are listed by the Food and Drug Administration for use as food without risk to human health, when cultivated in a proper environment, with hygiene and good manufacturing practices [51],[52].

In this context, Kuntzler et al. [12] developed a pH indicator based on PLA/PEO nanofibers with microalgal biomass. In this study, the biomass of the microalgae *Spirulina* sp. LEB 18 changes color according to the pH values tested (1–10). The presence of pigments inside the cell, such as phycocyanin and chlorophyll, induces color change. The indicator developed with 2% biomass and a thickness of 99.0 ± 7.7 μm showed the variation from brown to green color between pH 5 and 7 with ΔE values greater than 21 and statistically different. In this pH range, various foods are found such as beef, pork, and fish. Therefore, the colorimetric pH indicator developed with *Spirulina* sp. LEB 18 biomass can be a convenient and visual method to estimate quality changes in meat products since changes in pH values indicate meat deterioration.

13.4.2 Chlorophylls and Carotenoids

Chlorophylls are green pigments essential for photosynthesis in plants, microalgae, and cyanobacteria. Green microalgae, such as *Chlorella* sp., are primary sources of chlorophyll production and can present about 7% in its biomass [53]. Due to the high green pigmentation and the growing consumer demand for natural products, chlorophylls are applied as colorants in the food, pharmaceutical, and cosmetic industries [54],[55]. They can also act as biocompounds on indicators and sensors on food packaging.

Chlorophylls a and b are abundant compounds in the *Spirulina platensis*, which various between 3 and 10 mg/g of dry biomass [56]. The green coloration of chlorophylls, under acidic conditions, turns to olive green relative to pheophytin due to the loss of magnesium ions from the structure [16]. The rate of chlorophyll degradation accelerates as the acidic pH indicating the loss of green color occurs at lower pH levels [57]. Medina-Jaramillo et al. [58] observed that chlorophyll and carotenoids in basil and green tea extracts changed color under different acidic and alkaline pH. The film containing basil extract at acidic pH changed from greenish yellow to white. At the alkaline pH medium, the color of the tea extract changed to bright yellow. The film with green tea extract at basic pH showed the darkening of coloration.

Carotenoids are essential to protect microalgae cells of reactive species from oxygen generated during photosynthesis and from the high light intensity. The principal carotenoids of commercial interest for microalgae are β-carotene, lutein, and astaxanthin [59]. Carotenoids are susceptible to isomerization and oxidation due to conjugated double bonds, which present high sensitivity to light, oxygen, acids, and alkalis [60].

The principal sources of carotenoids are the microalgae belonging to the class Chlorophyceae. The carotenoids vary in color from yellow to red, and under adverse conditions of cultivation, these pigments can be produced to 12% in the phylum Chlorophyta [61],[62]. Some of the carotenoids produced by species of the genus *Chlorella*, such as astaxanthin and lutein, are exploited in the food, feed, cosmetic, and pharmaceutical industries [54],[63]-[65].

13.4.3 Phycocyanin

Phycocyanin is a pigment extracted mainly from the microalga *Spirulina* located in the thylakoid system or photosynthetic lamellae in the cytoplasmic membrane. This pigment constitutes up to 14% of the protein content of the microalgal biomass. Phycocyanin, composed of an apoprotein and a non-protein component known as phycocyanobilin (chromophore), is an open chain tetrapyrrole responsible for the blue color of phycocyanin. The apoprotein is linked to a phycocyanobilin by a thioether linkage. α and β subunits of molecular weights in the range of 18 and 20 kDa constitute the protein portion [66].

Phycocyanin has an intense blue color, is water-soluble, and has non-toxic properties (GRAS certified by the FDA) [66]. This pigment also has known antioxidant [40],[67], anti-inflammatory [68], and antitumor [69] properties. The degradation of phycocyanin and consequent color change occurs according to the state of protein aggregation when exposed to different environmental factors, mainly temperature and pH [2],[66]. According to Chaiklahan et al. [70], the maximum stability of the phycocyanin pigment was obtained in the range of pH 5.5 to 6 when applied at room temperature. Duangsee et al. [71] found that temperature variation affects the stability of the phycocyanin molecule, with a consequent change in color to variations in the blue hue. The loss of molecular stability occurred mainly in solutions at pH > 5 and pH < 3 with heating of approximately 10°C and could have been caused by the partial or total unfolding of the protein. Since this pigment is sensitive to pH variations with a consequent color change, it presents sensory properties for producing colorimetric indicators to apply in intelligent packaging [8].

Recent studies have investigated phycocyanin as an intelligent component for developing indicator nanofibers to apply in food packaging. Terra et al. [13] developed nanofibers with natural dyes (curcumin, quercetin, and/or phycocyanin). The authors monitored the color change for 24 hours using buffer solutions (pH 2 to 7). The best results were found with phycocyanin. When adding 2% of this microalgae pigment, a color variation ($\Delta E \geq 5$) was observed in the pH range of 3–6 in 5 hours of analysis. This result indicates a color change that is perceptible to the naked eye. Thus, the study found that systems of nanotechnological indicators added with phycocyanin can detect possible food degradation in less than 24 hours caused by inadequate storage conditions. Terra et al. [3] also produced nanofibers with microalgal pigment (phycocyanin). They observed that incorporating 1 and 2% of phycocyanin in nanofibers resulted in 75% and 71% of $\Delta E > 5$ values, respectively. When added 2% phycocyanin, the most significant color changes ($\Delta E \geq 8.5$) were at pH 3→4, 3→5, and 5→6. In addition, the work carried out irreversibility tests that proved the reliability of nanofiber indicators of phycocyanin to monitor the viability of foods. Another study evaluated the thickness of phycocyanin nanofibers in response concerning color change. Moreira et al. [2] found that from 3% phycocyanin, the highest thickness (68.7 μm) provided a value of $\Delta E = 18.85$ for pH 3 to 4 and $\Delta E = 18.66$ for pH 5 to 6, also proving the potential of this pigment to be applied in nanofibers to develop intelligent indicators in food packaging.

13.5 CHALLENGES

The shelf life of food is the period that ensures the product is safe, maintaining the sensory, chemical, and physical characteristics and fulfilling the nutritional claim. The shelf life estimation has become increasingly important due to technological developments and consumer interest in safe foods [72]. According to Codex Alimentarius [73], the shelf life of food is defined as the period during which a food preserves microbiological safety at a given storage temperature. However, the challenge is to maintain food safety in large-scale production processes and increasingly long distribution networks. Variations in temperature during food storage can influence the deterioration kinetics, accelerating the growth of deteriorating and pathogenic microorganisms. This problem leads to economic losses and compromises the quality and safety of food [74].

According to the World Health Organization [75], 600 million – almost 1 in 10 people worldwide – fall ill after eating contaminated food, and 420,000 die per year. In the Americas, around 77 million people suffer a foodborne illness each year. Available data indicate that foodborne diseases generate $700,000 to $19 million in annual healthcare costs in the Caribbean and more than $77 million in the United States [76]. Furthermore, food insecurity affects the image of companies and food segments, with consequent national economic losses [74]. Monitoring and controlling the production and distribution chain to the final consumer is essential for producing safe food [77]. Nanotechnology offers solutions for manufacturing, processing, distribution, and storage. Besides, nanomaterials contribute to the quality and safety of food, contributing to the health benefits that food provides [25].

13.6 CONCLUDING REMARKS

The food industry is increasing demand for products or processes that aim to improve food quality and safety, using environmentally friendly technologies or additives of natural origin. A relevant segment in the food area is intelligent packaging, which has been encouraging new alternatives to improve food monitoring. In this way, nanofiber engineering emerges to enhance and replace current food packaging materials. The applications of nanofibers produced by electrospinning have grown markedly in the last decade, with several research studies and reviews on intelligent food packaging.

Moreover, microalgae are relevant sources of natural pigments, such as phycocyanin, carotenoids, and chlorophyll. They can be explored individually or using total biomass for developing electrospun nanofibers. However, there are just a few studies on applying microalgae as intelligent food packaging, mainly *Chlorella.* Therefore, microalgae and their pigments were explored in this review to fill this gap and present these microorganisms as a source of research and innovation.

ACKNOWLEDGMENT

This study was financed in part by the Coordenação de Aperfeiçoamento de Pessoal de Nível Superior – Brasil (CAPES) – Finance Code 001. This research was developed within the scope of the Capes-PrInt Program (Process #88887.310848/2018–00).

REFERENCES

[1] Kalpana, S.; Priyadarshini, S.R.; Leena, M.M.; Moses, J.A.; Anandharamakrishnan, C. Intelligent packaging: Trends and applications in food systems. *Trends Food Sci. Technol.*2019, *93*, 145–157. https://doi.org/10.1016/j.tifs.2019.09.008.

[2] Moreira, J.B.; Terra, A.L.M.; Costa, J.A.V.; Morais, M.G. Development of pH indicator from PLA/PEO ultrafine fibers containing pigment of microalgae origin. *Int. J. Biol. Macromol.*2018, *118*(B):1855–1862. https://doi.org/10.1016/j.ijbiomac.2018.07.028.

[3] Terra, A.L.M.; Moreira, J.B.; Costa, J.A.V.; Morais, M.G. Development of pH indicators from nanofibers containing microalgal pigment for monitoring of food quality. *Food Biosci.*2021, *44*, 101387. https://doi.org/10.1016/j.fbio.2021.101387.

[4] Moreira, J.B., Kuntzler, S.G., Terra, A.L.M., Costa, J.A.V., Morais, M.G. Electrospun nanofibers: Fundamentals, food packaging technology, and safety. In Food packaging advanced materials, technologies, and innovations (pp. 223–254). Taylor and Francis, London, 2020.

[5] Sohail, M.; Sun, D.W.; Zhu, Z. 2018. Recent developments in intelligent packaging for enhancing food quality and safety. *Crit. Rev. Food Sci. Nut.*2018, *58*(15), 2650–2662. https://doi.org/10.1080/10408398.2018.1449731.

[6] Kumar, T.S.K.; Kumar, K.S.; Rajini, N.; Siengchin, S.; Ayrilmis, N.; Rajulu, A.V. A comprehensive review of electrospun nanofibers: Food and packaging perspective. *Compos. B. Eng. Compos. Part. B-Eng.*2019, *175*, 107074. https://doi.org/10.1016/j.compositesb.2019.107074.

[7] Leidy, R.; Ximena, Q.-C.M. Use of electrospinning technique to produce nanofibres for food industries: A perspective from regulations to characterisations. *Trends Food Sci. Technol.*2019, *85*, 92–106. https://doi.org/10.1016/j.tifs.2019.01.006.
[8] Moreira, J.B.; Morais, M.G.; Morais, E.G.; Vaz, B.S.; Costa, J.A.V. Electrospun polymeric nanofibers in food packaging. In Impact of nanoscience if food industry (pp. 387–417). Academic Press, London, 2018.
[9] Topuz, F.; Uyar, T. Antioxidant, antibacterial and antifungal electrospun nanofibers for food packaging applications. *Food Res. Int.*2020, *130*, 108927. https://doi.org/10.1016/j.foodres.2019.108927.
[10] Lehmkuhler, A.L.; Miller, M.D.; Bradman, A.; Castorina, R.; Mitchell, A.E. Dataset of certified food dye levels in over the counter medicines and vitamins intended for consumption by children and pregnant women. *Data Br.*2020, *32*, 106073. https://doi.org/10.1016/j.dib.2020.106073.
[11] Alizadeh-Sani, M.; Mohammadian, E.; Rhim, J.-W.; Jafari, S.M. pH-sensitive (halochromic) smart packaging films based on natural food colorants for the monitoring of food quality and safety. *Trends Food Sci. Technol.*2020 *105*, 93–144. https://doi.org/10.1016/j.tifs.2020.08.014.
[12] Kuntzler, S.G.; Costa, J.A.V.; Brizio, A.P.D.R.; Morais, M.G. Development of a colorimetric pH indicator using nanofibers containing *Spirulina* sp. LEB 18. *Food Chem.*2020, *328*, 126768. https://doi.org/10.1016/j.foodchem.2020.126768.
[13] Terra, A.M.; Moreira, J.B.; Costa, J.A.V.; Morais, M.G. Development of time-pH indicator nanofibers from natural pigments: An emerging processing technology to monitor the quality of foods. *LWT.*2021 *142*, 111020. https://doi.org/10.1016/j.lwt.2021.111020.
[14] Baidya, A.; Akter, T.; Islam, M.R.; Shah, A.K.M.A.; Hossain, M.A.; Salam, M.A.; Paul, S.I. Effect of different wavelengths of LED light on the growth, chlorophyll, β-carotene content and proximate composition of *Chlorella ellipsoidea*. *Heliyon.*2021 *7*(12), e08525. https://doi.org/10.1016/j.heliyon.2021.e08525.
[15] Bortolini, D.G.; Maciel, G.M.; Fernandes, I.A.A.; Pedro, A.C.; Ruio, F.T.V.; Branco, I.G.; Haminiuk, C.W.I. Functional properties of bioactive compounds from *Spirulina* spp.: Current status and future trends. *Food Chem.*2022, *5*, 100134. https://doi.org/10.1016/j.fochms.2022.100134.
[16] Yilmaz C., Gökmen V. *Chlorophyll*. In Encyclopedia of Food and Health (pp. 37–41). Elsevier, Amsterdam, 2016.
[17] Koeijer, B.; Lange, J.; Wever, R. Desired, perceived, and achieved sustainability: Trade-offs in strategic and operational packaging development. *Sustainability.*2017, *9*(10), 1923. https://doi.org/10.3390/su9101923.
[18] McMillin, K.W. Advancements in meat packaging. *Meat Sci.*2017, *132*, 153–162. https://doi.org/10.1016/j.meatsci.2017.04.015.
[19] Grundey, D. Functionality of product packaging: surveying consumers' attitude towards selected cosmetic brands. *Econ. Sociol.*2010, *3*, 87–103.
[20] Bucci, D.Z.; Tavares, L.B.B.; Sell, I. PHB packaging for the storage of food products. *Polym. Test.*2005, *24*(5), 564–571. https://doi.org/10.1016/j.polymertesting.2005.02.008.
[21] Schaefer, D.; Cheung, W.M. Smart packaging: opportunities and challenges. *Procedia CIRP.*2018, *72*, 1022–1027. https://doi.org/10.1016/j.procir.2018.03.240.
[22] Müller, P.; Schmid, M. Intelligent packaging in the food sector: a brief overview. *Foods.*2019, *8*(1), 16. https://doi.org/10.3390/foods8010016.
[23] Sharma, C.; Dhiman, R.; Rokana, N.; Panwar, H. Nanotechnology: an untapped resource for food packaging. *Front Microbiol.*2017, *8*, 1735. https://doi.org/10.3389/fmicb.2017.01735.
[24] Yildirim, S.; Röcker, B.; Pettersen, M.K.; Nilsen-Nygaard, J.; Ayhan, Z.; Rutkaite, R.; Radusin, T.; Suminska, P.; Marcos, B.; Coma, V. Active packaging applications for food. *Compr. Rev. Food Sci.*2017, *17*, 165–199. https://doi.org/10.1111/1541-4337.12322.

[25] Drago, E.; Campardelli, R.; Pettinato, M., Perego, P. Innovations in smart packaging concepts for food: An extensive review. *Foods*.2020, *9*(11), 1628. https://doi.org/10.3390/foods9111628.

[26] Pal, M.; Devrani, M.; Hadush, A. Recent developments in food packaging technologies. *Beverage & Food World*.2019, *46*(1), 21–25.

[27] Morais, M.G.; Kuntzler, S.G.; Almeida, A.C.A.; Alvarenga, A.G.P.; Costa, J.A.V. *Encapsulation of Bioactive Compounds in Electrospun Nanofibers for Food Packaging*. In Electrospun Nanofibers (pp. 473–490). Springer, Cham, 2022.

[28] Silva, C.K.; Mastrantonio, D.J.S.; Costa, J.A.V.; Morais, M.G. Innovative pH sensors developed from ultrafine fibers containing açaí (*Euterpe oleracea*) extract. *Food Chem*.2019, *294*, 397–404. https://doi.org/10.1016/j.foodchem.2019.05.059

[29] Duan, M.; Yu, S.; Sun, J.; Jiang, H.; Zhao, J.; Tong, C.; Hu, Y.; Pang, J.; Wu, C. Development and characterization of electrospun nanofibers based on pullulan/chitin nanofibers containing curcumin and anthocyanins for active-intelligent food packaging. *Int. J. Biol. Macromol*.2021, *187*, 332–340. https://doi.org/10.1016/j.ijbiomac.2021.07.140.

[30] Forghani, S.; Zeynali, F.; Almasi, H.; Hamishehkar, H. Characterization of electrospun nanofibers and solvent-casted films based on *Centaurea arvensis* anthocyanin-loaded PVA/κ-carrageenan and comparing their performance as colorimetric pH indicator. *Food Chem*.2022, *388*, 133057. https://doi.org/10.1016/j.foodchem.2022.133057.

[31] Duan, M.; Sun, J.; Huang, Y.; Jiang, H.; Hu, Y.; Pang, J.; Wu, C. Electrospun gelatin/chitosan nanofibers containing curcumin for multifunctional food packaging. *Food Sci. Hum. Wellness*.2023, *12*(2), 614–621. https://doi.org/10.1016/j.fshw.2022.07.064.

[32] Yildiz, E.; Sumnu, G.; Kahyaoglu, L. N. Monitoring freshness of chicken breast by using natural halochromic curcumin loaded chitosan/PEO nanofibers as an intelligent package. *Int. J. Biol. Macromol*.2021, *170*, 437–446. https://doi.org/10.1016/j.ijbiomac.2020.12.160.

[33] Luo, X.; Lim, L.-T. Curcumin-loaded electrospun nonwoven as a colorimetric indicator for volatile amines. *LWT*.2020, *128*, 109493. https://doi.org/10.1016/j.lwt.2020.109493.

[34] Jhuang, J.-R.; Lin, S.-B.; Chen, L.-C.; Lou, S.-N.; Chen, S.-H.; Chen, H.-H. Development of immobilized laccase-based time temperature indicator by electrospinning zein fiber. *Food Packag. Shelf Life*.2020, *23*, 100436. https://doi.org/10.1016/j.fpsl.2019.100436.

[35] Guo, M.; Wang, H.; Wang, Q.; Chen, M.; Li, L.; Li, X.; Jiang, S. Intelligent double-layer fiber mats with high colorimetric response sensitivity for food freshness monitoring and preservation. *Food Hydrocoll*.2020, *101*, 105468. https://doi.org/10.1016/j.foodhyd.2019.105468.

[36] Maftoonazad, N.; Ramaswamy, H. Design and testing of an electrospun nanofiber mat as a pH biosensor and monitor the pH associated quality in fresh date fruit (Rutab). *Polym. Test*.2019, *75*, 76–84. https://doi.org/10.1016/j.polymertesting.2019.01.011.

[37] Mercante, L.A.; Scagion, V.P.; Migliorini, F.L.; Mattoso, L.H.; Correa, D.S. Electrospinning-based (bio) sensors for food and agricultural applications: A review. *TrAC, Trends Anal. Chem*.2017, *91*, 91–103. https://doi.org/10.1016/j.trac.2017.04.004.

[38] Yang, T.; Zhan, L.; Huang, C.Z. Recent insights into functionalized electrospun nanofibrous films for chemo-/bio-sensors. *TrAC Trends Anal. Chem*.2020, *124*, 115813. https://doi.org/10.1016/j.trac.2020.115813.

[39] Moreira, J. B.; Kuntzler, S. G.; Alvarenga, A. G. P.; Costa, J. A. V.; Morais, M. G.; Lim, L. T. *Electrospinning and Electrospraying in Polylactic Acid/Cellulose Composites*. In Polylactic Acid-Based Nanocellulose and Cellulose Composites (pp. 277–292). CRC Press, 2022.

[40] Moreira, J.B.; Lim, L.-T.; Zavareze, E.R.; Dias, A.R.G.; Costa, J.A.V.; Morais, M.G. Antioxidant ultrafine fibers developed with microalga compounds using a free surface electrospinning. *Food Hydrocol*.2019, *93*, 131–136. doi:10.1016/j.foodhyd.2019.02.015.

[41] Xiao, Q.; Lim, L.-T. Pullulan-alginate fibers produced using free surface electrospinning. *Int. J. Biol. Macromol.*2018, *112*, 809–817. https://doi.org/10.1016/j.ijbiomac.2018.02.005.

[42] Yalcinkaya, F. Preparation of various nanofiber layers using wire electrospinning system. *Arab. J. Chem.*2019, *12*, 5162–5172. https://doi.org/10.1016/j.arabjc.2016.12.012.

[43] Olas, B.; Białecki, J.; Urbańska, K.; Bryś, M. The effects of natural and synthetic blue dyes on human health: A review of current knowledge and therapeutic perspectives. *Adv. Nutri.*2021, *12*(6), 2301–2311. https://doi.org/10.1093/advances/nmab081.

[44] Morais, M. G.; Vaz, B. S.; Morais, E. G.; Costa, J. A. V. Biologically active metabolites synthesized by microalgae. *Biomed. Res. Int.*2015, *2015*, 1–15. http://dx.doi.org/10.1155/2015/835761.

[45] Beijerinck, M.W. Kulturversuche mit Zoochlorellen, Lichenengonidien und anderen niederen Algen. *Botanische Ztg.*1980, *48*, 725–772.

[46] Lortou, U.; Gkelis, S. Polyphasic taxonomy of green algae strains isolated from Mediterranean freshwaters. *J. Biol. Res. Thessalon.*2019, *26*(1), 1–12. https://doi.org/10.1186/s40709–019–0105-y.

[47] Gohara-Beirigo, A.K.; Matsudo, M.C.; Cezare-Gomes, E.A.; Carvalho, J.C.M.; Danesi, E.D.G. Microalgae trends toward functional staple food incorporation: Sustainable alternative for human health improvement. *Trends Food Sci. Technol.*2022, *125*, 185–199. https://doi.org/10.1016/j.tifs.2022.04.030.

[48] Dragone, G. Challenges and opportunities to increase economic feasibility and sustainability of mixotrophic cultivation of green microalgae of the genus *Chlorella*. *Renewable Sustainable Energy Rev.*2022, *160*, 112284. https://doi.org/10.1016/j.rser.2022.112284.

[49] García, J. L.; Vicente, M.; Galán, B. Microalgae, old sustainable food and fashion nutraceuticals. *Microb. Biotechnol.*2017, 100, 1017–1024. https://doi.org/10.1111/1751-7915.12800.

[50] Chaiyasitdhi, A.; Miphonpanyatawichok, W.; Riehle, M.O.; Phatthanakun, R.; Surareungchai, W.; Kundhikanjana, W.; Kuntanawat, P. The biomechanical role of overall-shape transformation in a primitive multicellular organism: A case study of dimorphism in the filamentous cyanobacterium *Arthrospira platensis*. *Plos One.*2018, *13*(5), e0196383. https://doi.org/10.1371/journal.pone.0196383.

[51] Nabavi, S. M.; Silva, A. S. Nonvitamin and nonmineral nutritional supplements (1st ed.). Elsevier, 2019.

[52] Torres-Tiji, Y.; Fields, F.J.; Mayfield, S.P. Microalgae as a future food source. *Biotechnol. Adv.*2020, *41*, 107536. https://doi.org/10.1016/j.biotechadv.2020.107536.

[53] Khanra, S.; Mondal, M.; Halder, G.; Tiwari, O.N.; Gayen, K.; Bhowmick, T.K. Downstream processing of microalgae for pigments, protein and carbohydrate in industrial application: A review. *Food Bioprod. Process.*2018, *110*, 60–84. https://doi.org/10.1016/j.fbp.2018.02.002.

[54] Odjadjare, E.C.; Mutanda, T.; Olaniran, A.O. Potential biotechnological application of microalgae: a critical review. *Crit. Rev. Biotechnol.*2017, *370*, 37–52. https://doi.org/10.3109/07388551.2015.1108956.

[55] Sun, H.; Wang, Y.; He, Y.; Liu, B.; Mou, H.; Chen, F.; Yang, S. Microalgae-derived pigments for the food industry. *Mar. Drugs.*2023, *21*(2), 82. https://doi.org/10.3390/md21020082.

[56] Hynstova, V.; Sterbova, D.; Klejdus, B.; Hedbavny, J.; Huska, D.; Adam, V. Separation, identification and quantification of carotenoids and chlorophylls in dietary supplements containing *Chlorella vulgaris* and *Spirulina platensis* using high performance thin layer chromatography. *J. Pharm. Biomed. Anal.*2018, *148*, 108–118. https://doi.org/10.1016/j.jpba.2017.09.018.

[57] Chandra, R.D.; Prihastyanti, M.N.U.; Lukitasari, D.M. Effects of pH, high pressure processing, and ultraviolet light on carotenoids, chlorophylls, and anthocyanins of

fresh fruit and vegetable juices. *EFood*.2021, *2*(3), 113–124. https://doi.org/10.2991/efood.k.210630.001.

[58] Medina-Jaramillo, C.; Ochoa-Yepes, O.; Bernal, C.; Famá, L. Active and smart biodegradable packaging based on starch and natural extracts. *Carbohydr. Polym*.2017, *176*, 187–194. https://doi.org/10.1016/j.carbpol.2017.08.079.

[59] Sirohi, P.; Verma, H.; Singh, S.K.; Singh, V.K.; Pandey, J.; Khusharia, S., Kumar, D.; Kaushalendra; Teotia, P.; Kumar, A. Microalgal Carotenoids: Therapeutic application and latest approaches to enhance the production. *Curr. Issues Mol. Biol*.2022, *44*(12), 6257–6279. https://doi.org/10.3390/cimb44120427.

[60] Rodriguez-Amaya, D.B.; Carle, R. *Alterations of natural pigments*. In Chemical changes during processing and storage of foods (pp. 265–327). Academic Press, 2021.

[61] Berthon, J.-Y.; Nachat-Kappes, R.; Bey, M.; Cadoret, J.-P.; Renimel, I.; Filaire, E. Marine algae as attractive source to skin care. *Free Radic. Res*.2017, *510*, 555–567. https://doi.org/10.1080/10715762.2017.1355550.

[62] Lee, R.E. *Phycology*, 5th Edition. Cambridge University Press, New York, 2018.

[63] Bhalamur, G.L.; Valerie, O.; Mark, L. Valuable bioproducts obtained from microalgal biomass and their commercial applications: A review. *Environ. Eng. Res*.2018, *230*(3), 229–241. www.dbpia.co.kr/Article/NODE07417103.

[64] Molino, A.; Iovine, A.; Casella, P.; Mehariya, S.; Chianese, S.; Cerbone, A.; Rimauro, J.; Musmarra, D. Microalgae characterization for consolidated and new application in human food, animal feed and nutraceuticals. *Int. J. Environ. Res. Public Health*.2018, *150*, 2436. https://doi.org/10.3390/ijerph15112436.

[65] Siqueira, S.F.; Queiroz, M.I.; Zepka, L.Q.; Jacob-Lopes, E. *Microalgae biotechnology—A Brief Introduction*. In Microalgal biotechnology. Intech, 2018.

[66] Morais, M.G.; Prates, D.F.; Moreira, J.B.; Duarte, J.H.; Costa, J.A.V. Phycocyanin from Microalgae: Properties, extraction and purification, with some recent applications. *Ind. Biotechnol*.2018, *14*, 30–37. https://doi.org/10.1089/ind.2017.0009.

[67] Grover, P.; Bhatnagar, A.; Kumari, N.; Bhatt, A.N.; Nishad, D.K.; Purkayastha, J. C-Phycocyanin-a novel protein from *Spirulina platensis*- In vivo toxicity, antioxidant and immunomodulatory studies. *Saudi J. Biol. Sci*.2021, *28*(3), 1853–1859. https://doi.org/10.1016/j.sjbs.2020.12.037.

[68] Liu, R.; Qin, S.; Li, W. Phycocyanin: Anti-inflammatory effect and mechanism. *Biomed. Pharmacother*.2022, *153*, 113362. https://doi.org/10.1016/j.biopha.2022.113362.

[69] Silva, E.F.; FigueirA, F.S.; Lettnin, A.P.; Carrett-Dias, M.; Filgueira, D.M.V.B.; Kalil, S.; Trindade, G.S.; Votto, A.P.S. C-Phycocyanin: Cellular targets, mechanisms of action and multi drug resistance in cancer. *Pharmacol. Rep*.2018, *70*, 75–80. https://doi.org/10.1016/j.pharep.2017.07.018.

[70] Chaiklahan R.; Chirasuwan, N.; Bunnag, B. Stability of phycocyanin extracted from *Spirulina* sp.: Influence of temperature, pH and preservatives. *Process Biochem*.2012, *47*(4), 659–664. https://doi.org/10.1016/j.procbio.2012.01.010.

[71] Duangsee, R.; Phoopat. N.; Ningsanond, S. Phycocyanin extraction from *Spirulina platensis* and extract stability under various pH and temperature. *Asian J. Food Agro-Ind*.2009, 2, 819–826

[72] Giménez, A.; Ares, F.; Ares, G. Sensory shelf-life estimation: A review of current methodological approaches. *Food Res. Int*.2012, *49*(1), 311–325. https://doi.org/10.1016/j.foodres.2012.07.008.

[73] Codex Alimentarius. 1999. Code of hygienic practice for refrigerated packaged foods with extended shelf life—CAC/RCP 46. www.fao.org/input/download/standards/347/CXP_046e.pdf. Accessed March 2023

[74] Chisti, Y. Producing safe processed foods. *Biotechnol Adv*.2018, *36*(5), 1555. https://doi.org/10.1016/j.biotechadv.2018.06.005.

[75] World Health Organization. Food safety. www.who.int/news-room/fact-sheets/detail/food-safety#:~:text=An%20estimated%20600%20million%20%E2%80%93%20almost,healthy%20life%20years%20(DALYs). Accessed March 2023
[76] RETS—International Network of Education of Health Technicians. Food safety is everyone's business. www.rets.epsjv.fiocruz.br/en/news/food-safety-everyones-business. Accessed March 2023
[77] Siddh, M.M.; Soni, G.; Jain, R.; Sharma, M.K. Structural model of perishable food supply chain quality (PFSCQ) to improve sustainable organizational performance. *Benchmark Int J.*2018, *25*(7), 2272–2317. https://doi.org/10.1108/BIJ-01-2017-0003.

14 Electrospinning of Nanofibers for Stem Cell-Based Wound Healing and Tissue Engineering Applications

Suelen Goettems Kuntzler, Lívia da Silva Uebel, Daiane Angelica Schmatz, Bruna da Silva Vaz, Jorge Alberto Vieira Costa, Michele Greque de Morais, and Juliana Botelho Moreira

14.1 INTRODUCTION

Scaffolds or matrices based on electrospun nanofibers work as a support to cells reproducing in an extracellular matrix (ECM) in tissue engineering [1]. These nanomaterials provide an appropriate physical environment for the cells to attach, proliferate, migrate, and differentiate to form new tissues or cells [2]. Moreover, electrospun nanofiber facilitates the delivery of oxygen and nutrients to the tissue. Thus, they promote cell adhesion, proliferation, migration, differentiation, and tissue organization [3].

Nanofibers can be applied in wound healing and tissue engineering because their diameter is in the nanometer range, which provides physical properties such as high porosity, high surface area relative to volume, and fiber interconnectivity. Thus, this biomaterial becomes suitable for transporting nutrients, cellular communication, and efficient cellular responses. In addition, several synthetic and natural polymers and biopolymers are utilized to produce nanofiber scaffolds. They must have sufficient structural integrity and mechanical properties like the native tissue to mimic the microenvironment [4],[5].

There are several methods for developing nanofibers, such as self-assembly [6], phase separation [7], melt blowing [8], polymeric foaming [9], and electrospinning [10]. Due to the growing interest in nanotechnology, the electrospinning technique has received attention for biomedical applications. This process develops nanofibers with unique and suitable properties for reproducing the human biological environment because they are in the same size range as biological molecules [11]. Besides, electrospinning is the most versatile and practical method used to manufacture scaffold

 DOI: 10.1201978103333814-14

nanofibers due to its ability to control composition, structure, and functional properties using a relatively simple and low-cost approach [12]. Thus, this chapter addresses the electrospun nanofibers as a support for cell growth, highlighting the main polymers and bioactive compounds in applying scaffolds in wound healing and tissue engineering.

14.2 ELECTROSPUN NANOFIBERS

The electrospun nanofibers are highly porous with a large surface/volume ratio. These nanostructures have relatively high processability and formability into the desired shapes. They exhibit mechanical resistance and structural stability [13]. The interconnected porous structure of nanofibers provides a large surface area for cell adhesion and sufficient space to transport nutrients, oxygen, growth factors, and waste products [14]. The nanofiber scaffolds should also be biodegradable to allow complete tissue regeneration. They also must be biocompatible to avoid negative immune responses [15].

In general, the basic principles of electrospinning are based on uniaxial elongation of the polymer solution. Continuous nanofibers are obtained by applying an electrostatic repulsion force driven by an electric field while stretching a viscous polymer solution. The recipient discharges the polymer solution through the feed pump at a constant rate. High voltage is then applied to induce an electrified jet at the tip of the capillary. An electrically charged droplet is ejected when the applied voltage is sufficient for the surface tension to be less than the repulsive force. Elongation of the solution immediately occurs from the charged jet. At the same time, there is the evaporation of the solvent, and the polymer nanofibers are deposited on the collector [16].

Several factors affect the shape and size of nanofibers produced using the electrospinning technique, including equipment, solution, and environmental parameters. Equipment parameters comprise the applied electric field, the distance between the capillary and collector, the solution feed rate, and the capillary diameter. The solution parameters include the type of solvent, the polymer molecular weight, the polymer concentration, and the viscosity and conductivity of the solution. The environmental parameters include relative humidity, temperature, and airflow. All these parameters directly affect the production of uniform nanofibers [17].

The application of nanofibers as a biomaterial is one of the key factors in wound healing and tissue engineering (Figure 14.1). An increasing trend is to obtain bioactive scaffolds through combination with biomolecules, growth factors, and therapeutic genes, which present biological signals to modulate tissue regeneration [18]. Differentiation and behavior of stem cells are influenced by biochemical parameters emitted by the nanofiber support. In addition, cell development on scaffolds can result in increased facilitation of neural function and provide trophic support [19].

According to Morais et al. [20], scaffolds developing with polyhydroxybutyrate (PHB) and added microalgae biomass will provide a breakthrough in tissue engineering because that material may be reproduced of the ECM. Nanofibers contribute to reducing rejection during tissue and organ restructuring due to the use of a biopolymer compatible with cells and tissues. Nanofiber-based scaffolds stimulate cell

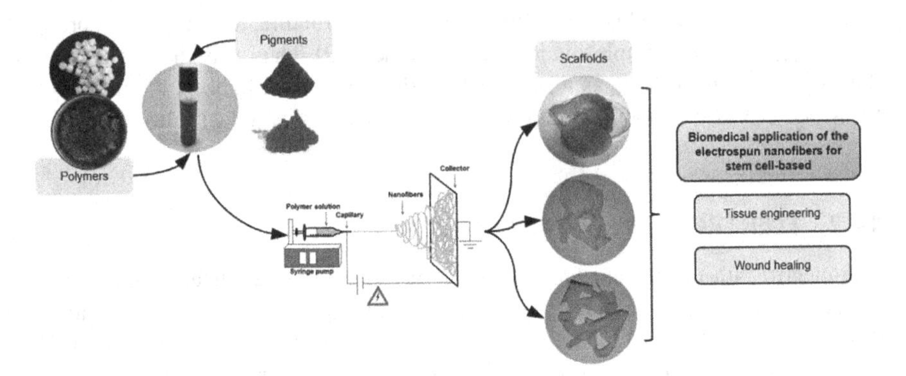

FIGURE 14.1 Electrospun nanofibers produced from polymers and natural pigments for biomedical applications.

growth and increase nutrient diffusion. In addition, they promote specific cellular interactions through the activities of microalgae biomass [20].

Schmatz et al. [21] produced polymer nanofiber scaffolds containing microalgae biomass. They concluded that the supports have the potential to be applied in the regeneration of tissues as biomaterials with pharmacological properties. The tensile strength and elongation of the nanofibers and the biomass presence in the scaffolds yielded the best results compared with those produced only with polymers. In an *in vitro* biodegradation experiment, enzymatic degradation of the biomaterial was observed, simulating the behavior of natural tissue. According to Morais et al. [22], microalgae bioactivity can endow the scaffolds with the ability to reduce wound healing times and stimulate cell growth.

14.3 NANOFIBER SCAFFOLDS FOR STEM CELL

Scaffolds are three-dimensional structures used as ECM for the development of tissues at the site of interest. The function of the ECM is to provide structural and functional support cellular. Besides, it directs the morphological organization and biochemical and biomechanical signals for morphogenesis, differentiation, and homeostasis during tissue formation [23].

Shafiei et al. [24] used nanofiber scaffolds produced by electrospinning with polycaprolactone (PCL) polymer made of layered double hydroxide (LDH). Their study evaluated the effect of the addition of LDH on the mechanical, physical, and chemical properties of the scaffold. Besides, biocompatibility and the ability to induce adipogenic differentiation of stem cells derived from adipose tissue of mice have been evaluated. The results of the mechanical analysis demonstrated that the addition of LDH to the nanofiber scaffolds produced a higher tensile strength and elongation compared with PCL nanofibers alone. Furthermore, the adherence, proliferation, and adipogenic differentiation of stem cells derived from murine adipose tissue were significantly improved with LDH.

Bagó et al. [25] described a nanofiber scaffold implant produced from poly (L-lactic acid) that was able to deliver and maintain cytotoxic human mesenchymal stem cells (hMSCs), which suppress the recurrence of postsurgical glioblastoma. They found that the nanofiber scaffold implant increased hMSC retention in the surgical cavity by 5 times and prolonged the persistence of these cells by 3 times compared with the standard direct injection approach. Another analysis showed that treatment with scaffold nanofibers and cytotoxic hMSCs destroyed cocultured human glioblastoma cells. Furthermore, an *in vivo* assay showed that this biomaterial with hMSCs and the release of an antitumor protein reduced the established glioblastoma volume by 3 times. Therefore, these data suggest that nanofiber scaffolds are an effective alternative to cytotoxic stem cell therapy for postsurgical glioblastoma.

Zhu et al. [26] produced nanofiber scaffolds by the electrospinning technique using poly (ether carbonate urethane) urea (PECUU) as support for annulus fibrous stem cells (AFSCs). The study explored the use of nanofiber scaffolds with AFSCs to promote diverse cell differentiation in the annulus fibrous since this differentiation depends on the elasticity of the polymer used to produce nanofibers. When AFSCs were grown on PECUU nanofiber scaffolds, the type I collagen gene in these cells increased with the elasticity of the polymer. Therefore, the results of this study indicate that AFSCs tend to differentiate into various cell types depending on the elasticity of the biomaterial, thus providing an application in tissue engineering.

Another relevant aspect is the arrangement of the nanofibers in the scaffolds. Electrospinning can produce these biomaterials with aligned or random nanofibers, interfering with cell adhesion. Han et al. [27] investigated whether an aligned or random orientation of PCL nanofiber scaffolds could induce anisotropic cell alignment. They also evaluated whether there could be increased maturation of cardiomyocytes derived from human pluripotent stem cells. In this sense, the study evaluated the structural, functional, and molecular properties of cells after two weeks of culture in nanofiber scaffolds. The results showed that cells grew on scaffolds according to the anisotropic and isotropic orientation of the nanofibers. In addition, the scaffolds with aligned nanofibers induced an anisotropic cell appearance but did not enhance the maturation of cardiomyocytes derived from human pluripotent stem cells. Therefore, the results suggest that anisotropic scaffolds have a limited effect on the improvement of cell maturation.

Zhong et al. [28] studied the possibility of formation of muscle cell agglomerates oriented on PCL nanofiber scaffolds and poly (lactic-co-glycoside acid). The nanofiber scaffolds were produced by the electrospinning technique using a static and rotating collector for the formation of random and aligned nanofibers, respectively. The shape and proliferation of human vascular muscle cells on this biomaterial were analyzed. The morphology of *in vivo* human vascular muscle cells is spindle shaped, and the aligned nanofiber scaffolds maintained this morphology throughout the culture process. However, random nanofiber scaffolds altered the cell shape from spindle to stellate. Therefore, the results suggest that the aligned scaffold may be suitable for *in vitro* cell culture and regeneration of *in vivo* cells.

Canadas et al. [29] used a polyhydroxyalkanoate extracted from bacteria to produce nanofiber scaffolds by the electrospinning process. Stem cells were cultured in the nanofiber scaffolds and on biopolymer film, which showed an ability to support the growth of cells at acceptable levels of proliferation. The comparative results showed that the topology of the nanofiber biomaterial aided cell adhesion and proliferation, whereas, on the polymeric films, such adhesion was not observed.

Chen et al. [30] produced nanofiber scaffolds from poly (butylene succinate) polymers and copolyesters containing butylene thioglycolate (BTDG) or butylene diglycolate (BDG) sequences. Osteogenic and chondrogenic-type stem cells were grown in these biomaterials. Besides, the BTDG copolyester was more favorable to chondrogenic cells than the BDG. Therefore, these new functional supports have the potential for stem cell applications that promote the recovery of native tissue.

14.4 POLYMERS USED IN THE PRODUCTION OF ELECTROSPUN NANOFIBERS

14.4.1 Chitosan

Chitosan is a partially deacetylated form of chitin. It is a natural polymer that is bioactive, biodegradable, biocompatible with cells and tissues, and safe. The applications and characteristics of chitosan depend on the deacetylation degree and size of the polymer chain. The deacetylation degree influences conformation, deformation, rupture tension, biodegradability, and immunological activity [31],[32]. Chitosan scaffolds (CS), silk fibroin (SF), and mixtures of CS/SF were prepared via electrospinning. The study evaluated the growth and osteogenic differentiation of human mesenchymal stem cells from bone marrow. The results showed that the CS/SF blend produced a promising scaffold for the development of bone tissue, with the CS inducing cell differentiation and SF promoting cell proliferation [33].

Trinca et al. [34] developed double-layer scaffolds produced by the electrospinning technique for application on the skin. One layer of PCL or PCL/cellulose acetate blend acts as mechanical support. The other layer, designed to be in direct contact with the lesion, was composed of a chitosan/poly (ethylene oxide) (PEO) blend that plays the role of a primary wound dressing. Scaffolds were not cytotoxic to L929 cells because they presented 84% of the remaining viable cells. The cell adhesion and increased cell count were observed after 48 hours, inoculating cells on the surface of the structures. Thus, the scaffolds can be used as a wound dressing to treat skin lesions.

14.4.2 Collagen

Collagen is abundant in nature and constitutes about 30% of animal protein [35]. This polymer is one of the native components of ECM in the skin, bones, tendons, and other connective tissues. The fibrillar structure of collagen has diameters of 50 to 500 nm. These fibrils have essential structural functions in the mechanical properties of tissues and are involved in numerous biological characteristics from the early stages of development to tissue repair[36]. Besides, collagen has low antigenicity, low inflammatory properties, and excellent biocompatibility with cells and tissues.

Collagen presents a three-dimensional network structure in native ECM, contributing to cell fixation and proliferation [37].

Luo et al. [38] produced nanofibers with type I collagen by electrospinning and evaluating the growth of MC3T3-El cells in scaffolds. The authors verified that the cells that adhered to the structures did not cause a modification in the fiber texture, and they resulted in scattering morphologies with cell extensions representative of the growth of MC3T3-El. Moreover, collagen nanofibers supported the adhesion, proliferation, and dissemination of MC3T3-El. Therefore, the scaffolds can mimic the biological and structural properties of native ECM proteins.

14.4.3 Alginate

Alginate polysaccharide is found in brown algae, consisting of monomers β-D-mannuronic acid and α-L-guluronic acid [39]. This polysaccharide, approved by the Food and Drug Administration (FDA), is recommended for biomedical uses, including wound dressings, dental impression materials, and cosmetics. This polymer is used by food additives, therapeutics, textiles, and agricultural industries since it is biodegradable, non-toxic, relatively low cost, and cytocompatible [40],[41]. Alginate is applied to the tissue engineering field due to its biocompatibility and non-immunogenicity [42]. Besides, alginate presents similar properties to glycosaminoglycan, optimal moisture vapor transfer, and exudate absorption. Thus, the production of alginate nanofibers combined with other biopolymers, such as collagen, silk fibroin, and chitosan, are promising alternative materials for wound dressings and tissue engineering [41],[43].

Ghalei et al. [44] investigated the production of a bioactive dressing composed of silk fibroin nanofibers with alginate hydrogel to provide amniotic fluid capable of promoting cellular response and healing of lesions. The viability and secretion of collagen were evaluated in the bioactive nanocomposite by seeding the dressings with L929 fibroblasts. Cell viability was higher in the dressing compared to the control group, suggesting that silk fibroin nanofibers with alginate hydrogel are non-toxic.

14.4.4 Poly(caprolactone)

Poly(caprolactone) (PCL) is the most used synthetic polymer for the manufacture of scaffolds. This semi-crystalline polymer has a glass transition temperature of −62°C and a low melting point (55–60 °C), which depends on crystallinity [45]. PCL is a biodegradable, flexible, and biocompatible polymer approved by the FDA. PCL has attracted considerable attention in biomedical applications such as materials for dressings, medical devices, degradable medical implants, drug delivery systems, and scaffolds for tissue engineering [46].

PCL promotes cytocompatibility, biodegradability, and mechanical resistance in materials for wound healing and tissue engineering [47]. PCL exhibits stability within the living body, representing some biological limitations, such as small cellular adhesion/proliferation and a long time to complete degradation [48]. Thus, the blending of PCL with natural polymers such as collagen [49], gelatin [50], and chitosan [51] can improve biocompatibility and modulate the rate of degradation [52].

The amino-functionalized tannin (TN) polymer was used by Martins et al. [53] for the development of blends with PCL and the formation of cytocompatible hydrophilic scaffolds. The authors performed cytocompatibility assay with scaffolds for seven days. They found that the PCL-TN membrane that supported binding, adhesion, and proliferation of stem cells was produced in the PCL:TN ratio of 78:22. Thus, these structures showed excellent characteristics for application in wound healing and tissue repair, aiming at new devices in tissue engineering.

14.4.5 Poly(lactic acid)

Poly(lactic acid) (PLA) is an FDA-approved synthetic polymer studied and applied in wound healing and tissue engineering. This polymer has biodegradability, biocompatibility, desirable mechanical characteristics, and excellent thermal stability [54],[55]. PLA is polyester from naturally occurring lactic acid produced by fermenting sugars from renewable resources [54].

Some characteristics of PLA limit its application in tissue engineering and long-term compatibility. The generation of by-products during the degradation of this polymer and the low stability induce tissue inflammation and cell death [56]. In addition, PLA is hydrophilic and has impaired mechanical properties, which reduce its overall strength and weaken cell adhesion, proliferation, and differentiation. Therefore, blends of PLA with other synthetic polymers or biopolymers can increase the mechanical strength of this polymer and are effective in extending its use in biomedical applications [57],[58].

Naghieh et al. [59] studied PLA and gelatin for scaffold production using fused deposition modeling and electrospinning techniques. The authors evaluated the ability of PLA/gelatin scaffolds to form apatite under the surface of the material. Furthermore, the scaffolds coated with hydroxyapatite crystals can be a substrate for the deposition. Thus, the authors suggested their use as a bioactive bone substitute in maxillofacial applications.

14.5 NANOFIBERS WITH ADDED BIOACTIVE COMPOUNDS FOR CELL GROWTH

Adding bioactive compounds to electrospun nanofibers to produce scaffolds improves cellular growth and adhesion. Also, these compounds incorporated in a polymer matrix can result in an increase in the mechanical strength of the biomaterial, bioactivity, and the ability to promote cell proliferation and differentiation [33],[60],[61] (Table 14.1).

Another study evaluated PLA nanofibers containing polyphenols and vitamin B6 for controlled release of the embedded compounds in tissue engineering applications. The attachment of cells on the scaffolds was not affected by antioxidants addition. In contrast, cell proliferation increased with antioxidant activity. The scaffolds protected cells against oxidative stress and provided innovative, fibrous 3D platforms for

TABLE 14.1
Studies with Bioactive Compounds Added into Electrospun Nanofibers for Stem Cell Application

Bioactive compound	Polymeric material	Cell culture	Potential application	Reference
Curcumin	PLA/hyperbranched polyglycerol	3T3 fibroblast	Patch for healing acute/chronic diabetic wound	[62]
	Poly(3-hydroxy butyric acid-co-3-hydroxy valeric acid)	L929 mouse fibroblasts	Wound dressing	[63]
	Poly (DL-lactic-co-glycolic) acid (PLGA)	Skin cancer (A431)	Controlled delivery of curcumin in squamous carcinoma	[64]
Hyaluronic acid	Chitosan/PCL	Vero cell (epithelial cell extracted from African Green Monkey, *Chlorocebus* sp.,)	Tissue engineering	[65]
Hyaluronic acid and collagen	poly(L-lactide-co-ε-caprolactone)	Adipose tissue and human umbilical vein endothelial cells	Vascular network formation (tissue engineering)	[66]
Quercetin	PCL/PLA; PCL/PLGA; PCL/poly (ethylene oxide)	Mammary carcinoma cell line (MCF-7)	Tissue engineering and drug delivery system as an anticancer carrier	[67]
Spirulina biomass	PDLLA	Mesenchymal stem cells	Wound healing	[68]
Hydroxyapatite (HA)	PCL	Human mesenchymal stem cells	Regeneration of bone tissue	[69]
HA and vitamin D3	PCL/gelatin	Human adipose tissue-derived stem cells	Tissue engineering	[70]
HA and vitamin D3	PLLA/gelatin	Bone mesenchymal stem cells	Bone regeneration	[71]

tissue growth and cell proliferation [72]. Moreover, nanofibers have been shown to work in drug delivery. Wang et al. [73] produced polyvinylpyrrolidone nanofibers to enhance the bioavailability of curcumin. Integrating curcumin into nanofibers' interstices solved the low bioavailability of the compound, increasing its effectiveness. Furthermore, *in vivo* assays revealed that curcumin-containing nanofibers helped to reduce toxicity to healthy cells [73].

The association of stem cells with biomaterials promises to be the future of regenerative medicine for treating tissue and organ injury. According to Steffens et al. [74], stem cells have been grown on poly-D, L-lactic acid (PDLLA) scaffolds produced by electrospinning without or with *Spirulina* biomass. Uniform nanofibers were developed with mechanical and morphological properties like natural ECM. Stem cells exhibited higher adhesion and viability in PDLLA scaffolds with Spirulina biomass compared to PDLLA scaffolds without biomass. In addition, the scaffolds were biocompatible with stem cells, presenting suitable characteristics for tissue engineering use because of the biological properties of the *Spirulina* biomass: antioxidant, antibacterial, antiviral, anticancer, anti-inflammatory, and antiallergic activities.

Compounds extracted from *Spirulina* biomass have been explored to produce nanofiber scaffolds. C-phycocyanin pigment extracted from microalgae is a blue fluorescent dye with nutritional and nutraceutical properties with potential applications in foods and cosmetics [20]. Nanofibers containing phycocyanin were developed by Figueira et al. [75] and Braga et al. [76]. These works showed that the nanofibers protect this bioactive compound against degradation by external conditions, such as light and temperature.

Bioactive glass is an amorphous structure consisting of elements in the body, including silicon dioxide, sodium dioxide, calcium oxide, and phosphorus pentoxide. The biocompatibility and bioactivity of this material have led to its use in the human body [77]. Talebian et al. [78] developed nanofiber scaffolds from a mixture of chitosan, poly (ethylene oxide), and bioactive glass using the electrospinning technique. Tensile strength tests and contact angle assessment suggested that the bioactive glass addition into nanofibers improved scaffolds' mechanical properties and hydrophobicity. The results indicated that the chitosan/poly (ethylene oxide)/bioactive glass scaffolds did not affect cell viability and therefore were not cytotoxic to human mesenchymal stem cells.

14.6 CONCLUSION

Several types of diseases are treated with stem cells since these cells are constantly renewing and can differentiate and generate any other cell in the body. Therefore, researchers seek alternative methods to grow and manipulate these cells and induce specific functions of interest. Studies using electrospun nanofiber scaffolds for stem cell culture have proved promising. Besides, nanofiber scaffolds have several advantages over polymeric films or liquid culture media, primarily due to their three-dimensional structure and porosity. These scaffold characteristics facilitate cell adhesion and allow the addition of bioactive substances or compounds. The fixation of these biocompounds into nanofibers' interstices protects them against degradation and contributes to the controlled release at the target site. Thus, rather than just being inert physical structures, scaffolds provide a suitable environment for cell proliferation, facilitate the distribution of nutrients, and ensure protection against undesired biochemical reactions, thus allowing the replacement of damaged tissue healthily and safely.

ACKNOWLEDGMENT

This study was financed in part by the Coordenação de Aperfeiçoamento de Pessoal de Nível Superior – Brasil (CAPES) – Finance Code 001. This research was developed within the scope of the Capes-PrInt Program (Process #88887.310848/2018–00).

REFERENCES

[1] Sridhar, R.; Sundarrajan, S.; Venugopar, J. R.; Ravichandran, R.; Ramakrishna, S. Electrospun inorganic and polymer composite nanofibers for biomedical applications. *J. Biomater. Sci. Polym. Ed.*2013, *24*, 4. https://doi.org/10.1080/09205063.2012.690711.

[2] Qian, Y.; Lin, H.; Yan, Z.; Shi, J.; Fan, C. Functional nanomaterials in peripheral nerve regeneration: scaffold design, chemical principles and microenvironmental remodeling. *Mater. Today.*2021, *51*, 165–187. https://doi.org/10.1016/j.mattod.2021.09.014

[3] Edalat, F.; Sheu, I.; Manoucheri, S.; Khademhosseini, A. Material strategies for creating artificial cell-instructive niches. *Curr. Opin. Biotechnol.*2012, *23*, 820–825. https://doi:10.1016/j.copbio.2012.05.007.

[4] Dahlin, R. L.; Kasper, F. K.; Mikos, A.G. Polymeric Nanofibers in Tissue Engineering. *Tissue. Eng. Part. B. Rev.*2011, *17*, 349–364. https://doi: 10.1089/ten.TEB.2011.0238.

[5] Stella, J. A.; D'Amore, A.; Wagner, W. R.; Sacks, M. S. On the biomechanical function of scaffolds for engineering load bearing soft tissues. *Acta. Biomater.*2010, *6*, 2365–2381. https://doi: 10.1016/j.actbio.2010.01.001.

[6] Sun, Y.; Li, X.; Zhao, M.; Chen, Y.; Xu, Y.; Wang, K.; Bian, S.; Jiang, Q.; Fan, Y.; Zhang, X. Bioinspired supramolecular nanofiber hydrogel through self-assembly of biphenyl-tripeptide for tissue engineering. *Bioact. Mater.*2022, *8*, 396–408. https://doi.org/10.1016/j.bioactmat.2021.05.054.

[7] Zhao, J.; Han, W.; Tu, M.; Huan, S.; Zeng, R.; Wu, H.; Cha, Z.; Zhou, C. Preparation and properties of biomimetic porous nanofibrous poly (l-lactide) scaffold with chitosan nanofiber network by a dual thermally induced phase separation technique. *Mater. Sci. Eng. C.*2012, *32*(6), 1496–1502. https://doi.org/10.1016/j.jddst.2021.102623

[8] Zhang, H.; Zhen, Q.; Liu, Y.; Liu, R.; Zhang, Y. One-step melt blowing process for PP/PEG micro-nanofiber filters with branch networks. *Results in Phys.*2019, *12*, 1421–1428. https://doi.org/10.1016/j.rinp.2019.01.012

[9] Ghorbankhani, A.; Zahedi, A. R. Micro-cellular polymer foam supported polyaniline-nanofiber: Eco-friendly tool for petroleum oil spill cleanup. *J. Clean. Prod.*2022, *368*, 133240. https://doi.org/10.1016/j.jclepro.2022.133240

[10] Qamar, Z.; Khan, T. M.; Abideen, Z. U.; Shahzad, K.; Hassan, A.; Khan, S. U.; Haider, S.; Akhtar, M. S. Optical, morphological, and impedance characteristics of $Ni_{(x)}$ –$(CdO)_{(1-x)}$ nanofibers fabricated by electrospinning technique. *Mater Sci Eng B.*2022, *282*, 115779. https://doi.org/10.1016/j.mseb.2022.115779

[11] Kai, D.; Liowa, S. S.; Loh, X. J. Biodegradable polymers for electrospinning: Towards biomedical applications. *Mater. Sci. Eng. C. Mater. Biol. Appl.*2014, *45*, 659–670. https://doi: 10.1016/j.msec.2014.04.051.

[12] Arida, I. A.; Ali, I. H.; Nasr, M.; El-Sherbiny, I. M. Electrospun polymer-based nanofiber scaffolds for skin regeneration. *J. Drug Deliv. Sci. Technol.*2021, *64*, 102623. https://doi.org/10.1016/j.jddst.2021.102623

[13] Haider, A.; Haider, S.; Kang, I. K. A comprehensive review summarizing the effect of electrospinning parameters and potential applications of nanofibers in biomedical and biotechnology. *Arab. J. Chem.*2018, *11*(8), 1165–1188. https://doi.org/10.1016/j.arabjc.2015.11.015

[14] Karuppuswamy, P.; Venugopal, J. R.; Navaneethan, B.; Laiva, A. L.; Sridhar, S.; Ramakrishna, S. Functionalized hybrid nanofibers to mimic native ECM for tissue engineering applications. *Appl. Surf. Sci.*2014, *322*, 162–168. https://doi.org/10.1016/j.apsusc.2014.10.074.

[15] Jaganathan, S. K.; Mani, M. P.; Nageswaran, G.; Krishnasamy, N. P.; Ayyar, M. The potential of biomimetic nanofibrous electrospun scaffold comprising dual component for bone tissue engineering. *Int. J. Polym. Anal.*2019, *24*(3), 204–218. https://doi.org/10.1080/1023666X.2018.1564127

[16] Moreira, J. B.; Kuntzler, S. G.; Alvarenga, A. G. P.; Costa, J. A. V.; Morais, M. G.; Lim, L. T. *Electrospinning and Electrospraying in Polylactic Acid/Cellulose Composites.* In Polylactic Acid-Based Nanocellulose and Cellulose Composites (pp. 277–292). CRC Press, 2022.

[17] Morais, M. G.; Kuntzler, S. G.; Almeida, A. C. A.; Alvarenga, A. G. P.; Costa, J. A. V. *Encapsulation of Bioactive Compounds in Electrospun Nanofibers for Food Packaging. In Electrospun Nanofibers* (pp. 473–490). Springer, Cham, 2022.

[18] Afsharian, Y. P.; Rahimnejad, M. Bioactive electrospun scaffolds for wound healing applications: A comprehensive review. *Polym. Test.*2021, *93*, 106952. https://doi.org/10.1016/j.polymertesting.2020.106952

[19] Leena, M. M.; Silvia, M. G.; Moses, J. A.; Anandharamakrishnan, C. *Nanofiber-integrated hydrogel as nanocomposites for tissue engineering.* In Bionanocomposites in Tissue Engineering and Regenerative Medicine (pp. 119–147). Woodhead Publishing, 2021.

[20] Morais, M. G.; Vaz, B. S.; Morais, E. G.; Costa, J. A. V. Biologically Active Metabolites Synthesized by Microalgae. *Biomed. Res. Int.*2015, *2015*, 1–a15. http://dx.doi.org/10.1155/2015/835761.

[21] Schmatz, D. A.; Uebel, L. S.; Kuntzler, S. G.; Dora, C. L.; Costa, J. A. V.; Morais, M. G. Scaffolds Containing *Spirulina* sp. LEB 18 Biomass: Development, Characterization and Evaluation of *In Vitro* Biodegradation. *J. Nanosci. Nanotechnol.*2016, *16*, 1050–1059. https://doi: 10.1166/jnn.2016.12331.

[22] Morais, M. G.; Stillings, C.; Dersch, R.; Rudisile, M.; Pranke, P.; Costa, J. A. V.; Wendorff, J. Preparation of nanofibers containing the microalga *Spirulina* (*Arthrospira*). *Bioresour. Technol.*2010, *101*, 2872–2876. https://doi.org/10.1016/j.biortech.2009.11.059.

[23] Fattahi, R.; Chamkhorami, F. M.; Taghipour, N.; Keshel, S. H. The effect of extracellular matrix remodeling on material-based strategies for bone regeneration. *Tissue Cell.*2022, *76*, 101748. https://doi.org/10.1016/j.tice.2022.101748

[24] Shafiei, S. S.; Shavandi, M.; Ahangari, G.; Shoklohari, F. Electrospun layered double hydroxide/poly (ε-caprolactone) nanocomposite scaffolds for adipogenic differentiation of adipose-derived mesenchymal stem cells. *Appl. Clay. Sci.*2016, *127–128*, 52–53. https://doi.org/10.1016/j.clay.2016.04.004.

[25] Bagó, J. R.; Pegna, G. J.; Okolie, O.; Mohiti-Asli, M.; Loboa, E. G.; Hingtgen, S. D. Electrospun nanofibrous scaffolds increase the efficacy of stem cell-mediated therapy of surgically resected glioblastoma. *Biomaterials.*2016, *90*, 116–125. https://doi:10.1016/j.biomaterials.2016.03.008.

[26] Zhu, C.; Li, J.; Liu, C.; Zhou, P.; Yang, H.; Li, B. Modulation of the gene expression of annulus fibrosus-derived stem cells using poly (ether carbonate urethane) urea scaffolds of tunable elasticity. *Acta Biomater.*2016, *29*, 228–238. https://doi:10.1016/j.actbio.2015.09.039.

[27] Han, J.; Wu, Q.; Xia, Y.; Wagner, M. B.; Xu, C. Cell alignment induced by anisotropic electrospun fibrous scaffolds alone has limited effect on cardiomyocyte maturation. *Stem. Cell. Res.*2016, *16*, 740–750. https://doi: 10.1016/j.scr.2016.04.014.

[28] Zhong, J.; Zhang, H.; Yan, J.; Gong, X. Effect of nanofiber orientation of electrospun nanofibrous scaffolds on cell growth and elastin expression of muscle cells. *Colloids. Surf. B. Biointerfaces*.2015, *136*, 772–778. https://doi: 10.1016/j.colsurfb.2015.10.017.
[29] Canadas, R. F.; Cavalheiro, J. M. B. T.; Guerreiro, J. D. T.; Almeida, M. C. M. D.; Pollet, E.; Silva, C. L.; Fonseca, M. M. R.; Ferreira, F.C. Polyhydroxyalkanoates: Waste glycerol upgrade into electrospun fibrous scaffolds for stem cells culture. *Int. J. Biol. Macromol*.2014, *71*, 131–140. https://doi: 10.1016/j.ijbiomac.2014.05.008.
[30] Chen, H.; Gigli, M.; Gualandi, C.; Trukenmüller, R.; Blitterswijk, C. V.; Lotti, N.; Murani, A.; Focarete, M. L.; Moroni, L. Tailoring chemical and physical properties of fibrous scaffolds from block copolyesters containing ether and thio-ether linkages for skeletal differentiation of human mesenchymal stromal cells. *Biomaterials*.2016, *76*, 261–272. https://doi: 10.1016/j.biomaterials.2015.10.071.
[31] Safdar, R.; Omar, A. A.; Arunagiri, A.; Regupathi, I.; Thanabalan, M. Potential of Chitosan and its derivatives for controlled drug release applications–A review. *J. Drug Deliv. Sci. Technol*.2019, *49*, 642–659. https://doi.org/10.1016/j.jddst.2018.10.020.
[32] Vunain, E.; Mishra, A. K.; Mamba, B. B. Fundamentals of chitosan for biomedical applications. In *Chitosan Based Biomaterials* Volume 1 (pp. 3–30). Woodhead Publishing, 2017.
[33] Lai, G.; Shalumon, K. T.; Chen, S. H.; Chen, J. P. Composite chitosan/silk fibroin nanofibers for modulation of osteogenic differentiation and proliferation of human mesenchymal stem cells. *Carbohydr. Polym*.2014, *111*, 2014. https://doi: 10.1016/j.carbpol.2014.04.094.
[34] Trinca, R. B.; Westin, C. B.; Silva, J. A. F.; Moraes, A. M. Electrospun multilayer chitosan scaffolds as potential wound dressings for skin lesions. *Eur. Polym. J*.2017, *88*, 161–170. https://doi.org/10.1016/j.eurpolymj.2017.01.021.
[35] Felician, F. F.; Yu, R. H.; Li, M. Z.; Li, C. J.; Chen, H. Q.; Jiang, Y.; Tang, T.; Qi, W. Y.; Xu, H. M. The wound healing potential of collagen peptides derived from the jellyfish *Rhopilema esculentum*. *Chin. J. Traumatol*.2019, *22*, 12.20. https://doi.org/10.1016/j.cjtee.2018.10.004
[36] Addad, S.; Exposito, J.-Y.; Faye, C.; Ricard-Blum, S.; Lethias. C. Isolation, Characterization and Biological Evaluation of Jellyfish Collagen for Use in Biomedical Applications. *Mar. Drugs*.2011, 9, 967–983. https://doi:10.3390/md9060967.
[37] Wang, J.; Windbergs, M. Functional electrospun fibers for the treatment of human skin wounds. *Eur J Pharm Biopharm*.2017, *119*, 283–299. https://doi:10.1016/j.ejpb.2017.07.001.
[38] Luo, X.; Guo, Z.; He, P.; Chen, T.; Li, L.; Ding, S.; Li, H. Study on structure, mechanical property and cell cytocompatibility of electrospun collagen nanofibers crosslinked by common agents. *Int. J. Biol. Macromol*.2018, *113*, 476–486. https://doi:10.1016/j.ijbiomac.2018.01.179s.
[39] Dekamin, M. G.; Karimi, Z.; Latifidoost, Z.; Ilkhanizadeh, S.; Daemi, H.; Naimi-Jamal, M. R.; Barikani, M. Alginic acid: A mild and renewable bifunctional heterogeneous biopolymeric organocatalyst for efficient and facile synthesis of polyhydroquinolines. *Int. J. Biol. Macromol*.2018, *108*, 1273–1280. https://doi.org/10.1016/j.ijbiomac.2017.11.050.
[40] Ching, S. H.; Bansal, N.; Bhandari, B. Alginate gel particles–A review of production techniques and physical properties. *Crit. Rev. Food Sci. Nutr*.2017, *57*(6), 1133–1152. https://doi.org/10.1080/10408398.2014.965773
[41] Taemeh, M. A.; Shiravandi, A.; Korayem, M. A.; Daemi, H. Fabrication challenges and trends in biomedical applications of alginate electrospun nanofibers. *Carbohydr. Polym*.2020, *228*, 115419. https://doi.org/10.1016/j.carbpol.2019.115419
[42] Silva, R.; Singh, R.; Sarker, B.; Papageorgiou, D. G.; Juhasz-Bortuzzo, J. A.; Roether, J.A.; Cicha, I.; Kaschta, J.; Schubert, D. W.; Chrissafis, K.; Detsch, R.; Boccaccini, A. R. Hydrogel matrices based on elastin and alginate for tissue engineering applications. *Int. J. Biol. Macromol*.2018, *114*, 614–625. https://doi.org/10.1016/j.ijbiomac.2018.03.091.

[43] Sobhanian, P.; Khorram, M.; Hashemi, S. S.; Mohammadi, A. Development of nanofibrous collagen-grafted poly (vinyl alcohol)/gelatin/alginate scaffolds as potential skin substitute. *Int. J. Biol. Macromol.*2019, *130*, 977–987. https://doi.org/10.1016/j.ijbiomac.2019.03.045

[44] Ghalei, S.; Nourmohammadi, J.; Solouk, A.; Mirzadeh, H. Enhanced cellular response elicited by addition of amniotic fluid to alginate hydrogel-electrospun silk fibroin fibers for potential wound dressing application. *Colloids. Surf. B. Biointerfaces.*2018, *172*, 82–89. https://doi: 10.1016/j.colsurfb.

[45] Bartnikowski, M.; Dargaville, T. R.; Ivanovski, S.; Hutmacher, D. W. Degradation mechanisms of polycaprolactone in the context of chemistry, geometry and environment. *Prog. Polym. Sci.*2019, *96*, 1–20. https://doi.org/10.1016/j.progpolymsci.2019.05.004

[46] Abdal-hay, A.; Sheikh, F. A.; Gómez-Cerezo, N.; Alneairi, A.; Luqman, M.; Pant, H. R.; Ivanovski, S. A review of protein adsorption and bioactivity characteristics of poly ε-caprolactone scaffolds in regenerative medicine. *Eur. Polym. J.*2022, *162*, 110892. https://doi.org/10.1016/j.eurpolymj.2021.110892

[47] Backes, E. H.; Harb, S. V.; Beatrice, C. A. G.; Shimomura, K. M. B.; Passador, F. R.; Costa, L. C.; Pessan, L. A. Polycaprolactone usage in additive manufacturing strategies for tissue engineering applications: A review. *J. Biomed. Mater. Res. Part B Appl. Biomater.*2022, *110*(6), 1479–1503. https://doi.org/10.1002/jbm.b.34997

[48] Zhao, X.; Li, J.; Liu, J.; Zhou, W.; Peng, S. Recent progress of preparation of branched poly(lactic acid) and its application in the modification of polylactic acid materials. *Int. J. Biol. Macromol.*2021, *193*, 874–892. https://doi:10.1016/j.ijbiomac.2021.10.154

[49] Chong, C.; Wang, Y.; Fathi, A.; Parungao, R.; Maitz, P. K.; Li, Z. Skin wound repair: Results of a pre-clinical study to evaluate electropsun collagen–elastin–PCL scaffolds as dermal substitutes. *Burns.*2019, *45*(7), 1639–1648. https://doi.org/10.1016/j.burns.2019.04.014

[50] Ren, K.; Wang, Y.; Sun, T.; Yue, W.; Zhang, H. Electrospun PCL/gelatin composite nanofiber structures for effective guided bone regeneration membranes. *Mater. Sci. Eng. C.*2017, *78*, 324–332. https://doi.org/10.1016/j.msec.2017.04.084

[51] Shaltooki, M.; Dini, G.; Mehdikhani, M. Fabrication of chitosan-coated porous polycaprolactone/strontium-substituted bioactive glass nanocomposite scaffold for bone tissue engineering. *Mater. Sci. Eng. C Mater. Biol. Appl.*2019, *105*, 110138. https://doi.org/10.1016/j.msec.2019.110138

[52] Arif, Z. U.; Khalid, M. Y.; Noroozi, R.; Sadeghianmaryan, A.; Jalalvand, M.; Hossain, M. Recent advances in 3D-printed polylactide and polycaprolactone-based biomaterials for tissue engineering applications. *Int. J. Biol. Macromol.*2022, *218*, 930–968. https://doi.org/10.1016/j.ijbiomac.2022.07.140

[53] Martins, A. F.; Facchi, S. P.; Câmara, P. C. F.; Camargo, S. E. A.; Camargo, C. H. R.; Popat, K. C.; Kipper, M. J. Novel poly(e-caprolactone)/amino-functionalized tannin electrospun membranes as scaffolds for tissue engineering. *J. Colloid Interface Sci.*2018, *525*, 21–30. https://doi.org/10.1016/j.jcis.2018.04.060.

[54] Albuquerque, T. L.; Júnior, J. E. M.; Queiroz, L. P.; Ricardo, A. D. S.; Rocha, M. V. P. Polylactic acid production from biotechnological routes: A review. *Int. J. Biol. Macromol.*2021, *186*, 933–951. https://doi:10.1016/j.ijbiomac.2021.07.074

[55] Khalid, M.Y.; Arif, Z.U. Novel biopolymer-based sustainable composites for food packaging applications: a narrative review. *Food Packag. Shelf Life.*2022, *33*, 100892. https://doi:10.1016/j.fpsl.2022.100892

[56] Marra, A.; Silvestre, C.; Duraccio, D.; Cimmino, S. Polylactic acid/zinc oxide biocomposite films for food packaging application. *Int. J. Biol. Macromol.*2016, *88*, 254–262. https://doi.org/10.1016/j.ijbiomac.2016.03.039

[57] Mehrpouya, M.; Vahabi, H.; Janbaz, S.; Darafsheh, A.; Mazur, T. R; Ramakrishna, S. 4D printing of shape memory polylactic acid (PLA). *Polymer.*2021, *230*, 124080. https://doi.org/10.1016/J.POLYMER.2021.124080

[58] Rosli, N. A.; Karamanlioglu, M.; Kargarzadeh, H; Ahmad, I. Comprehensive exploration of natural degradation of poly(lactic acid) blends in various degradation media: a review. *Int. J. Biol. Macromol.*2021, *187*, 732–741. https://doi.org/10.1016/j.ijbiomac.2021.07.196
[59] Naghieh, S.; Foroozmehr, E.; Badrossamay, M.; Kharaziha, M. Combinational processing of 3D printing and electrospinning of hierarchical poly (lactic acid)/gelatin-forsterite scaffolds as a biocomposite: Mechanical and biological assessment. *Mater. Des.*2017, *133*, 128–135. https://doi.org/10.1016/j.matdes.2017.07.051.
[60] Nune, M.; Krishnan, U. M.; Sethuraman, S. PLGA nanofibers blended with designer self-assembling peptides for peripheral neural regeneration. *Mater. Sci. Eng. C. Mater. Biol. Appl.*2016, *62*, 329–337. https://doi: 10.1016/j.msec.2016.01.057.
[61] Uebel, L. S.; Schmatz, D. A.; Kuntzler, S. G.; Dora, C. L.; Muccillo-Baisch, A. L.; Costa, J. A. V.; Morais, M. G. Quercetin and curcumin in nanofibers of polycaprolactone and poly(hydroxybutyrate-co-hydroxyvalerate): Assessment of *in vitro* antioxidant activity. *J. Appl. Polym. Sci.*2016, *133*, 30. https://doi.org/10.1002/app.43712.
[62] Perumal, G.; Pappuru, S.; Chakraborty, D.; Nandkumar, A. M.; Chand, D. K.; Doble, M. Synthesis and characterization of curcumin loaded PLA—Hyperbranched polyglycerol electrospun blend for wound dressing applications. *Mater. Sci. Eng. C.*2017, *76*, 1196–1204. https://doi: 10.1016/j.msec.2017.03.200.
[63] Mutlu, G.; Calamak, S.; Ulubayram, K.; Guven, E. Curcumin-loaded electrospun PHBV nanofibers as potential wound dressing material. *J. Drug. Deliv. Sci. Technol.*2018, *43*, 185–193. https://doi.org/10.1016/j.jddst.2017.09.017.
[64] Sampath, M.; Lakra, R.; Korrapati, P. S.; Sengottuvelan, B. Curcumin loaded poly (lactic-co-glycolic) acid nanofiber for thetreatment of carcinoma. *Colloids. Surf. B. Biointerfaces.*2014, 117, 128–134. https://doi.org/10.1016/j.colsurfb.2014.02.020.
[65] Chanda, A.; Adhikari, J.; Ghosh, A.; Chowdhury, S. R.; Thomas, S.; Datta, P; Saha, P. Electrospun chitosan/polycaprolactone-hyaluronic acid bilayered scaffold for potential wound healing applications. *Int. J. Biol. Macromol.*2018, *116*, 774–785. https://doi:10.1016/j.ijbiomac.2018.05.099.
[66] Kenar, H.; Ozdogan, C. Y.; Dumlu, C.; Doger, E.; Kose, G. T.; Hasirci, V. Microfibrous scaffolds from poly(L-lactide-co-ε-caprolactone) blended with xeno-free collagen/hyaluronic acid for improvement of vascularization in tissue engineering applications. *Mater. Sci. Eng. C.*2019, *97*, 31–44. https://doi.org/10.1016/j.msec.2018.12.011.
[67] Eskitoros-Togay, Ş. M.; Bulbul, Y. E.; Dilsiz, N. Quercetin-loaded and unloaded electrospun membranes: Synthesis, characterization and in vitro release study. *J. Drug. Deliv. Sci. Technol.*2018, *47*, 22–30. https://doi.org/10.1016/j.jddst.2018.06.017.
[68] Steffens, D.; Leonardi, D.; Soster, P. R. L.; Lersch, M.; Rosa, A.; Crestani, T.; Scher, C.; Morais, M. G.; Costa, J. A. V.; Pranke, P. Development of a new nanofiber scaffold for use with stem cells in a third degree burn animal model. *Burns.*2014, *40*, 710–718. https://doi: 10.1016/j.burns.2014.03.008.
[69] Surmenev, R. A.; Shkarina, S.; Syromotina, D. S.; Melnik, E. V.; Shkarin, R.; Selezneva, I. I.; Ermakov, A. M.; Ivlev, S. I.; Cecilia, A.; Weinhardt, V.; Baumbach, T.; Rijavec, T.; Lapanje, A.; Chaikina, M. V.; Surmeneva, M. A. Characterization of biomimetic silicate- and strontium-containing hydroxyapatite microparticles embedded in biodegradable electrospun polycaprolactone scaffolds for bone regeneration. *Eur. Polym. J.*2019, *113*, 67–77. https://doi.org/10.1016/j.eurpolymj.2019.01.042.
[70] Sattary, M.; Rafienia, M.; Kazemi, M.; Salehi, H.; Mahmoudzadeh, M. Promoting effect of nano hydroxyapatite and vitamin D3 on the osteogenic differentiation of human adipose-derived stem cells in polycaprolactone/gelatin scaffold for bone tissue engineering. *Mater. Sci. Eng. C. Mater. Biol. Appl.*2019, *97*, 141–155. https://doi: 10.1016/j.msec.2018.12.030.
[71] Ye, K.; Liu, D.; Kuang, H.; Cai, J.; Chen, W.; Sun, B.; Xia, L.; Fang, B.; Morsi, Y.; Mo, X. Three-dimensional electrospun nanofibrous scaffolds displaying bone morpho-

genetic protein-2-derived peptides for the promotion of osteogenic differentiation of stem cells and bone regeneration. *J. Colloid. Interface. Sci.*2019, *534*, 625–636. https://doi:10.1016/j.jcis.2018.09.071.

[72] Llorens, E.; Valle, L. J.; Díaz, A.; Casas, M. T.; Puiggalí, J. Polylactide nanofibers loaded with vitamin B6 and polyphenols as bioactive platform for tissue engineering. *Macromol. Res.*2013, *21*, 775–787. https://doi.org/10.1007/s13233-013-1090-x.

[73] Wang, C.; Ma, C.; Wu, Z.; Liang, H.; Yan, P.; Song, J.; Ma, N.; Zhao, Q. Enhanced Bioavailability and Anticancer Effect of Curcumin-Loaded Electrospun Nanofiber: *In Vitro* and *In Vivo* Study. *Nanoscale. Res. Lett.*2015, *10*, 439. https://doi: 10.1186/s11671-015-1146-2.

[74] Steffens, D.; Lersch, M.; Rosa, A.; Scher, C.; Crestani, T.; Morais, M. G; Costa, J. A. V.; Pranke, P. A New Biomaterial of Nanofibers with the Microalga *Spirulina* as Scaffolds to Cultivate with Stem Cells for Use in Tissue Engineering. *J. Biomed. Nanotechnol.*2013, *9*, 710–718. https://doi.org/10.1166/jbn.2013.1571.

[75] Figueira, F.S.; Gettens, J.G.; Costa, J. A. V.; Morais, M. G.; Moraes, C. C.; Kalil, S. J. Production of Nanofibers Containing the Bioactive Compound C-Phycocyanin. *J Nanosci Nanotechnol.*2016, *16*, 944–949. https://doi: 10.1166/jnn.2016.10906.

[76] Braga, A. R. C.; Figueira, F. S.; Silveira, J. T.; Morais, M. G.; Costa, J. A. V.; Kalil, S. J. Improvement of Thermal Stability of C-Phycocyanin by Nanofiber and Preservative Agents. *J. Food. Process. Preserv.*2016, *40*, 1264–1269. https://doi.org/10.1111/jfpp.12711.

[77] Yadav, V. S.; Sankar, M. R.; Pandey, L. M. Coating of bioactive glass on magnesium alloys to improve its degradation behavior: Interfacial aspects. *J. Magnes. Alloy.*2020, *8*(4), 999–1015. https://doi.org/10.1016/j.jma.2020.05.005

[78] Talebian, S.; Mehrali, M; Mohan, S; Raghavendran, H. B; Mehrali, M.; Khanlou, H. M; Kamarul, T.; Afifi, A. M.; Abass, A. A. Chitosan (PEO)/bioactive glass hybrid nanofibers for bone tissue engineering. *RSC Adv.*2014, *4*, 49144–49152. https://doi:10.1039/C4RA06761D.

15 Green Nanofibers via Electrospinning for Tissue Regrowth

Trishna Bal, Aditya Dev Rajora, Anima Pandey, Biplob De, Anant Nag, Adrika Maji, Mrinal Kanti Pradhan, Samsur Ali Dafadar, and Sauvik Mazumdar

15.1 INTRODUCTION

Cosmetic science has developed a lot in the modern era where major organs and tissues can be replaced by the use of prosthetic substrates. Such substrates can be fabricated from versatile biomaterials prepared from isolates of human or animal or plant origin in the form of a specific system, which can ultimately mimic the extracellular matrix of the parent tissues, either as implantable systems or as scaffolds to serve a major role in tissue engineering and regenerative medicine. The extracellular matrix (ECM) is the key component in the creation of cells, tissues or organs and scaffolds to mimic this ECM to regenerate new cells and tissues which can be considered as a major platform for new organ development. The ECM comprises of many growth factors and other information which are necessary for cell regulation and their sustenance [1]. It is a difficult task to completely mimic ECM by virtue of its complex structure and function and thus researchers attempt to prepare different types of scaffolds which should be able to target the ECM of the native tissue, at least partially for tissue regrowth. Tissue grafts or organ transplants are not always successful as they are sometimes rejected by the recipient and such opportunities become a platform for fabrication of biomaterials systems which can replace the normal tissues and thus fortify the parent cells by triggering them to proliferate in a very rapid manner and ultimately instigate the formation of new cells to create tissues and organs. Such concepts of utilization of biomaterials work magically for external wound healing as well as other forms of tissue regeneration related to bones, nerves, vaginal endometrium repair and so on. When taking into consideration of biomaterials for tissue regrowth, they are classified into either first generation type which are basically those materials which do not show any biological reactions and are used as prosthetics and implants, or second generation materials which help in creating or mimicking the host tissue and thus promote tissue proliferation, or third generation materials which are biodegradable in nature and lastly fourth generation materials which are ideally the smart biomaterials to mimic the

DOI: 10.1201978100333814-15

ECM for tissue regrowth[2]. Many polymers like gelatin, collagen, alginate, dextran, chitosan, cellulose, etc. serve the purpose of playing the role of scaffolds and are highly efficacious in accelerating healing and have superior biocompatibility properties, yet they suffer from poor mechanical strength [3] which can be improved by either preparing modified polymer compositions viz blending natural polymers with synthetic moieties or fabricating a more robust system which will have high mechanical strength to sustain tissue proliferation. Mechanical strength plays a major role for any material to be utilized as a scaffold, and nanofiber systems are one such mode giving ample opportunity for healing and accelerating other biological functions on application. Nanofibers prepared from blends of natural polymers with synthetic water-soluble polymers is the most desired combination which can achieve both biocompatibility and hemocompatibility parameters making them tissue and environment friendly. The nanofibers being lighter in weight are manageable with enhanced porosity making them appropriate for cell adherence with high folding endurance and tensile strength [3,4]. Nanofibers as devices for drug delivery as well as tissue scaffolds fulfil most of the desirable properties for an ideal tissue scaffold and thus are appropriate for tissue regrowth platforms. The current chapter describes the wide and recent applications of green nanofiber mats and its modified forms in different fields of tissue engineering and their principles.

15.2 MECHANISM OF ACTION OF NANOFIBERS IN TISSUE REGROWTH

Tissue regeneration via scaffolds demands many requirements like mechanical strength, cell compatibility, and tissue mimicking architecture to perform as a perfect stage for growth. Nanofibers possess inbuilt properties of being highly porous which makes sufficient room for the cells to adapt and accommodate for their redevelopment and proliferation [4]. The mechanism of cell proliferation within the strands of nanofiber are illustrated in Figure 15.1.

The pores of the fiber mat are usually within the range in nanometer from 2–500nm [5] and mainly depend on the type of solvents used for preparation. As seen from different research, highly volatile solvents which require less spinning time form porous nanofiber mats and there are five different mechanisms involved in the pore formation of nanofiber sheets like in a Breath figures mechanism where the pore formed is between 20 and 200nm, in Vapor induced phase separation mechanism pore formed is 50–300nm, in Nonsolvent induced phase separation mechanism pore formed is 20–100nm, in Thermally induced phase separation mechanism the pore formed is 2–50nm, and in Selective removal mechanism the pore formed is 5–30nm [5]. The porosity of nanofibers makes them appropriate for any active site generation, shortening of ion transport path, or increase in ion transport channels, or increase in surface area which makes the surface more viable for cell expansion. These pores as well as the pores created within the individual nanofiber strands facilitate the easy dispersal of cells and enhanced drug loading when used as a delivery device.

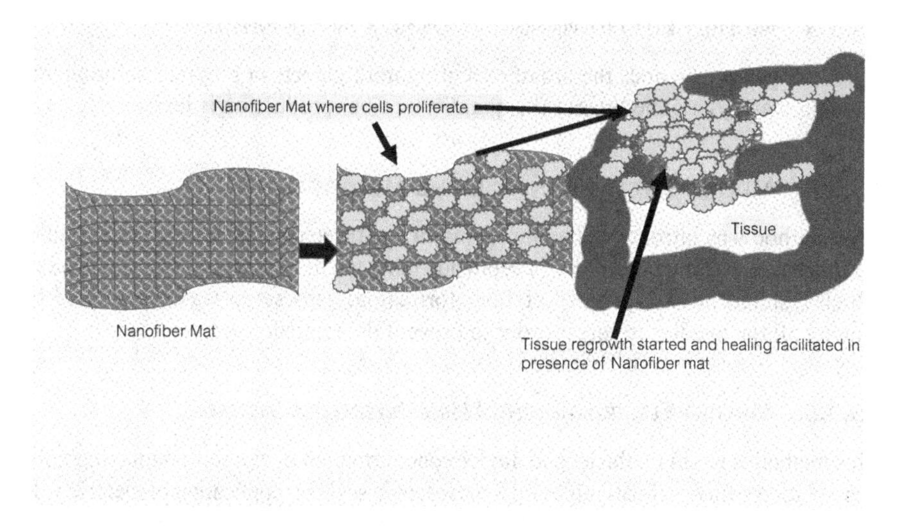

FIGURE 15.1 Mechanism of nanofiber in facilitation of cell proliferation.

15.3 ADVANCED METHODS USED FOR GREEN NANOFIBER FORMATION VIA ELECTROSPINNING

When considering green electrospinning, solvent selection is most important and green electrospinning aims for a compatible mode of fabrication using green solvents and biodegradable polymers which should have nil negative impact on society. With time, a variety of methods has been tried to increase the efficiency of the standard electrospinning method, and some of the most prominent methods which are used for green electrospinning are outlined as follows.

15.3.1 Emulsion Electrospinning

This method is also called two-phase electrospinning by Sanders et al [6,7]. This method encourages the incorporation of both hydrophilic and lipophilic drugs via by the formulation of both water in oil emulsion or oil in water emulsion strategy. Research revealed this method exhibiting better drug release with increased impact and bioavailability [7]. Also, this method proved better when essential oils are to be incorporated within the nanofiber matrix.

15.3.2 Suspension Electrospinning

This method also called as Green electrospinning, involves the use of an aqueous polymer dispersion in association with a polymer template, where for electrospinning of hydrophobic polymer substrates, water is used as the electrospinning medium resulting in effective dispersion and enhanced productivity of nanofibers [8].

15.3.3 Multijet Electrospinning by Using a Single Needle

Here a single needle does the job of executing multiple jets of polymer solution and forming the Taylor cone [9], but the quantity of fiber production is less.

15.3.4 Multijet Electrospinning Using Multiple Needles

This method was introduced by employing two to three needles for the formation of jets but there was the possibility of repulsion between the jets from different needles which resulted in less quantity of fiber formation. This setup was remodelled by placing all the needles in linear order to prevent the repulsion.

15.3.5 Multijet Electrospinning Using Needleless System

This method is based on the jet and Taylor cone formation by the self-conducting polymer solution which self-assembles on a mesoscopic scale on application of electric voltage at or above the critical value and this process runs without any needle ejection [10].

15.3.6 Electrospinning by Porous Hollow Tube

This is another modified form of needleless electrospinning where a hollow tube with pores is used through which the conductive polymer solution is forced under high pressure and the solution forces out through the pores and gets stretched in the presence of an electric field to form fibers[10].

15.3.7 Bubble Electrospinning

This method involves the formation of bubbles over a conductive polymer solution by application of air pressure, and the formed bubbles instantly get converted to Taylor cones as they are stretched in the presence of an electric field resulting in nanofiber mat formation [10].

15.3.8 Coaxial Electrospinning

Here two different solutions are driven from two syringes and later form a composite and then they are spun under the influence of electric voltage [34].

15.3.9 Electroblowing Electrospinning

In this method, the polymer solution is blown by the air under an electric voltage combining the use of air and electric voltage for the formation of Taylor cone and fiber formation [10].

15.3.10 Nanospider Electrospinning

This method uses a roller spinning electrode placed in the polymer solution and with the aid of an electric field, the solution is stretched and collected on a collector plate [10].

15.4 APPLICATIONS OF GREEN POLYMERIC NANOFIBERS IN TISSUE REGROWTH

Polymers from natural and renewable sources are in demand for preparation of different types of polymeric systems as they are abundant and biocompatible in nature as being from natural origin. Synthetic polymers are biocompatible too, and their esterified modified forms enhance their biodegradation by hydrolysis of the polymer chains. The amalgamation of natural and synthetic polymers acquires the blended properties of both categories making them achieve the ideal qualities of a biomaterial for tissue regrowth which are mechanically stronger and more cost efficient. Green polymers prepared by chemical-free methods [11] have desired cytocompatibility to serve the purpose of tissue regrowth. Moreover, green electrospinning mainly relies on suitable solvent selection as well as polymer which do not pose any threat neither to the environment nor to the human body [7,12]. Polymers which are widely used in green electrospinning are like a combination of polycaprolactone with polyethylene oxide, silk fibroin, collagen, Silk fibroin with polyethylene oxide, polyvinyl alcohol with polyethylene oxide, polystyrene and many more [7]. Biocompatible polymers are in demand as they are easily metabolized by the body fluids and can serve as effective scaffolds. The application of different green polymers in different tissue regrowth is discussed.

15.4.1 Skin Wound Healing

Burns and skin injury lead to scars, and prolonged infections if not attended properly can lead to serious issues. Numerous research has been conducted on skin wound healing utilizing different green polymers which are outlined as follows.

Cellulose acetate with soyprotein hydrolysate was used for skin wound healing in the form of nanofiber scaffolds as soya protein almost replicates the ECM proteins and estrogens and this scaffold helped in re-epithialization and reducing scars [13].

In another study **polyvinyl alcohol** combined with soyprotein isolate and formulated nanofiber mat was used for skin wound healing [14].

Diabetic foot ulcers are one of the major medical issues for which many researchers worked and in this aspect **gelatin** combined with **cellulose acetate** in the form of nanofiber wound dressings embedding berberine was used for healing diabetic foot ulcers in streptozotocin induced murine model [15].

In another study, **polylactic acid** was fabricated in the form of multilayered nanofiber patch where Phenytoin, Sildenafil Citrate and Simvastatin were incorporated in different layers of nanofibers separately to for wound healing of diabetic foot ulcers [16].

In another study, Baicalin isolated from a Chinese herb having the efficacy for wound healing was incorporated in **polycaprolactone** combined with transition metals like titanium, molybdenum etc. with carbides, nitrites or carbonitrites as antibacterial wound dressings [17].

In another study, **chitosan** combined with **gelatin** nanofiber mat reinforced with different quantity of graphene nanosheets to enhance the wound healing process and act as antibacterial scaffolds [18].

In another study, **Gum Ghatti** combined with polyvinyl alcohol in the form of nanofiber mat was used as scaffold for wound healing, and it was represented that the material itself was efficient enough to mimic the ECM and accelerate the wound healing of skin [19].

Natural polysaccharides play a very important role in skin wound healing and in this aspect, in a study **Hibiscus leaves mucilage** blended with pectin and polyvinyl alcohol in the form of nanofiber scaffold was effective for wound healing within a span of only six days when compared to that of control murine model. The pectin along with hibiscus leaves mucilage was effective in tissue proliferation as both are biocompatible [20].

Similarly another natural polysaccharide, **neem gum** when blended with polyvinyl alcohol in the form of nanofiber mat very effectively showed skin wound healing within 10 days and that too instantly hair started developing which was absent in case of control mice and thus material can be explored as antibacterial coating material as well as in cosmetic burns applications [4]. Also, it was found that silk fibroin nanofiber patch very effectively was used for cell adhesion blood-derived fibroblast type of cells for accelerated wound healing [21].

15.4.2 Tympanic Membrane Reconstruction

Perforations in the tympanic membrane is a very prominent issue leading to hearing loss, and to solve this clinical problem, many scientists are working on the development of polymer membranes which can help in reconstruction of the tympanic membrane in its presence. Next are discussed some of the important research related to the tympanic membrane reconstruction by using tissue proliferation in presence of nanofiber patch as a wound healing platform.

Bacterial cellulose derived from *Gluconacetobacter xylinus* was used for fabrication of nanofiber patch for the repairing of wounded tympanic membrane [22].

In another study, **bacterial cellulose** isolated from *Komagataeibacter xylinus* was used for tympanic membrane repair. Due to slight differences in crystal and microfibril structure of vegetable cellulose, bacterial cellulose is more superior as it transforms to 3D network of 1.5nm with greater strength and durability and this material was utilized for repair of wounded tympanic membrane [23]. Electrospin nanofiber of silk fibroin blended with polycaprolactone served a very good version of tympanic membrane with good mechanical strength with good acoustic vibrations when the pore size of the nanofibers were less than 50% [24]. Another study was conducted using gelatin which was blended with the crosslinking agent genipin and electrospun to form nanofibers which showed increased mechanical strength and sufficient biocompatibility was used for healing the tympanic membrane [25].

Chitosan combined with polyvinyl alcohol nanofiber mat was studied for the tympanic membrane repair and the nanofiber scaffolds were crosslinked with glutaraldehyde solution and the scaffolds showed good biocompatibility with endothelial cells, mesenchymal stem cells and fibroblasts [26].

15.4.3 Myocardial Tissue Regrowth

Cardiac diseases are another major concern gripping the world and valve failures and non-functionality of other myocardial tissues aggravate such situations. To reduce cardiac failure associated with valves, prosthetics materials are tried for and research is being conducted on different materials which can help in the repair of valves and other myocardial tissue regeneration. Some research related to myocardial tissue regrowth is discussed later.

Cellulose acetate polymer in the form of nanofiber mat showed very promising results in triggering the ECM and preventing the heart valve thrombosis when applied on the surface of aortic heart valve as scaffolds [27]. Also chitosan 3D nanofiber scaffold served as a good platform for regrowth of rat cardiomyocytes and fibroblasts [28].

15.4.4 Ocular Tissue Regrowth

Burn injury causes major damage to ocular tissues and these tissues cannot be regained back and thus many people lose their vision due to chemical attack or other injuries. To solve this problem, many polymers are tried which can effectively help in tissue regrowth; they are discussed later.

Silk fibroin nanofiber scaffolds help in mimicking corneal stroma and showed no body reactions and biocompatibility with the host cornea when tested in Japanese white male rabbits [29].

In another study, nonmulberry silk from *A. mylitta* was used for the preparation of silk fibroin nanofiber sheet which was tested for the regrowth of epithelial cells and keratocytes from rat corneal explants. The material was effective in maintaining tear film in rabbit eyes, and histology examination of the tissue showed absence of any inflammation and neovascularization [30].

Polycaprolcatone nanofiber surface was functionalized with helium-oxygen plasma discharge and proved their transparency which was required for optical clarity as well as helped in construction of damaged ocular surface [31].

15.4.5 Bone Tissue Regeneration

Numerous research is being carried out on bone tissue and their alternatives to improve the lifestyle of people suffering from different bone defects.

Silk fibroin with kappa carrageenan blend prepared in the form of nanofiber and crosslinked with genipin showed improved growth of bone tissues as both silk fibroin and carrageenan helps in stimulating and mimicking the ECM of bone tissues for regrowth, and genipin applications made the scaffold with increased mechanical strength [32].

In another study, polycaprolactone combined with gelatin and to this added Zinc(II) to increase the mechanical strength of the scaffold and nanofibers prepared and this scaffold was used for bone tissue regeneration [33]. Also in another study, polycaprolactone nanofiber scaffold was prepared and incubated with three different types of mesenchymal stem cells from three sources like umbilical, bone marrow and adipose tissue to check the viability of cells in the scaffold medium, and it was found that these mesenchymal stem cells sufficiently proliferated and also their osteogenic activity got enhanced in the presence of polycaprolactone nanofiber scaffold which stated that the polymer in the form of nanofiber was biocompatible [34].

Another on polycaprolactone was performed by fabricating it in nanofiber form with pomegranate peel extract for bone tissue regrowth and it was found that natural waste like pomegranate peel extract not only increased the mechanical strength of the nanofiber mat but also made it more biodegradable with huge antioxidant potential which was essential in case of bone defects [35].

Also, a study was conducted with polycaprolactone nanofiber mat which was prepared by blending nanohydroxyapatite and incorporated zinc oxide in different concentrations to increase the biocompatibility as well as its durability too for bone tissue regrowth. The material showed very good biocompatibility with huge biomineralization and higher alkaline phosphatases activity and more cell viability in comparison to pristine polycaprolactone scaffolds [36].

Also, a 3D nanofibrous scaffold was designed with polycaprolactone/polylactic acid and gelatin combined with taurine to behave in a bioactive and biodegradable form for bone tissue regrowth, and this material provoked 10% more woven bone tissues with angiogenesis in test animal than in control group [37].

Also another study was conducted with polylactic acid as nanofibers, and multiwalled carbon nanotubes were added to it to increase the mechanical strength as well as polyethylene glycol was added to modulate the release of Dexamethasone as the drug which induced the osteogenesis activity and also the nanofibrous scaffold so obtained attained very good mechanical strength and biocompatibility [38].

Another 3D nanofibrous scaffold was prepared with polycaprolactone and gelatin incorporating berberine, but it was found that the mechanical strength of the material decreased due to berberine but hydrophilicity of the material enhanced which increased the degradation rate when implanted invivo and thus nanofiber scaffold showed very good bone cell proliferation and healing and proved to be a good platform for bone tissue regrowth [39].

15.4.6 Urethral Tissue Regeneration

A study was conducted using hyaluronic acid functionalized collagen nanofibers for regeneration of wounded urethral tissues in male beagle puppies and the material showed tremendous impact in tissue growth as hyaluronic acid can induce macrophage elongation and this activity helps in inducing tissue proliferation [40].

In another study, fibrinogen-poly(l-lactide-co-caprolactone) copolymer nanofiber scaffold was fabricated for urethral restoration by seeding the scaffold with epithelial cells on the surface of the scaffold and there was an increased growth of cytokeratin and actin filaments in presence of the prepared nanofiber scaffold and when the material was loaded in rabbit urethral replacement model, it showed rapid luminal epithelization, urethral smooth muscle cell remodelling and capillary development, which indicated the material to be appropriate for the purpose [41].

15.4.7 Dura Mater Regeneration

Chitosan nanofiber when implanted with human dermal fibroblasts and human umbilical vein endothelial cells showed growth and the prepared mat showed cerebrospinal fluid leakage prevention when tested in rabbit duraplasty model [42].

Also Poly(L-lactic acid) was grafted with Tetra calcium phosphate and prepared a stereo complex and electrospun as nanofiber scaffold and applied for proliferation of mesenchymal stem cells over the nanofiber scaffold which showed that mesenchymal stem cells could easily differentiate into neuron like cells and thus material could be used for dura mater substitute [43].

In another study, chitosan along with polycaprolactone was used for nanofibers preparation which were used for immobilization of nerve growth factor and by conjugating dopamine onto the nanofiber scaffold the cell attachment onto the scaffold increased and these activities showed enhanced growth of PC12 cell over the scaffold [44].

15.4.8 Myelin Membrane Regeneration

Blends of poly (3-hydroxybutyrate) (PHB) and poly (3-hydroxy butyrate-co-3-hydroxyvalerate) (PHBV) were investigated for myelin membrane and also this blend was electrospun with and without type I collagen and it was found that Schwann cells when seeded in these nanofiber scaffolds showed promising attachment properties and cell proliferation till 14 days and it was observed that on sixth day, neuron growth factor was secreted and thus proved that the nanofiber scaffold was effective for nerve cell differentiation and this was more prominent in the presence of type I collagen conjugated PHB-PHBV nanofiber scaffolds [45].

In another study, nanofibers were electrospun using polycaprolactone-gelatin blends with and without cerium oxide nanoparticles and later these nanofiber scaffolds of polycaprolactone-gelatin blend both containing cerium oxide nanoparticles and without cerium oxide nanoparticles were tested in rat model for spinal cord injury for reduction of pain and it was observed that with nanofiber with cerium oxide nanoparticles there was a marked reduction in pain which was confirmed by behavioural motor test [46].

In another study, polycaprolactone was electrospun for nanofiber preparation and tissue plasminogen activator was deposited onto the nanofiber scaffold by placing the nanofiber mat in the plasma chamber and by using argon discharge plasma for 240 minutes, the nanofiber sheet treated with tissue plasminogen activator was placed for reaction at 4°C overnight. The prepared mat was tested for nerve tissue regrowth after sciatic nerve transection in male rats [47].

15.5 CONCLUSION

Tissue engineering and regenerative therapy leads to solving many issues related to organ transplant and cosmetic surgery and paves a role in accidental cases where patients lose their major organs. Such emergency leads to a search of potential donors for organs and other molecular modes where stem cells or body tissues are utilized for grafting for tissue engineering, but sometimes, a recipient's body rejects these contributions due to biological incompatibilities. Many systems are fabricated which can trigger the recipient body cells by mimicking the ECM and replicate and restructure tissues and organs which reduces the burden of finding donors. In this aspect, green nanofibers prepared by green electrospinning methods, where no harmful chemicals are used, are appropriate to serve as tissue regrowth media proving their potential in accelerating the healing process for wounds and reconstruction of damaged organs. The green nanofibers even help in regrowth of new organs when such nanofibrous platforms are seeded with mesenchymal stem cells having the potential of exactly replicating new cells and tissues. Thus, a lot of new unexplored green polymers can be tried for nanofiber fabrication for tissue regrowth and reconstruction helping in making lives easier.

REFERENCES

[1] Chan, B. P., Leong, K. W. Scaffolding in tissue engineering: general approaches and tissue-specific considerations. Eur Spine J. 2008, 17 (Suppl. 4), S467–S479, DOI 10.1007/s00586-008-0745-3

[2] Simionescu, B.C., Ivanov, D. Natural and synthetic polymers for designing composite materials. In Handbook of Bioceramics and Biocomposites, Ed. I.V. Antoniac. Cham: Springer International Publishing, 2015, pp. 1–54, DOI: 10.1007/978-3-319-09230-0_11–1.

[3] Graça, M. F. P., Miguel, S. P., Cabral, C. S. D., Correia, I. J. Hyaluronic acid—based wound dressings: a review. Carbohydr. Polym. 2020, 241, 116364.

[4] Rajora, A.D, Bal, T. Evaluating neem gum-polyvinyl alcohol (NGP-PVA) blend nanofiber mat as a novel platform for wound healing in murine model. Int. J. Biol. Macromol. 2023, 226, 760–771. DOI: 10.1016/j.ijbiomac.2022.12.014.

[5] Cao, X., Chen, W., Zhao, P., Yang, Y., Yu, D.-G. Electrospun porous nanofibers: Pore–forming mechanisms and applications for photocatalytic degradation of organic pollutants in wastewater. Polymers, 2022, 14, 3990. DOI: 10.3390/polym14193990

[6] Sanders, E. H. et al. Two-phase electrospinning from a single electrified jet: microencapsulation of aqueous reservoirs in poly (ethylene-co-vinyl acetate) fibers. Macromolecules, 2003, 36(11), 3803–3805.

[7] Çallioğlu, F. C., Güler, H. K. 7. Natural nanofibers and applications. Green Electrospinning, Eds. N. Horzum, M. M. Demir, R. Muñoz-Espí and D. Crespy. Berlin and Boston: De Gruyter, 2019, 157–188. Doi: 10.1515/9783110581393-007

[8] Gonzalez, E., Barquero, A., Munoz Sanchez, B., Paulis, M., Leiza, JR. Green Electrospinning of Polymer Latex: A systematic study of the effect of Latex Properties on Fiber Morphology. Nanomaterials (Basel). 2021, 11(3), 706, Doi:10.3390/nano11030706.

[9] Yamashita, Y., Tanaka, A., Ko, F. Characteristics of elastomeric nanofiber membranes produced by electrospinning. J Text Eng. 2007, 53, 137–142

[10] Alghoraibi, I., Alomar, S. Different methods for nanofiber design and fabrication. Handbook of Nanofibers, Eds. A. Barhoum et al. Cham: Springer International Publishing AG, 2018. DOI: 10.1007/978-3-319-42789-8_11-2

[11] Cheng, H. N., Gross, R. A., Smith, P. B. Green polymer chemistry: some recent developments and examples. Green Polymer Chemistry: Biobased Materials and Biocatalysis ACS Symposium Series. Eds. Cheng et al. Washington, DC: American Chemical Society, 2015

[12] Selvakumar, K., Madhan, R., Venkat Kumar, G. Biodegradable polymers for nanofibre production. Biol Forum Int J. 2020, 12(2), 68–73

[13] Ahn, S., Chantre, C. O., Gannon, A. R., Lind, J. U., Campbell, P. H., Grevesse, T., O'Connor, B. B., Parker, K. K. Soy protein/cellulose nanofiber scaffolds mimicking skin extracellular matrix for enhanced wound healing. Adv Healthc Mater. 2018, 7(9), e1701175. DOI: 10.1002/adhm.201701175.
[14] Khabbaz, B., Solouk, A., Mirzadeh, H. Polyvinyl alcohol/soy protein isolate nanofibrous patch for wound-healing applications. Prog Biomat 2019, 8, 185–196.
[15] Samadian, H., Zamiri, S., Ehterami, A. et al. Electrospun cellulose acetate/gelatin nanofibrous wound dressing containing berberine for diabetic foot ulcer healing: in vitro and in vivo studies. Sci Rep 2020, 10, 8312. DOI: 10.1038/s41598-020-65268-7.
[16] Ali, I. H., Khalil, I. A., El-Sherbiny, I. M. Design, development, in-vitro and in-vivo evaluation of polylactic acid-based multifunctional nanofibrous patches for efficient healing of diabetic wounds. Sci Rep 2023, 13, 3215. DOI: 10.1038/s41598-023-29032-x.
[17] Zeng, W., Cheng, N. M., Liang, X., et al. Electrospun polycaprolactone nanofibrous membranes loaded with baicalin for antibacterial wound dressing. Sci Rep 12, 10900 (2022). DOI: 10.1038/s41598-022-13141-0
[18] Ali, I. H., et al. Antimicrobial and wound-healing activities of graphene-reinforced electrospun chitosan/gelatin nanofibrous nanocomposite scaffolds. ACS Omega 2022, 7(2), 1838–1850. DOI: 10.1021/acsomega.1c05095
[19] Dey, P., Bal, T., Gupta, R. N. Fabrication and invitro evaluation of electrospun gum ghatti-polyvinyl alcohol polymeric blend green nanofibre mat (GG-PVA NFM) as a novel material for polymeric scaffolds in wound healing. Polymer Test, 2020, 91, 106826. DOI: 10.1016/j.polymertesting
[20] Sen, S., Bal, T., Rajora, A. D. Green nanofiber mat from HLM–PVA–Pectin (Hibiscus leaves mucilage–polyvinyl alcohol–pectin) polymeric blend using electrospinning technique as a novel material in wound-healing process. Appl Nanosci, 2022, 12, 237–250. DOI: 10.1007/s13204-021-02295-4
[21] Nikam, V. S., Punde, D. S., Bhandari, R. S. Silk fibroin nanofibers enhance cell adhesion of blood-derived fibroblast-like cells: A potential application for wound healing. Indian J Pharmacol. 2020, 52(4), 306–312. DOI: 10.4103/ijp.IJP_609_19
[22] Kim, J., et al. Bacterial cellulose nanofibrillar patch as a wound healing platform of tympanic membrane perforation. Adv Healthc Mat, 2023, 2, 1525–1531. DOI: 10.1002/adhm.201200368
[23] Azimi, B., Milazzo, M., Danti, S. Cellulose-based fibrous materials from bacteria to repair tympanic membrane perforations. Front Bioeng Biotechnol. 2021, 9, 669863. DOI: 10.3389/fbioe.2021.669863
[24] Benecke, L., et al. Development of electrospun, biomimetic tympanic membrane implants with tunable mechanical and oscillatory properties for myringoplasty. Biomater. Sci. 2022, 10, 2287–2301
[25] Li, L., Zhang, W., Huang, M., Li, J., Chen, J., Zhou, M., He, J. Preparation of gelatin/genipin nanofibrous membrane for tympanic member repair. J Biomat Sci. 2018, 17, 2154–2167. DOI: 10.1080/09205063.2018.1528519
[26] Huang, M., et al. Preparation of CS/PVA nanofibrous membrane with tunable mechanical properties for tympanic member repair. Macromol. Res. 2018, 26, 892–899. DOI: 10.1007/s13233-018-6127-8
[27] Chainoglou, E., Karagkiozaki, V., Choli-Papadopoulou, T., Mavromanolis, C., Laskarakis, A., Logothetidis, S. Development of biofunctionalized cellulose acetate nanoscaffolds for heart valve tissue engineering. World J Nano Sci Eng. 2016, 6, 129–152. DOI: 10.4236/wjnse.2016.64013

[28] Ding, F., Deng, H., Du, Y., Shi, X., Wang, Q. Emerging chitin and chitosan nanofibrous materials for biomedical applications. Nanoscale. 2014, 6(16), 9477–9493. doi:10.1039/c4nr02814g.

[29] Kobayashi, H., Hattori, S., Honda, T., Kameda, T., Tamada, Y. Long term implantation of silk fibroin nanofiber in rabbit cornea as a scaffold for corneal stromal regeneration. Front Bioeng Biotechnol, 2016. DOI: 10.3389/conf.FBIOE.2016.01.01881

[30] Hazra, S., Nandi, S., Naskar, D., et al. Non-mulberry silk fibroin biomaterial for corneal regeneration. Sci Rep, 2016, 6, 21840. DOI: 10.1038/srep21840

[31] Sharma, S., et al. Surface-modified electrospun poly(ε-Caprolactone) scaffold with improved optical transparency and bioactivity for damaged ocular surface reconstruction. Invest. Ophthalmol. Vis. Sci. 2014, 55(2), 899–907. DOI: 10.1167/iovs.13-12727

[32] Roshanfar, F., Hesaraki, S., Dolatshahi-Pirouz, A. Electrospun silk fibroin/kappa-carrageenan hybrid nanofibers with enhanced osteogenic properties for bone regeneration applications. Biology, 2022, 11, 751. DOI: 10.3390/biology11050751

[33] Preeth, D. R., et al. Bioactive Zinc(II) complex incorporated PCL/gelatin electrospun nanofiber enhanced bone tissue regeneration. Eur J Pharm Sci, 2021, 160, 105768. DOI: 10.1016/j.ejps.2021.105768.

[34] Xue, R., Qian, Y., Li, L., et al. Polycaprolactone nanofiber scaffold enhances the osteogenic differentiation potency of various human tissue-derived mesenchymal stem cells. Stem Cell Res Ther, 2017, 8, 148. DOI: 10.1186/s13287-017-0588-0

[35] Sadek, K. M., et al. In Vitro Biological Evaluation of a Fabricated Polycaprolactone/Pomegranate Electrospun Scaffold for Bone Regeneration. ACS Omega, 2021, 6 (50), 34447–34459. DOI: 10.1021/acsomega.1c04608

[36] Shitole, A. A., Raut, P. W., Sharma, N. et al. Electrospun polycaprolactone/hydroxyapatite/ZnO nanofibers as potential biomaterials for bone tissue regeneration. J Mater Sci: Mater Med, 2019, 30, 51. DOI: 10.1007/s10856-019-6255-5

[37] Samadian, H., Farzamfar, S., Vaez, A. et al. A tailored polylactic acid/polycaprolactone biodegradable and bioactive 3D porous scaffold containing gelatin nanofibers and Taurine for bone regeneration. Sci Rep, 2020, 10, 13366. DOI: 10.1038/s41598-020-70155-2

[38] Wang, S.-F., Wu, Y.-C., Cheng, Y.-C., Hu, W.-W. The development of polylactic acid/multi-wall carbon nanotubes/polyethylene glycol scaffolds for bone tissue regeneration application. Polymers 2021, 13, 1740. DOI: 10.3390/polym13111740

[39] Ehterami, A., Abbaszadeh-Goudarzi, G., Haghi-Daredeh, S., et al. Bone tissue engineering using 3-D polycaprolactone/gelatin nanofibrous scaffold containing berberine: In vivo and in vitro study. Polym Adv Technol. 2022; 33(2), 672–681. DOI: 10.1002/pat.5549

[40] Niu, Y., Stadler, F.J., Yang, X. et al. HA-coated collagen nanofibers for urethral regeneration via in situ polarization of M2 macrophages. J Nanobiotechnol 2021, 19, 283. DOI: 10.1186/s12951-021-01000-5

[41] Jiao, W., et al. Fibrinogen/poly(l-lactide-co-caprolactone) copolymer scaffold: A potent adhesive material for urethral tissue regeneration in urethral injury treatment. Regenerat Ther, 2023, 22, 136–147, DOI: 10.1016/j.reth.2022.12.004

[42] Dou, Y., Sun, X., Guo, G., Dong, J., Lu, M., Zhang, W. Electrospun pure chitosan nanofibrous mats with high structural stability for dura mater regeneration. Front. Bioeng. Biotechnol. 2016, DOI: 10.3389/conf.FBIOE.2016.01.01956

[43] Chuan, D., Wang, Y., Fan, R., Zhou, L., Chen, H., Xu, J., Guo, G. Fabrication and properties of a biomimetic dura matter substitute based on stereocomplex poly(Lactic Acid) nanofibers. Int J Nanomed. 2020, 15, 3729–3740. DOI: 10.2147/IJN.S248998.

[44] Afrash, H., Nazeri, N., Davoudi, P., Faridi Majidi, R., Ghanbar, H. Development of a bioactive scaffold based on NGF containing PCL/chitosan nanofibers for nerve regeneration. Biointerf Res Appl Chem, 2021, 11(5), 12606–12617. DOI: 10.33263/BRIAC115.1260612617.

[45] Masaeli, E., et al. Fabrication, characterization and cellular compatibility of poly(hydroxy alkanoate) composite nanofibrous scaffolds for nerve tissue engineering. PLoS One. 2013, 8(2), e57157. DOI: 10.1371/journal.pone.0057157. Epub 2013 Feb 27. PMID: 23468923; PMCID: PMC3584130.
[46] Rahimi, B., Behroozi, Z., Motamed, A. et al. Study of nerve cell regeneration on nanofibers containing cerium oxide nanoparticles in a spinal cord injury model in rats. J Mater Sci: Mater Med, 2023, 34, 9. DOI: 10.1007/s10856-023-06711-9
[47] Sajadi, E., Aliaghaei, A., Farahni, R.M. et al. Tissue plasminogen activator loaded pcl nanofibrous scaffold promoted nerve regeneration after sciatic nerve transection in male rats. Neurotox Res 2021, 39, 413–428. DOI: 10.1007/s12640-020-00276-z

Index

Note: Numbers in *italics* indicate figures on the corresponding page.

C

G

H

I

O

Q

R

S

T

U

Printed in the United States
by Baker & Taylor Publisher Services

Electrospinning is a versatile method to synthesize fiber materials. *Electrospun Nanofibres: Materials, Methods, and Applications* explores the technical aspects of electrospinning methods used to derive a wide range of functional fiber materials and their applications in various technical sectors. As electrospinning is a process that can be modified strategically to achieve different fibers of interest, this book covers the wide spectrum of electrospinning methodologies, such as coaxial, triaxial, emulsion, suspension, electrolyte and gas-assisted spinning processes. It:

- Discusses a broad range of materials, including synthetic polymers, biodegradable polymers, metals and their oxides, hybrid materials, nonpolymers, and more.
- Reviews different electrospinning methods and combined technologies.
- Describes process-related parameters and their influence on material properties and performance.
- Examines modeling of the electrospinning process.
- Highlights applications across different industries.

This book is aimed at researchers, professionals, and advanced students in materials science and engineering.

MATERIALS SCIENCE AND ENGINEERING

www.routledge.com

CRC Press titles are available as eBook editions in a range of digital formats